Solutions Guide for

CHEMISTRY

THIRD EDITION

by
Steven S. Zumdahl

Kenneth C. Brooks
Thomas J. Hummel
Steven S. Zumdahl

University of Illinois at Urbana - Champaign

D. C. Heath and Company
Lexington, Massachusetts Toronto

Published simultaneously in Canada.

Printed in the United States of America.

International Standard Book Number: 0-669-32869-3

10 9 8 7 6 5 4

TO THE STUDENT: HOW TO USE THIS GUIDE

Chemistry is an applied science. Chemistry is valuable because a collection of facts about chemical behavior can be dealt with in a systematic manner and applied to solving the new problems that chemists encounter daily. In your study of chemistry you should give a priority to solving problems over other activities.

Solutions to all of the odd end of chapter exercises are in this manual. In addition, all of the solutions for the review questions in Chapter 11 and 17 and the solution to Exercise 14.40 are included. The latter is the first occasion in Chapter 14 that successive approximations are used to solve an equilibrium problem. This "Solutions Guide" can be very valuable if you use it properly. The way NOT to use it is to look at an exercise in the book and then check the solution, often saying to yourself, "That's easy, I can do it." Chemistry is easy once you get the hang of it, but it takes work. Don't look up a solution to a problem until you have tried to work it on your own. If you are completely stuck, see if you can find a similar problem in the Sample Exercises in the chapter. Then look up the solution. After you do this, look for a similar problem in the end of chapter exercises and try working it. The more problems you do, the easier chemistry becomes. It is also in your self interest to try to work as many problems as possible. Most exams that you will take in chemistry will involve a lot of problem solving. If you have worked several problems similar to the ones on an exam you will do much better than if the exam is the first time you try to solve a particular type problem. No matter how much you read and study the text, or how well you think you understand the material, you don't really understand it until you have taken the information in the text and used it to solve a problem.

In this manual we have worked problems as in the textbook. We have shown intermediate answers to the correct number of significant figures and used that rounded answer in later calculations. Thus, some of your answers may differ slightly from ours. When we have not followed this convention, we have noted this in the solution.

We are grateful to Delores Wyatt for her outstanding effort in preparing the manuscript of this manual.

TABLE OF CONTENTS

CHAPTER ONE: CHEMICAL FOUNDATIONS

QUESTIONS

1. a. law: A law is a concise statement or equation that summarizes a great variety of observations.

   theory: A theory is an hypothesis that has been tested over a length of time.

   A law is less likely to be challenged or modified than a theory.

   b. theory versus experiment: A theory is our explanation of why things behave the way they do, while experiment is the process of observing that behavior. Theories attempt to explain the results of experiments and are, in turn, tested by further experiments.

   c. qualitative versus quantitative: A qualitative measurement only measures a quality while a quantitative measurement attaches a number to the observation. Examples:

   qualitative observations: The water was hot to the touch. Mercury was found in the drinking water. quantitative observations: The temperature of the water was 62°C. The concentration of mercury in the drinking water was 1.5 ppm.

   d. hypothesis versus theory: Both are explanations of experimental observation. A theory is an hypothesis that has been tested over time and found to still be valid, with, perhaps, some modifications.

3. a. No b. Yes c. Yes

5. accuracy: How close a measurement or series of measurements is to an accepted or true value.

   precision: How close a series of measurements of the same thing are to each other.

7. Experimental results are the facts that we deal with. Theories are our attempt to rationalize those facts. If the experiment is done properly and the theory can't account for the facts, then the theory is wrong.

9. Chemical changes involve the making and breaking of chemical forces (bonds). Physical changes do not. The identity of a substance changes after a chemical change, but not after a physical change.

EXERCISES

Uncertainty, Precision, Accuracy and Significant Figures

11. a. inexact b. exact c. exact

    For c, $\frac{36\ \text{in}}{\text{yd}} \times \frac{2.54\ \text{cm}}{\text{in}} \times \frac{1\ \text{m}}{100\ \text{cm}} = \frac{0.9144\ \text{m}}{\text{yd}}$ (All conversion factors used are exact.)

    d. inexact - Although this number appears to be exact, it probably isn't. The announced attendance may be tickets sold but not the number who were actually in the stadium. Some people who paid may not have gone, some may leave early, or arrive late, some may sneak in without paying, etc.

    e. exact f. inexact

13. a. 0.0012; 2 S.F., 1.2×10^{-3}

b. 437,000; 3 S.F., 4.37×10^{5}

c. 900.0; 4 S.F., 9.000×10^{2}

d. 106, 3 S.F.; 1.06×10^{2}

e. 125, 904,000; 6 S.F., 1.25904×10^{8}

f. 1.0012; 5 S.F., 1.0012×10^{0}

g. 2006; 4 S. F., 2.006×10^{3}

h. 3050; 3 S.F., 3.05×10^{3}

i. 0.001060; 4 S.F., 1.060×10^{-3}

15. a. 6×10^{8} b. 5.8×10^{8} c. 5.82×10^{8}

d. 5.8200×10^{8} e. 5.820000×10^{8}

17. a. 467; (25.27 - 24.16 = 1.11, only 3 significant figures)

b. 0.24; (8.925 - 8.904 = 0.021, 2 significant figures)

c. $(9.04 - 8.23 + 21.954 + 81.0) \div 3.1416 = 103.8 \div 3.1416 = 33.04$

d. $\dfrac{9.2 \times 100.65}{8.321 + 4.026} = \dfrac{9.2 \times 100.65}{12.347} = 7.5 \times 10^{1}$

e. $0.1654 + 2.07 - 2.114 = 0.12$

Uncertainty begins to appear in the second decimal place. Numbers were added as written and the answer was rounded off to 2 decimal places at the end. If you round to 2 decimal places and add you get 0.13.

f. $8.27(4.987 - 4.962) = 8.27(0.025) = 0.21$

g. $\dfrac{9.5 + 4.1 + 2.8 + 3.175}{4} = 4.9$ Uncertainty appears in the first decimal place.

h. $\dfrac{9.025 - 9.024}{9.025} \times 100 = \dfrac{0.001}{9.025} \times 100 = 0.01$

Units and Unit Conversions

19. a. $1\ \text{km} = 10^{3}\ \text{m} = 10^{6}\ \text{mm} = 10^{15}\ \text{pm}$

b. $1\ \text{g} = 10^{-3}\ \text{kg} = 10^{3}\ \text{mg} = 10^{9}\ \text{ng}$

c. $1\ \text{mL} = 10^{-3}\ \text{L} = 10^{-3}\ \text{dm}^{3} = 1\ \text{cm}^{3}$

d. $1\ \text{mg} = 10^{-6}\ \text{kg} = 10^{-3}\ \text{g} = 10^{3}\ \mu\text{g} = 10^{6}\ \text{ng} = 10^{9}\ \text{pg} = 10^{12}\ \text{fg}$

e. $1\ \text{s} = 10^{3}\ \text{ms} = 10^{9}\ \text{ns}$

21. $1\ \text{A} \times \dfrac{10^{-8}\ \text{cm}}{\text{Å}} \times \dfrac{1\ \text{m}}{100\ \text{cm}} \times \dfrac{10^{9}\ \text{nm}}{\text{m}} = 1 \times 10^{-1}\ \text{nm}$

$1 \times 10^{-1}\ \text{nm} \times \dfrac{10^{-9}\ \text{m}}{\text{nm}} \times \dfrac{1\ \text{pm}}{10^{-12}\ \text{m}} = 1 \times 10^{2}\ \text{pm}$

23. a. Since 1 in = 2.54 cm, then an uncertainty of $\pm \frac{1}{2}$ in is about $\pm$ 1 cm after we convert from in to cm. Thus, we should express the heights to the nearest centimeter.

7'6" = 7(12) + 6 in = 90 in

$$90 \text{ in} \times \frac{2.54 \text{ cm}}{\text{in}} = 229 \text{ cm} = 229 \times 10^{-2} \text{ m} = 2.29 \text{ m}$$

5'2" = 5(12) + 2 in = 62 in

$$62 \text{ in} \times \frac{2.54 \text{ cm}}{\text{in}} = 157 \text{ cm} = 157 \times 10^{-2} \text{ m} = 1.57 \text{ m}$$

b. $$25{,}000 \text{ mi} \times \frac{1.609 \text{ km}}{\text{mi}} = 4.0 \times 10^4 \text{ km}$$

We can derive this conversion factor from information in the chapter.

$$1 \text{ mi} \times \frac{5280 \text{ ft}}{\text{mi}} \times \frac{1 \text{ yd}}{3 \text{ ft}} = 1760 \text{ yd and}$$

$$1760 \text{ yd} \times \frac{1 \text{ m}}{1.094 \text{ yd}} \times \frac{1 \text{ km}}{1000 \text{ m}} = 1.609 \text{ km}$$

or 1 mi = 5280 ft = 1760 yd = 1.609 km

$$4.0 \times 10^4 \text{ km} \times \frac{1000 \text{ m}}{\text{km}} = 4.0 \times 10^7 \text{ m}$$

c. $V = l \times w \times h$

$$V = 1.0 \text{ m} \times \left(5.6 \text{ cm} \times \frac{1 \text{ m}}{100 \text{ cm}}\right) \times \left(2.1 \text{ dm} \times \frac{1 \text{ m}}{10 \text{ dm}}\right) = 1.2 \times 10^{-2} \text{ m}^3$$

$$1.2 \times 10^{-2} \text{ m}^3 \times \left(\frac{10 \text{ dm}}{\text{m}}\right)^3 \left(\frac{1 \text{ L}}{\text{dm}^3}\right) = 12 \text{ L}$$

$$12 \text{ L} \times \frac{1000 \text{ cm}^3}{\text{L}} \times \left(\frac{1 \text{ in}}{2.54 \text{ cm}}\right)^3 = 730 \text{ in}^3$$

$$730 \text{ in}^3 \times \left(\frac{1 \text{ ft}}{12 \text{ in}}\right)^3 = 0.42 \text{ ft}^3$$

25. a. $$928 \text{ mi} \times \frac{5280 \text{ ft}}{\text{mi}} \times \frac{1 \text{ fathom}}{6 \text{ ft}} \times \frac{1 \text{ cable length}}{100 \text{ fathoms}} \times \frac{1 \text{ nautical mi}}{10 \text{ cable lengths}}$$

$$\times \frac{1 \text{ league}}{3 \text{ nautical miles}} = 272 \text{ leagues}$$

$$928 \text{ mi} \times \frac{5280 \text{ ft}}{\text{mi}} \times \frac{1 \text{ yd}}{3 \text{ ft}} \times \frac{1 \text{ m}}{1.094 \text{ yd}} \times \frac{1 \text{ km}}{1000 \text{ m}} = 1.49 \times 10^3 \text{ km}$$

b. $1.0 \text{ cable length} \times \frac{100 \text{ fathom}}{\text{cable length}} \times \frac{6 \text{ ft}}{\text{fathom}} \times \frac{1 \text{ yd}}{3 \text{ ft}} \times \frac{1 \text{ m}}{1.094 \text{ yd}} \times \frac{1 \text{ km}}{1000 \text{ m}}$

$= 0.18 \text{ km}$

$1.0 \text{ cable length} = 0.18 \text{ km} \times \frac{1000 \text{ m}}{\text{km}} \times \frac{100 \text{ cm}}{\text{m}} = 1.8 \times 10^4 \text{ cm}$

c. $315 \text{ ft} \times \frac{12 \text{ in}}{\text{ft}} \times \frac{2.54 \text{ cm}}{\text{in}} \times \frac{1 \text{ m}}{100 \text{ cm}} = 96.0 \text{ m}$

$37 \text{ ft} \times \frac{12 \text{ in}}{\text{ft}} \times \frac{2.54 \text{ cm}}{\text{in}} \times \frac{1 \text{ m}}{100 \text{ cm}} = 11 \text{ m}$

$315 \text{ ft} \times \frac{1 \text{ fathom}}{6 \text{ ft}} \times \frac{1 \text{ cable length}}{100 \text{ fathoms}} = 5.25 \times 10^{-1} \text{ cable length}$

$37 \text{ ft} \times \frac{1 \text{ fathom}}{6 \text{ ft}} = 6.2 \text{ fathoms}$

27. a. $\frac{100.8 \text{ mi}}{\text{hr}} \times \frac{1760 \text{ yd}}{\text{mi}} \times \frac{1 \text{ m}}{1.0936 \text{ yd}} \times \frac{\text{hr}}{60 \text{ min}} \times \frac{1 \text{ min}}{60 \text{ s}} = \frac{45.06 \text{ m}}{\text{s}}$

b. 60 ft 6 in = 60(12) + 6 in = 726 in

$726 \text{ in} \times \frac{2.54 \text{ cm}}{\text{in}} \times \frac{1 \text{ m}}{100 \text{ cm}} \times \frac{1 \text{ s}}{45.06 \text{ m}} = 0.409 \text{ s}$

29. a. $1 \text{ lb. troy} \times \frac{12 \text{ oz tr}}{1 \text{ lb troy}} \times \frac{20 \text{ pw}}{1 \text{ troy oz}} \times \frac{24 \text{ gr}}{\text{pw}} \times \frac{0.0648 \text{ g}}{\text{gr}} \times \frac{1 \text{ kg}}{1000 \text{ g}} = 0.373 \text{ kg}$

$1 \text{ troy lb} = 0.373 \text{ kg} \times \frac{2.205 \text{ lb}}{\text{kg}} = 0.822 \text{ lb}$

b. $1 \text{ oz tr} \times \frac{20 \text{ pw}}{\text{oz tr}} \times \frac{24 \text{ gr}}{\text{pw}} \times \frac{0.0648 \text{ g}}{\text{gr}} = 31.1 \text{ g}$

$1 \text{ oz tr} = 31.1 \text{ g} \times \frac{1 \text{ carat}}{0.200 \text{ g}} = 156 \text{ carats}$

c. 1 lb tr = 0.373 kg

$0.373 \text{ kg} \times \frac{1000 \text{ g}}{\text{kg}} \times \frac{1 \text{ cm}^3}{19.3 \text{ g}} = 19.3 \text{ cm}^3$

31. $2240 \text{ lb} \times \frac{1 \text{ kg}}{2.205 \text{ lb}} = 1016 \text{ kg} = 1 \text{ long ton}$

$1016 \text{ kg} \times \frac{1 \text{ tonne}}{1000 \text{ kg}} = 1.016 \text{ tonne}$

1 long ton = 1.016 metric tonne

Temperature

33. $°\text{C} = \frac{5}{9}(°\text{F} - 32) = \frac{5}{9}(102.5 - 32) = 39.2°\text{C}$, $\text{K} = °\text{C} + 273.2 = 312.4 \text{ K}$ (Note: 32 is exact)

35. $°\text{F} = \frac{9}{5}°\text{C} + 32 = \frac{9}{5}(25) + 32 = 77°\text{F}$, $\text{K} = 25 + 273 = 298 \text{ K}$

37. We can do this two ways.

First, we calculate the high and low temperature and get the uncertainty from the range.

20.6 °C ± 0.1 °C means the temperature can range from 20.5 °C to 20.7 °C.

$T_F = \frac{9}{5} T_c + 32$ ⟵ (exact)

$T_F\,(\text{min}) = \frac{9}{5}(20.5) + 32 = 68.9\ °F$; $T_F\,(\text{max}) = \frac{9}{5}(20.7) + 32 = 69.3\ °F$

So, the temperature ranges from 68.9 °F to 69.3 °F which we can express as 69.1 °F ± 0.2 °F.

An alternative way is to treat the uncertainty and the temperature in °C separately.

$T_F = \frac{9}{5} T_c + 32 = \frac{9}{5}(20.6) + 32 = 69.1\ °F$

and $\pm 0.1\ °C \times \frac{9\ °F}{5\ °C} = \pm 0.18\ °F \approx \pm 0.2\ °F$

Combining the two calculations: $T_F = 69.1\ °F \pm 0.2\ °F$

Density

39. $\frac{2.70\ g}{cm^3} \times \frac{1\ kg}{1000\ g} \times \left(\frac{100\ cm}{m}\right)^3 = \frac{2.70 \times 10^3\ kg}{m^3}$

$\frac{2.70\ g}{cm^3} \times \frac{1\ lb}{453.6\ g} \times \left(\frac{2.54\ cm}{in}\right)^3 \times \left(\frac{12\ in}{ft}\right)^3 = \frac{169\ lb}{ft^3}$

41. $D = \frac{\text{mass}}{\text{volume}}$; mass $= 1.67 \times 10^{-24}$ g; $r = d/2 = 5.0 \times 10^{-4}$ pm

$V = \frac{4}{3}\pi r^3 = \frac{4}{3}(3.14) \times \left(5.0 \times 10^{-4}\ pm \times \frac{10^{-12}\ m}{pm} \times \frac{100\ cm}{m}\right)^3 = 5.2 \times 10^{-40}\ cm^3$

$D = \frac{1.67 \times 10^{-24}\ g}{5.2 \times 10^{-40}\ cm^3} = \frac{3.2 \times 10^{15}\ g}{cm^3}$

43. $5.0\ \text{carat} \times \frac{0.200\ g}{\text{carat}} \times \frac{1\ cm^3}{3.51\ g} = 0.28\ cm^3$

45. a. both are the same mass

b. 1.0 mL of mercury. Mercury has a greater density than water.

Mass of mercury $= 1.0\ mL \times \frac{13.6\ g}{mL} = 13.6$ g of mercury

mass of water $= 1.0\ mL \times \frac{1.0\ g}{mL} = 1.0$ g of water.

c. same, 19.32 g

d. 1.0 L of benzene

Classification and Separation of Matter

47. Solid: own volume, own shape, does not flow

Liquid: own volume, takes shape of container, flows

Gas: takes volume and shape of container, flows

49. a. pure b. mixture c. mixture d. pure e. mixture

f. pure g. mixture h. mixture i. pure

ADDITIONAL EXERCISES

51. a. 8.41 (2.16 has only three significant figures)

b. 16.1 (uncertainty appears in the first decimal place, 8.1)

c. 52.5 d. 5 (2 contains one significant figure)

e. 0.009 f. 429.59 (uncertainty appears in 2nd decimal, 2.17, 4.32)

53. a. Density = $\frac{\text{mass}}{\text{volume}}$; Volume of a sphere = $\frac{4}{3}\pi r^3$

$$D = \frac{2 \times 10^{36}\ \text{kg}}{\frac{4}{3} \times 3.14 \times (6.96 \times 10^5\ \text{km})^3} = \frac{2 \times 10^{36}\ \text{kg}}{1.41 \times 10^{18}\ \text{km}^3}$$

$$= \frac{1.4 \times 10^{18}\ \text{kg}}{\text{km}^3} \approx \frac{1 \times 10^{18}\ \text{kg}}{\text{km}^3}$$

b. $$\frac{1 \times 10^{18}\ \text{kg}}{\text{km}^3} \times \left(\frac{1\ \text{km}}{1000\ \text{m}}\right)^3 = \frac{1 \times 10^9\ \text{kg}}{\text{m}^3}$$

c. $$\frac{1 \times 10^{18}\ \text{kg}}{\text{km}^3} \times \frac{1000\ \text{g}}{\text{kg}} \times \left(\frac{1\ \text{km}}{1000\ \text{m}}\right)^3 \times \left(\frac{1\ \text{m}}{100\ \text{cm}}\right)^3 = \frac{1 \times 10^6\ \text{g}}{\text{cm}^3}$$

55. $$\frac{36.1\ \text{miles}}{\text{gallon}} \times \frac{1760\ \text{yd}}{\text{mi}} \times \frac{1\ \text{m}}{1.094\ \text{yd}} \times \frac{1\ \text{km}}{1000\ \text{m}} \times \frac{1\ \text{gal}}{4\ \text{qt}} \times \frac{1.057\ \text{qt}}{\text{L}} = 15.3\ \text{km/L} = 15\ \text{km/L}$$

57. a. For $\frac{103 \pm 1}{101 \pm 1}$: Maximum = $\frac{104}{100}$ = 1.04; Minimum = $\frac{102}{102}$ = 1.00

So $\frac{103 \pm 1}{101 \pm 1} = 1.02 \pm 0.02$

b. For $\frac{101 \pm 1}{99 \pm 1}$: Maximum = $\frac{102}{98}$ = 1.04; Minimum = $\frac{100}{100}$ = 1.00

So $\frac{101 \pm 1}{99 \pm 1} = 1.02 \pm 0.02$

c. For $\frac{99 \pm 1}{101 \pm 1}$: Maximum = $\frac{100}{100}$ = 1.00; minimum = $\frac{98}{102}$ = 0.96

So $\frac{99 \pm 1}{101 \pm 1} = 0.98 \pm 0.02$

Considering the error limits, (a) and (b) should be expressed to three significant figures and (c) to two significant figures. The division rule differs in (b). The rule says (b) should be expressed to two significant figures. If we do this for (b) we imply that the answer is 1.0 or between 0.95 and 1.05. The actual range is less than this, so we should use the more precise way of expressing uncertainty. The significant figure rules only give us guidelines for estimating uncertainty. When we have a better handle on the uncertainty we should use it in precedence of the significant figure guidelines.

59. $126 \text{ gal} \times \frac{4 \text{ qt}}{\text{gal}} \times \frac{1 \text{ L}}{1.057 \text{ qt}} = 477 \text{ L}$

61. a. $85 \text{ crowns} \times \frac{1 \text{ royal}}{20 \text{ crowns}} = 4.25 \text{ royals}$

$85 \text{ crowns} \times \frac{100 \text{ weights}}{\text{crown}} = 8.5 \times 10^3 \text{ weights}$

$50 \text{ royals} \times \frac{1 \text{ horse}}{4.25 \text{ royals}} = 11.8 \text{ horses.}$ So, 11 horses can be bought.

$11 \text{ horses} \times \frac{4.25 \text{ royals}}{\text{horse}} = 46.75 \text{ royals}, \; 50 - 46.75 = 3.25$ royals will be left

b. $\frac{13 \text{ crowns}}{\text{bundle}} \times \frac{100 \text{ weights}}{\text{crown}} \times \frac{1 \text{ bundle}}{25 \text{ haigus hides}} = \frac{52 \text{ weights}}{\text{haigus hide}}$

$288 \text{ bundles} \times \frac{13 \text{ crowns}}{\text{bundle}} \times \frac{1 \text{ royal}}{20 \text{ crowns}} = 187.2 \text{ royals}$

63. $D_{max} = \frac{M_{max}}{V_{min}}$; V = V(final) - V(initial)

We get V_{min} from $9.7 \text{ cm}^3 - 6.5 \text{ cm}^3 = 3.2 \text{ cm}^3$

$$D_{max} = \frac{28.93 \text{ g}}{3.2 \text{ cm}^3} = \frac{9.0 \text{ g}}{\text{cm}^3}$$

$$D_{min} = \frac{M_{min}}{V_{max}} = \frac{28.87 \text{ g}}{9.9 \text{ cm}^3 - 6.3 \text{ cm}^3} = \frac{8.0 \text{ g}}{\text{cm}^3}$$

The density is $\frac{8.5 \text{ g}}{\text{cm}^3} \pm \frac{0.5 \text{ g}}{\text{cm}^3}$

65. $\frac{100. \text{ m}}{9.86 \text{ s}} = 10.1 \text{ m/s}$

$$\frac{10.1 \text{ m}}{\text{s}} \times \frac{1 \text{ km}}{1000 \text{ m}} \times \frac{60 \text{ s}}{\text{min}} \times \frac{60 \text{ min}}{\text{hr}} = 36.4 \text{ km/hr}$$

$$\frac{100. \text{ m}}{9.86 \text{ s}} \times \frac{1.094 \text{ yd}}{\text{m}} \times \frac{3 \text{ ft}}{\text{yd}} = 33.3 \text{ ft/s}$$

$$\frac{33.3 \text{ ft}}{\text{s}} \times \frac{1 \text{ mi}}{5280 \text{ ft}} \times \frac{60 \text{ s}}{\text{min}} \times \frac{60 \text{ min}}{\text{hr}} = 22.7 \text{ mi/hr}$$

$$1.00 \times 10^2 \text{ yds} \times \frac{1 \text{ m}}{1.094 \text{ yd}} \times \frac{9.86 \text{ s}}{100. \text{ m}} = 9.01 \text{ s}$$

67. $K = °C + 273.15; 0 = °C + 273.15; °C = -273.15$

$°F = \frac{9}{5}°C + 32; °F = \frac{9}{5}(-273.15) + 32 = -459.67\ °F$ (32 exact)

69. The object that sinks has the greater density. Since it sinks its density is greater than that of water, 1.0 g/cm^3. The second object must have a density less than water since it floats. Both objects have the same mass. Thus, the sphere that sinks has the smaller volume. The sphere that floats has the larger diameter.

71. $D_{cube} = \frac{140.4\ g}{(3.00\ cm)^3} = \frac{5.20\ g}{cm^3}$

If this is correct to 1.00% then the density is: $\frac{5.20\ g}{cm^3} \pm \frac{0.05\ g}{cm^3}$

$V_{sphere} = \frac{4}{3}\pi r^3 = \frac{4}{3}\pi(1.42\ cm)^3 = 12.0\ cm^3$

$D_{sphere} = \frac{61.6\ g}{12.0\ cm^3} = \frac{5.13\ g}{cm^3}$; $D_{sphere} = \frac{5.13\ g}{cm^3} \pm \frac{0.05\ g}{cm^3}$

D_{cube} is between 5.15 g/cm^3 and 5.25 g/cm^3

D_{sphere} is between 5.08 g/cm^3 and 5.18 g/cm^3

We can't decisively say if they are the same or different. The data are not precise enough to say.

CHALLENGE PROBLEMS

73. In a subtraction, the results get smaller, but the uncertainties add. If the two numbers are very close together, the uncertainty may be larger than the result. For example, let us assume we want to take the difference of the following two measured quantities, 999,999 ± 2 and 999,996 ± 2. The difference is 3 ± 4.

75. Heavy pennies (old): mean mass = 3.08 ± 0.05 g

Light pennies (new): mean mass = $\frac{(2.467 + 2.545 + 2.518)}{3}$

= 2.51 ± 0.04 g

Average Density of old pennies:

$$D_{old} = \frac{\frac{95 \times 8.96\ g}{cm^3} + \frac{5 \times 7.14\ g}{cm^3}}{100} = \frac{8.9\ g}{cm^3}$$

Average Density of new pennies:

$$D_{new} = \frac{\frac{2.4 \times 8.96\ g}{cm^3} + \frac{97.6 \times 7.14\ g}{cm^3}}{100} = \frac{7.18\ g}{cm^3}$$

Since $D = \frac{mass}{volume}$ and the volume of old and new pennies are the same,

then $\frac{D_{new}}{D_{old}} = \frac{Mass_{new}}{Mass_{old}}$; $\frac{D_{new}}{D_{old}} = \frac{7.18}{8.9} = 0.807 = 0.81$

$\frac{Mass_{new}}{Mass_{old}} = \frac{2.51}{3.08} = 0.81$

To two decimal places the ratios are the same. We can reasonably conclude that yes, the difference in mass is accounted for by the difference in the alloy used.

77. a. One possibility is that rope B is not attached to anything and rope A and rope C are connected via a pair of pulleys and/or gears.

b. Try to pull rope B out of the box. Measure the distance moved by C for a given movement of A. Hold either A or C firmly while pulling on the other.

CHAPTER TWO: ATOMS, MOLECULES, AND IONS

QUESTIONS

1. There should be no difference. It does not matter how a substance is produced, it is still that substance.

3. Some elements exist as molecular substances. That is, hydrogen normally exists as H_2 molecules, not single hydrogen atoms.

5. We now know that some atoms of the same element have different masses. We have had to include the existence of isotopes in our models.

7. β-particles are electrons. A cathode ray is a stream of electrons.

9. The atomic number of an element is equal to the number of protons in the nucleus of an atom of that element. The mass number is the sum of the number of protons plus neutrons in the nucleus. The atomic weight is the actual mass of a particular isotope (including electrons). As we will see in chapter three, the average mass of an atom is taken from a measurement made on a large number of atoms and it is this average value we see on the periodic table.

11. A compound will always contain the same numbers (and types) of atoms. A given amount of hydrogen will react only with a specific amount of oxygen. Any excess oxygen will remain unreacted.

EXERCISES

Development of the Atomic Theory

13. a. The composition of a substance depends on the numbers of atoms of each element making up the compound (i.e., the formula of the compound) and not on the composition of the mixture from which it was formed.

    b. $H_2 + Cl_2 \rightarrow 2\ HCl$. The volume of HCl produced is twice the volume of H_2 (or Cl_2) used.

15. $\frac{1.188}{1.188} = 1.000; \frac{2.375}{1.188} = 1.999; \frac{3.563}{1.188} = 2.999$

    The masses of fluorine are simple ratios of whole numbers to each other, 1:2:3.

17. To get the atomic mass of H to be 1.00, we divide the mass that reacts with 1.00 g of oxygen by 0.1260. $\frac{0.1260}{0.1260} = 1.00$

    To get Na and Mg on the same scale, we do the same division.

    Na: $\frac{2.8750}{0.1260} = 22.8$; Mg: $\frac{1.5000}{0.1260} = 11.9$

For O: $\frac{1.00}{0.1260} = 7.94$

	H	O	Na	Mg
Scale	1.00	7.94	22.8	11.9
Accepted Value	1.01 (1.008)	16.00	22.99	24.31

The atomic masses of O and Mg are incorrect. The atomic masses of H and Na are close. Something must be wrong about the assumed formulas of the compounds. It turns out the correct formulas are H_2O, Na_2O, and MgO. The smaller discrepancies result from the error in the atomic mass of H.

The Nature of the Atom

19. Density of nucleus:

$$V_{nucleus} = \frac{4}{3}\pi r^3 = \frac{4}{3}(3.14)(5 \times 10^{-14}\ \text{cm})^3 = 5 \times 10^{-40}\ \text{cm}^3$$

$$D = \frac{1.67 \times 10^{-24}\ \text{g}}{5 \times 10^{-40}\ \text{cm}^3} = 3 \times 10^{15}\ \text{g/cm}^3$$

Density of H-atom:

$$V_{atom} = \frac{4}{3}(3.14)(1 \times 10^{-8}\ \text{cm})^3 = 4 \times 10^{-24}\ \text{cm}^3$$

$$D = \frac{1.67 \times 10^{-24} + 9 \times 10^{-28}\ \text{g}}{4 \times 10^{-24}\ \text{cm}^3} = 0.4\ \text{g/cm}^3$$

21. $5.93 \times 10^{-18}\ \text{C} \times \frac{1\ \text{neg charge}}{1.602 \times 10^{-19}\ \text{C}}$ = 37 negative charges on the oil drop

23. gold - Au; silver - Ag; mercury - Hg; potassium - K: iron - Fe; antimony - Sb; tungsten - W

25. fluorine - F; chlorine - Cl; bromine - Br; sulfur - S; oxygen - O; phosphorus -P

27. Sn - tin; Pt - platinum; Co - cobalt; Ni - nickel; Mg - magnesium; Ba - barium; K - potassium

29. The noble gases are He, Ne, Ar, Kr, Xe, and Rn (helium, neon, argon, krypton, xenon, and radon). Radon has only radioactive isotopes. In the periodic table the whole number enclosed in paranthesis is the mass number of the longest lived isotope of the element.

31. a. Eight, Li to Ne b. Eight, Na to Ar c. Eighteen, K to Kr d. Five, N, P, As, Sb, Bi

33. a. $^{238}_{94}Pu$; 94 protons, 238 - 94 = 144 neutrons
    d. $^{4}_{2}He$; 2 protons, 2 neutrons

    b. $^{65}_{29}Cu$; 29 protons, 65 - 29 = 36 neutrons
    e. $^{60}_{27}Co$; 27 protons, 33 neutrons

    c. $^{52}_{24}Cr$; 24 protons, 28 neutrons
    f. $^{54}_{24}Cr$; 24 protons, 30 neutrons

35. 9 protons means the atomic number is 9. The mass number is 9 + 10 = 19; Symbol: $^{19}_{9}F$

37. Atomic number = 63 (Eu); Charge = +63 - 60 = +3; Mass number = 63 + 88 = 151; Symbol: ${}^{151}_{63}Eu^{3+}$

39. Atomic number = 16 (S); Charge = +16 - 18 = -2; Mass number = 16 + 18 = 34; Symbol: ${}^{34}_{16}S^{2-}$

41.

Symbol	Number of protons in nucleus	Number of neutrons in nucleus	Number of electrons	Net charge
${}^{75}_{33}As^{3+}$	33	42	30	3+
${}^{128}_{52}Te^{2-}$	52	76	54	2-
${}^{32}_{16}S$	16	16	16	0
${}^{204}_{81}Tl^{+}$	81	123	80	1+
${}^{195}_{78}Pt$	78	117	78	0

43. Metals: Mg, Ti, Au, Bi, Ge, Eu, Am Non-metals: Si, B, At, Rn, Br

45. a and d

A group is a vertical column of elements in the periodic table. Elements in the same family (group) have similar chemical properties.

47. Carbon is a non-metal. Silicon and germanium are metalloids. Tin and lead are metals. Thus, metallic character increases as one goes down a family in the periodic table.

49. a. Na^+ lose one e^- d. I^- gain one e^-

b. Sr^{2+} lose two e^- e. Al^{3+} lose three e^-

c. Ba^{2+} lose two e^- f. S^{2-} gain two e^-

Nomenclature

51. a. chromium(VI) oxide d. selenium dioxide g. phosphorus trichloride
b. chromium(III) oxide e. selenium trioxide h. sulfur difluoride
c. aluminum oxide f. nitrogen triiodide

53. a. sodium chloride f. hydrogen iodide
b. magnesium chloride g. nitrogen monoxide or nitric oxide (common name)
c. rubidium bromide h. nitrogen trifluoride
d. cesium fluoride i. dinitrogen tetrafluoride k. silicon tetrafluoride
e. aluminum iodide j. dinitrogen dichloride l. dihydrogen selenide

55. a. nitric acid
b. nitrous acid
c. phosphoric acid
d. phosphorous acid
e. sodium hydrogen sulfate or sodium bisulfate (common name)
f. calcium hydrogen sulfite or calcium bisulfite (common name)
g. sodium bromate
h. iron(III) periodate
i. ruthenium(III) nitrate
j. vanadium(V) oxide
k. platinum(IV) chloride
l. iron(III) phosphate

57. a. CsBr
b. $BaSO_4$
c. NH_4Cl
d. ClO
e. $SiCl_4$
f. ClF_3
g. BeO
h. MgF_2

59. a. SF_2
b. SF_6
c. NaH_2PO_4
d. Li_3N
e. $Cr_2(CO_3)_3$
f. SnF_2
g. $(NH_4)(C_2H_3O_2)$
h. $(NH_4)(HSO_4)$
i. $Co(NO_3)_3$
j. Hg_2Cl_2: mercury(I) exists as Hg_2^{2+}
k. $KClO_3$
l. NaH

ADDITIONAL EXERCISES

61. There should be no difference. The composition of insulin from both sources will be the same and therefore, it will have the same activity regardless of the source. As a practical note, trace contaminants in the two types of insulin may be different. These trace components may be important.

63. a. $Pb(C_2H_3O_2)_2$: lead(II) acetate
b. $CuSO_4$: copper(II) sulfate
c. CaO: calcium oxide
d. $MgSO_4$: magnesium sulfate
e. $Mg(OH)_2$: magnesium hydroxide
f. $CaSO_4$: calcium sulfate
g. N_2O: dinitrogen monoxide or nitrous oxide

65. Yes, 1.0 g H would react with 37.0 g ^{37}Cl and 1.0 g H would react with 35.0 g ^{35}Cl.

No, the ratio of H/Cl would always be 1 g H/37 g Cl for ^{37}Cl and 1 g H/35 g Cl for ^{35}Cl.

As long as we had pure ^{35}Cl and pure ^{37}Cl the above ratios will always hold. If we have a mixture (such as the natural abundance of chlorine) the ratio will also be constant as long as the composition of the mixture of the two isotopes does not change.

CHALLENGE PROBLEMS

67. a. Ba^{2+} and O^{2-}: BaO, barium oxide
b. Li^+ and H^-: LiH, lithium hydride
c. In^{3+} and F^-: InF_3, indium(III) fluoride
d. As^{3+} and S^{2-}: As_2S_3, diarsenic trisulfide
e. B^{3+} and O^{2-}; B_2O_3, boron oxide or diboron trioxide

f. OF_2; oxygen difluoride

g. Al^{3+} and H^-; AlH_3, aluminum hydride

h. In^{3+} and P^{3-}; InP, indium(III) phosphide

69. The solid residue must have come from the flask.

71. The PO_4^{3-} ion is phosphate.

Na_3AsO_4 contain Na^+ ions and AsO_4^{3-} ions. So the name would be sodium arsenate.

H_3AsO_4: arsenic acid

$Mg_3(SbO_4)_2$: Mg^{2+} ions and SbO_4^{3-} ions, magnesium antimonate

CHAPTER THREE: STOICHIOMETRY

QUESTIONS

1. The two major isotopes of boron are ^{10}B and ^{11}B. The listed mass of 10.81 is the average mass of a very large number of boron atoms.

3. The molecular formula tells us the actual number of atoms of each element in a molecule (or formula unit) of a compound. The empirical formula tells only the simplest whole number ratio of atoms of each element in a molecule. The molecular formula is a whole number multiple of the empirical formula. If that multiplier is one, the molecular and empirical formulas are the same. For example, both the molecular and empirical formulas of water are H_2O. They are the same. For hydrogen peroxide, the empirical formula is OH; the molecular formula is H_2O_2.

EXERCISES

Atomic Masses and Mass Spectrometer

5. $$\begin{aligned} AW &= 0.7899(23.9850 \text{ amu}) + 0.1000(24.9858 \text{ amu}) + 0.1101(25.9826 \text{ amu}) \\ &= 18.95 \text{ amu} + 2.499 \text{ amu} + 2.861 \text{ amu} = 24.31 \text{ amu} \end{aligned}$$

7. Let x = % of ^{151}Eu and y = % of ^{153}Eu

$x + y = 100$ so $y = 100 - x$

$$151.96 = \frac{x(150.9196) + (100 - x)(152.9209)}{100}$$

$$15196 = 150.9196x + 15292.09 - 152.9209x$$

$$-96 = -2.0013x$$

$x = 48\%$; 48% ^{151}Eu and 52% ^{153}Eu

9. There are three peaks in the mass spectrum, each 2 mass units apart. This is consistent with two isotopes, differing in mass by two mass units. The peak at 157.84 corresponds to a Br_2 molecule composed of two atoms of the lighter isotope. This isotope has mass equal to 157.84/2 or 78.92. This corresponds to ^{79}Br. The second isotope is ^{81}Br with mass 161.84/2 = 80.92. The peaks in the mass spectrum correspond to $^{79}Br_2$, $^{79}Br^{81}Br$ and $^{81}Br_2$ in order of increasing mass. The intensities of the highest and lowest mass tell us the two isotopes are present at about equal abundance. The actual abundance is 50.69% ^{79}Br and 49.31% ^{81}Br. The calculation of the abundance from the mass spectrum is beyond the scope of this text.

11. $$\frac{9.123 \times 10^{-23} \text{ g}}{\text{atom}} \times \frac{6.022 \times 10^{23} \text{ atom}}{\text{mol}} = \frac{54.94 \text{ g}}{\text{mol}}$$

The atomic mass is 54.94. The element is manganese (Mn).

Moles and Molar Masses

13. a. $1.0 \text{ mol } NH_3 \times \frac{1 \text{ mol N}}{\text{mol } NH_3} = 1.0 \text{ mol N}$ $1.0 \text{ mol } NH_3 \times \frac{3 \text{ mol H}}{\text{mol } NH_3} = 3.0 \text{ mol H}$

b. $1.0 \text{ mol } N_2H_4 \times \frac{2 \text{ mol N}}{\text{mol } N_2H_4} = 2.0 \text{ mol N}$ $1.0 \text{ mol } N_2H_4 \times \frac{4 \text{ mol H}}{\text{mol } N_2H_4} = 4.0 \text{ mol H}$

c. $1.0 \text{ mol } (NH_4)_2Cr_2O_7 \times \frac{2 \text{ mol N}}{\text{mol } (NH_4)_2Cr_2O_7} = 2.0 \text{ mol N}$

$1.0 \text{ mol } (NH_4)_2Cr_2O_7 \times \frac{8 \text{ mol H}}{\text{mol } (NH_4)_2Cr_2O_7} = 8.0 \text{ mol H}$

$1.0 \text{ mol } (NH_4)_2Cr_2O_7 \times \frac{2 \text{ mol Cr}}{\text{mol } (NH_4)_2Cr_2O_7} = 2.0 \text{ mol Cr}$

$1.0 \text{ mol } (NH_4)_2Cr_2O_7 \times \frac{7 \text{ mol O}}{\text{mol } (NH_4)_2Cr_2O_7} = 7.0 \text{ mol O}$

15. a. 1.0 mol NH_3 contains 1.0 mol N and 3.0 mol H.

$1.0 \text{ mol N} \times \frac{6.02 \times 10^{23} \text{ atoms N}}{\text{mol N}} = 6.0 \times 10^{23} \text{ atoms N}$

$3.0 \text{ mol H} \times \frac{6.02 \times 10^{23} \text{ atoms H}}{\text{mol H}} = 1.8 \times 10^{24} \text{ atoms H}$

b. 1.0 mol N_2H_4 contains 2.0 mol N and 4.0 mol H.

$2.0 \text{ mol N} \times \frac{6.02 \times 10^{23} \text{ atoms N}}{\text{mol N}} = 1.2 \times 10^{24} \text{ atoms N}$

$4.0 \text{ mol H} \times \frac{6.02 \times 10^{23} \text{ atoms H}}{\text{mol H}} = 2.4 \times 10^{24} \text{ atoms H}$

c. 1.0 mol of $(NH_4)_2Cr_2O_7$ contains 2.0 mol N, 8.0 mol H, 2.0 mol Cr, and 7.0 mol O.

$2.0 \text{ mol N} \times \frac{6.02 \times 10^{23} \text{ atoms N}}{\text{mol N}} = 1.2 \times 10^{24} \text{ atoms N}$

$8.0 \text{ mol H} \times \frac{6.02 \times 10^{23} \text{ atoms H}}{\text{mol H}} = 4.8 \times 10^{24} \text{ atoms H}$

$2.0 \text{ mol Cr} \times \frac{6.02 \times 10^{23} \text{ atoms Cr}}{\text{mol Cr}} = 1.2 \times 10^{24} \text{ atoms Cr}$

$7.0 \text{ mol O} \times \frac{6.02 \times 10^{23} \text{ atoms O}}{\text{mol O}} = 4.2 \times 10^{24} \text{ atoms O}$

17. a. 1.0 mol NH_3 contains 1.0 mol N and 3.0 mol H.

$1.0 \text{ mol N} \times \frac{14.01 \text{ g N}}{\text{mol N}} = 14 \text{ g N}$

$3.0 \text{ mol H} \times \frac{1.008 \text{ g H}}{\text{mol H}} = 3.0 \text{ g H}$

b. 1.0 mol N_2H_4 contains 2.0 mol N and 4.0 mol H.

$$2.0 \text{ mol N} \times \frac{14.01 \text{ g N}}{\text{mol N}} = 28 \text{ g N}$$

$$4.0 \text{ mol H} \times \frac{1.008 \text{ g H}}{\text{mol H}} = 4.0 \text{ g H}$$

c. 1.0 mol $(NH_4)_2Cr_2O_7$ contains 2.0 mol N, 8.0 mol H, 2.0 mol Cr and 7.0 mol O.

$$2.0 \text{ mol N} \times \frac{14.01 \text{ g N}}{\text{mol N}} = 28 \text{ g N}$$

$$8.0 \text{ mol H} \times \frac{1.008 \text{ g H}}{\text{mol H}} = 8.0 \text{ g H}$$

$$2.0 \text{ mol Cr} \times \frac{52.00 \text{ g Cr}}{\text{mol Cr}} = 104 \text{ g Cr} = 1.0 \times 10^2 \text{ g Cr (2 S.F.)}$$

$$7.0 \text{ mol O} \times \frac{16.00 \text{ g O}}{\text{mol O}} = 112 \text{ g O} = 1.1 \times 10^2 \text{ g O (2 S.F.)}$$

19. a. For NH_3 the molar mass is 14.01 + 3(1.008) = 17.03 g/mol.

$$1.0 \text{ g } NH_3 \times \frac{1 \text{ mol } NH_3}{17.03 \text{ g } NH_3} \times \frac{1 \text{ mol N}}{\text{mol } NH_3} = 5.9 \times 10^{-2} \text{ mol N}$$

$$1.0 \text{ g } NH_3 \times \frac{1 \text{ mol } NH_3}{17.03 \text{ g } NH_3} \times \frac{3 \text{ mol H}}{\text{mol } NH_3} = 1.8 \times 10^{-1} \text{ mol H}$$

b. The molar mass of N_2H_4 is 2(14.01) + 4(1.008) = 32.05 g/mol.

$$1.0 \text{ g } N_2H_4 \times \frac{1 \text{ mol } N_2H_4}{32.05 \text{ g } N_2H_4} \times \frac{2 \text{ mol N}}{\text{mol } N_2H_4} = 6.2 \times 10^{-2} \text{ mol N}$$

$$1.0 \text{ g } N_2H_4 \times \frac{1 \text{ mol } N_2H_4}{32.05 \text{ g } N_2H_4} \times \frac{4 \text{ mol H}}{\text{mol } N_2H_4} = 1.2 \times 10^{-1} \text{ mol H}$$

c. The molar mass of $(NH_4)_2Cr_2O_7$ is 2(14.01) + 8(1.008) + 2(52.00) + 7(16.00) = 252.08 g/mol.

$$1.0 \text{ g } (NH_4)_2Cr_2O_7 \times \frac{1 \text{ mol } (NH_4)_2Cr_2O_7}{252.08 \text{ g}} \times \frac{2 \text{ mol N}}{\text{mol } (NH_4)_2Cr_2O_7} = 7.9 \times 10^{-3} \text{ mol N}$$

$$1.0 \text{ g } (NH_4)_2Cr_2O_7 \times \frac{1 \text{ mol } (NH_4)_2Cr_2O_7}{252.08 \text{ g}} \times \frac{8 \text{ mol H}}{\text{mol } (NH_4)_2Cr_2O_7} = 3.2 \times 10^{-2} \text{ mol H}$$

$$1.0 \text{ g } (NH_4)_2Cr_2O_7 \times \frac{1 \text{ mol } (NH_4)_2Cr_2O_7}{252.08 \text{ g}} \times \frac{2 \text{ mol Cr}}{\text{mol } (NH_4)_2Cr_2O_7} = 7.9 \times 10^{-3} \text{ mol Cr}$$

$$1.0 \text{ g } (NH_4)_2Cr_2O_7 \times \frac{1 \text{ mol } (NH_4)_2Cr_2O_7}{252.08 \text{ g}} \times \frac{7 \text{ mol O}}{\text{mol } (NH_4)_2Cr_2O_7} = 2.8 \times 10^{-2} \text{ mol O}$$

21. a. $1.0 \text{ g } NH_3 \times \frac{1 \text{ mol } NH_3}{17.03 \text{ g}} \times \frac{1 \text{ mol N}}{\text{mol } NH_3} \times \frac{6.02 \times 10^{23} \text{ atoms N}}{\text{mol N}} = 3.5 \times 10^{22} \text{ atoms N}$

The first two conversions are the same ones that we did in Exercise 3.19. So, we can take our answers from the previous exercise and convert the moles of each element in 1.0 g into the number of atoms of each element in 1.0 g.

$1.8 \times 10^{-1} \text{ mol H} \times \frac{6.02 \times 10^{23} \text{ atoms H}}{\text{mol H}} = 1.1 \times 10^{23} \text{ atoms H}$

b. 1.0 g of N_2H_4 contains (from Exercise 3.19):

$6.2 \times 10^{-2} \text{ mol N} \times \frac{6.02 \times 10^{23} \text{ atoms N}}{\text{mol N}} = 3.7 \times 10^{22} \text{ atoms N}$

$1.2 \times 10^{-1} \text{ mol H} \times \frac{6.02 \times 10^{23} \text{ atoms H}}{\text{mol H}} = 7.2 \times 10^{22} \text{ atoms H}$

But, by doing the complete calculation and carrying an extra significant figure:

$1.0 \text{ g } N_2H_4 \times \frac{1 \text{ mol } N_2H_4}{32.05 \text{ g}} \times \frac{4 \text{ mol H}}{\text{mol } N_2H_4} = 1.25 \times 10^{-1} \text{ mol H}$

$1.25 \times 10^{-1} \text{ mol H} \times \frac{6.02 \times 10^{23} \text{ atoms H}}{\text{mol H}} = 7.5 \times 10^{22} \text{ atoms H}$

c. 1.0 g of $(NH_4)_2Cr_2O_7$ contains:

$7.9 \times 10^{-3} \text{ mol N} \times \frac{6.02 \times 10^{23} \text{ atoms N}}{\text{mol N}} = 4.8 \times 10^{21} \text{ atoms N}$

$3.2 \times 10^{-2} \text{ mol H} \times \frac{6.02 \times 10^{23} \text{ atoms H}}{\text{mol H}} = 1.9 \times 10^{22} \text{ atoms H}$

$7.9 \times 10^{-3} \text{ mol Cr} \times \frac{6.02 \times 10^{23} \text{ atoms Cr}}{\text{mol Cr}} = 4.8 \times 10^{21} \text{ atoms Cr}$

$2.8 \times 10^{-2} \text{ mol O} \times \frac{6.02 \times 10^{23} \text{ atoms O}}{\text{mol O}} = 1.7 \times 10^{22} \text{ atoms O}$

23. Al_2O_3; 2(26.98) + 3(16.00) = 101.96 g/mol

Na_3AlF_6; 3(22.99) + 1(26.98) + 6(19.00) = 209.95 g/mol

25. Molar mass of $C_6H_8O_6$ = 6(12.01) + 8(1.008) + 6(16.00) = 176.12 g/mol

$500.0 \text{ mg} \times \frac{1 \text{ g}}{1000 \text{ mg}} \times \frac{1 \text{ mol}}{176.12 \text{ g}} = 2.839 \times 10^{-3} \text{ mol}$

$2.839 \times 10^{-3} \text{ mol} \times \frac{6.022 \times 10^{23} \text{ molecules}}{\text{mol}} = 1.710 \times 10^{21} \text{ molecules}$

27. a. $100 \text{ molecules } H_2O \times \frac{1 \text{ mol}}{6.022 \times 10^{23} \text{ molecules}} = 1.661 \times 10^{-22} \text{ mol}$

b. $100.0 \text{ g } H_2O \times \frac{1 \text{ mol } H_2O}{18.02 \text{ g } H_2O} = 5.549 \text{ mol}$

c. $500 \text{ atoms Fe} \times \frac{1 \text{ mol Fe}}{6.022 \times 10^{23} \text{ atoms}} = 8.303 \times 10^{-22} \text{ mol}$

d. $500.0 \text{ g Fe} \times \frac{1 \text{ mol Fe}}{55.85 \text{ g Fe}} = 8.953 \text{ mol}$

e. $150 \text{ molecules } O_2 \times \frac{1 \text{ mol } O_2}{6.022 \times 10^{23} \text{ molecules } O_2} = 2.491 \times 10^{-22} \text{ mol}$

29. a. $3.00 \times 10^{20} \text{ molecules } N_2 \times \frac{1 \text{ mol } N_2}{6.022 \times 10^{23} \text{ molecules}} \times \frac{28.02 \text{ g } N_2}{\text{mol } N_2} = 1.40 \times 10^{-2} \text{ g } N_2$

b. $3.00 \times 10^{-3} \text{ mol } N_2 \times \frac{28.02 \text{ g } N_2}{\text{mol } N_2} = 8.41 \times 10^{-2} \text{ g } N_2$

c. $1.5 \times 10^2 \text{ mol } N_2 \times \frac{28.02 \text{ g } N_2}{\text{mol } N_2} = 4.2 \times 10^3 \text{ g } N_2$

d. $1 \text{ molecule } N_2 \times \frac{1 \text{ mol } N_2}{6.022 \times 10^{23} \text{ molecules } N_2} \times \frac{28.02 \text{ g } N_2}{\text{mol } N_2} = 4.653 \times 10^{-23} \text{ g } N_2$

e. $2.00 \times 10^{-15} \text{ mol } N_2 \times \frac{28.02 \text{ g } N_2}{\text{mol } N_2} = 5.60 \times 10^{-14} \text{ g } N_2 = 56.0 \text{ fg } N_2$

f. $18.0 \text{ pmol } N_2 \times \frac{1 \text{ mol } N_2}{10^{12} \text{ pmol}} \times \frac{28.02 \text{ g } N_2}{\text{mol } N_2} = 5.04 \times 10^{-10} \text{ g } N_2 = 504 \text{ pg } N_2$

g. $5.0 \text{ nmol } N_2 \times \frac{1 \text{ mol } N_2}{10^9 \text{ nmol}} \times \frac{28.02 \text{ g } N_2}{\text{mol } N_2} = 1.4 \times 10^{-7} \text{ g } N_2 = 140 \text{ ng}$

31. a. $14 \text{ mol C} \left(\frac{12.01 \text{ g}}{\text{mol C}}\right) + 18 \text{ mol H} \left(\frac{1.008 \text{ g}}{\text{mol H}}\right) + 2 \text{ mol N} \left(\frac{14.01 \text{ g}}{\text{mol N}}\right) + 5 \text{ mol O} \left(\frac{16.00 \text{ g}}{\text{mol O}}\right)$

$= 294.30 \text{ g/mol}$

b. $10.0 \text{ g aspartame} \times \frac{1 \text{ mol}}{294.30 \text{ g}} = 3.40 \times 10^{-2} \text{ mol}$

c. $1.56 \text{ mol} \times \frac{294.30 \text{ g}}{\text{mol}} = 459 \text{ g}$

d. $5.0 \text{ mg} \times \frac{1 \text{ g}}{1000 \text{ mg}} \times \frac{1 \text{ mol}}{294.30 \text{ g}} \times \frac{6.02 \times 10^{23} \text{ molecules}}{\text{mol}} = 1.0 \times 10^{19} \text{ molecules}$

e. $1.2 \text{ g aspartame} \times \frac{1 \text{ mol aspartame}}{294.30 \text{ g aspartame}} \times \frac{2 \text{ mol N}}{\text{mol aspartame}} \times \frac{6.02 \times 10^{23} \text{ atoms N}}{\text{mol N}}$

$= 4.9 \times 10^{21} \text{ atoms of nitrogen}$

f. $1.0 \times 10^9 \text{ molecules} \times \frac{1 \text{ mol}}{6.02 \times 10^{23} \text{ molecules}} \times \frac{294.30 \text{ g}}{\text{mol}} = 4.9 \times 10^{-13} \text{ g or 490 fg}$

g. $1 \text{ molecule aspartame} \times \frac{1 \text{ mol}}{6.022 \times 10^{23} \text{ molecules}} \times \frac{294.30 \text{ g}}{\text{mol}} = 4.887 \times 10^{-22} \text{ g}$

Percent Composition

33. From Exercise 2.15

i. 1.188 g for every 1.000 g S

$\%F = \frac{1.188 \text{ g}}{1.188 \text{ g} + 1.000 \text{ g}} \times 100 = 54.30\% \text{ F};$ $\%S = 100.00 - 54.30 = 45.70\% \text{ S}$

ii. 2.375 g F for every 1.000 g S

$\%F = \frac{2.375 \text{ g}}{3.375 \text{ g}} \times 100 = 70.37\% \text{ F};$ $\%S = 100.00 - 70.37 = 29.63\% \text{ S}$

iii. 3.563 g F for every 1.000 g S

$\%F = \frac{3.563 \text{ g}}{4.563 \text{ g}} \times 100 = 78.08\% \text{ F};$ $\%S = 100.00 - 78.08 = 21.92\% \text{ S}$

35. In 1 mole of $YBa_2Cu_3O_7$ there are 1 mole of Y, 2 moles of Ba, 3 moles of Cu, and 7 moles of O.

$$\text{Molar mass} = 1 \text{ mol Y}\left(\frac{88.91 \text{ g Y}}{\text{mol Y}}\right) + 2 \text{ mol Ba}\left(\frac{137.3 \text{ g Ba}}{\text{mol Ba}}\right)$$

$$+ 3 \text{ mol Cu}\left(\frac{63.55 \text{ g Cu}}{\text{mol Cu}}\right) + 7 \text{ mol O}\left(\frac{16.00 \text{ g O}}{\text{mol O}}\right)$$

Molar mass = 88.91 + 274.6 + 190.65 + 112.00

Molar mass = 666.2 g/mol

$\%Y = \frac{88.91 \text{ g}}{666.2 \text{ g}} \times 100 = 13.35\% \text{ Y};$ $\%Ba = \frac{274.6 \text{ g}}{666.2 \text{ g}} \times 100 = 41.22\% \text{ Ba}$

$\%Cu = \frac{190.65 \text{ g}}{666.2 \text{ g}} \times 100 = 28.62\% \text{ Cu};$ $\%O = \frac{112.0 \text{ g}}{666.2 \text{ g}} \times 100 = 16.81\% \text{ O}$

or %O = 100.00 - (13.35 + 41.22 + 28.62) = 100.00 - (83.19) = 16.81% O

37. Na_3PO_4: molar mass = 3(22.99) + 30.97 + 4(16.00) = 163.94 g/mol

$\%P = \frac{30.97 \text{ g}}{163.94 \text{ g}} \times 100 = 18.89\% \text{ P}$

PH_3 : molar mass = 30.97 + 3(1.008) = 33.99 g/mol

$\%P = \frac{30.97 \text{ g}}{33.99 \text{ g}} \times 100 = 91.12\% \text{ P}$

P_4O_{10}: molar mass = 4(30.97) + 10(16.00) = 283.88 g/mol

$\%P = \frac{123.88 \text{ g}}{283.88 \text{ g}} \times 100 = 43.638\%$

$(NPCl_2)_3$: molar mass = 3(14.01) + 3(30.97) + 6(35.45) = 347.64 g/mol

$\%P = \frac{92.91 \text{ g}}{347.64 \text{ g}} \times 100 = 26.73\%$

The order from lowest to highest percentage of phosphorus is:

$$Na_3PO_4 < (NPCl_2)_3 < P_4O_{10} < PH_3$$

39. $\%Co = 4.34 = \frac{58.93 \text{ g Co}}{\text{molar mass}} \times 100$; molar mass = 1360 g/mol

41. Cds: $\%Cd = \frac{112.4 \text{ g Cd}}{144.5 \text{ g CdS}} \times 100 = 77.79\% \text{ Cd}$

CdSe: $\%Cd = \frac{112.4 \text{ g}}{191.4 \text{ g}} \times 100 = 58.73\% \text{ Cd}$

CdTe: $\%Cd = \frac{112.4 \text{ g}}{240.0 \text{ g}} \times 100 = 46.83\% \text{ Cd}$

43. a. $C_3H_4O_2$: Molar mass = 3(12.01) + 4(1.008) + 2(16.00) = 36.03 + 4.032 + 32.00 = 72.06 g/mol

$\%C = \frac{36.03 \text{ g C}}{72.06 \text{ g compound}} \times 100 = 50.00\% \text{ C}$

$\%H = \frac{4.032 \text{ g H}}{72.06 \text{ g compound}} \times 100 = 5.595\% \text{ H}$

$\%O = 100.00 - (50.00 + 5.595) = 44.41\% \text{ O}$

or $\%O = \frac{32.00 \text{ g}}{72.06 \text{ g}} \times 100 = 44.41\% \text{ O}$

b. $C_4H_6O_2$: Molar mass = 4(12.01) + 6(1.008) + 2(16.00) = 48.04 + 6.048 + 32.00 = 86.09 g/mol

$\%C = \frac{48.04 \text{ g}}{86.09 \text{ g}} \times 100 = 55.80\% \text{ C}$

$\%H = \frac{6.048 \text{ g}}{86.09 \text{ g}} \times 100 = 7.025\% \text{ H}$

$\%O = 100.00 - (55.80 + 7.025) = 37.18\% \text{ O}$

c. C_3H_3N: Molar mass = 3(12.01) + 3(1.008) + 1(14.01) = 36.03 + 3.024 + 14.01 = 53.06 g/mol

$\%C = \frac{36.03 \text{ g}}{53.06 \text{ g}} \times 100 = 67.90\% \text{ C}$; $\%H = \frac{3.024 \text{ g}}{53.06 \text{ g}} \times 100 = 5.699\% \text{ H}$

$\%N = \frac{14.01 \text{ g}}{53.06 \text{ g}} \times 100 = 26.40\% \text{ N}$; or $\%N = 100.00 - (67.90 + 5.699) = 26.40\%N$

Empirical and Molecular Formulas

45. From Exercise 2.15 we get

i. 1.188 g F for every 1.000 g S

$$1.188 \text{ g F} \times \frac{1 \text{ mol F}}{19.00 \text{ g F}} = 6.253 \times 10^{-2} \text{ mol F}$$

$$1.000 \text{ g S} \times \frac{1 \text{ mol S}}{32.07 \text{ g S}} = 3.118 \times 10^{-2} \text{ mol S}$$

$$\frac{6.253 \times 10^{-2} \text{ mol F}}{3.118 \times 10^{-2} \text{ mol S}} = \frac{2.005 \text{ mol F}}{\text{mol S}}; \text{ Empirical formula: } SF_2$$

ii. $$2.375 \text{ g F} \times \frac{1 \text{ mol F}}{19.00 \text{ g F}} = 1.250 \times 10^{-1} \text{ mol F}$$

$$\frac{1.250 \times 10^{-1} \text{ mol F}}{3.118 \times 10^{-2} \text{ mol S}} = \frac{4 \text{ mol F}}{\text{mol S}}; \text{ Empirical formula: } SF_4$$

iii. $$3.563 \text{ g F} \times \frac{1 \text{ mol F}}{19.00 \text{ g F}} = 1.875 \times 10^{-1} \text{ mol F}$$

$$\frac{1.875 \times 10^{-1} \text{ mol F}}{3.118 \times 10^{-2} \text{ mol S}} = \frac{6 \text{ mol F}}{\text{mol S}}; \text{ Empirical formula: } SF_6$$

47. a. SNH: Mass of a single SNH is 32.07 + 14.01 + 1.008 = 47.09 g

$\frac{188.35}{47.09} = 4$; So the molecular formula is $(SNH)_4$ or $S_4N_4H_4$.

b. $NPCl_2$: Mass of an $NPCl_2$ unit is 14.01 + 30.97 + 2(35.45) = 115.88 g

$\frac{347.64}{115.88} = 3$; Molecular formula is $(NPCl_2)_3$ or $N_3P_3Cl_6$.

c. CoC_4O_4: 58.93 + 4(12.01) + 4(16.00) = 170.97 g

$\frac{341.94}{170.97} = 2$; Molecular formula: $Co_2C_8O_8$

d. SN: 32.07 + 14.01 = 46.08 g

$\frac{184.32}{46.08} = 4$; Molecular formula: S_4N_4

49. a. Molar mass of CH_2O = $1 \text{ mol C}\left(\frac{12.01 \text{ g}}{\text{mol C}}\right) + 2 \text{ mol H}\left(\frac{1.008 \text{ g H}}{\text{mol H}}\right)$

$$+ 1 \text{ mol O}\left(\frac{16.00 \text{ g}}{\text{mol O}}\right) = 30.03 \text{ g/mol}$$

$$\%C = \frac{12.01 \text{ g C}}{30.03 \text{ g } CH_2O} \times 100 = 39.99\% \text{ C}$$

$$\%H = \frac{2.016 \text{ g H}}{30.03 \text{ g } CH_2O} \times 100 = 6.713\% \text{ H}$$

$$\%O = \frac{16.00 \text{ g O}}{30.03 \text{ g } CH_2O} \times 100 = 53.28\% \text{ O} \qquad \text{or } \%O = 100 - (39.99 + 6.713) = 53.30\%$$

b. Molar mass of $C_6H_{12}O_6$ = 6(12.01) + 12(1.008) + 6(16.00) = 180.16 g/mol

$$\%C = \frac{72.06 \text{ g C}}{180.16 \text{ g } C_6H_{12}O_6} \times 100 = 40.00\% \qquad \%H = \frac{12.096 \text{ g}}{180.16 \text{ g}} \times 100 = 6.7140\%$$

%O = 100 - (40.00 + 6.714) = 53.29%

c. Molar mass of $HC_2H_3O_2$ = 2(12.01) + 4(1.008) + 2(16.00) = 60.05 g/mol

$$\%C = \frac{24.02 \text{ g}}{60.05 \text{ g}} \times 100 = 40.00\% \quad \%H = \frac{4.032 \text{ g}}{60.05 \text{ g}} \times 100 = 6.714\%$$

%O = 100 - (40.00 + 6.714) = 53.29%

51. Out of 100 g of compound, there are:

$$48.38 \text{ g C} \times \frac{1 \text{ mol C}}{12.01 \text{ g C}} = 4.028 \text{ mol C}$$

$$8.12 \text{ g H} \times \frac{1 \text{ mol H}}{1.008 \text{ g H}} = 8.06 \text{ mol H}$$

%O = 100 - 48.38 - 8.12 = 43.50%

$$43.50 \text{ g O} \times \frac{1 \text{ mol O}}{16.00 \text{ g O}} = 2.719 \text{ mol O}$$

Look for common ratios.

$$\frac{4.028}{2.719} = 1.481 \approx \frac{3}{2}; \text{ So we try } \frac{2.719}{2} = 1.360$$

$$\frac{4.028}{1.36} \approx 3 \quad \frac{8.06}{1.36} \approx 6 \quad \frac{2.719}{1.36} \approx 2$$

So empirical formula is $C_3H_6O_2$.

53. $$0.979 \text{ g Na} \times \frac{1 \text{ mol Na}}{22.99 \text{ g Na}} = 4.26 \times 10^{-2} \text{ mol Na}$$

$$1.365 \text{ g S} \times \frac{1 \text{ mol S}}{32.07 \text{ g S}} = 4.256 \times 10^{-2} \text{ mol S}, \frac{\text{mol S}}{\text{mol Na}} = 1.00$$

$$1.021 \text{ g O} \times \frac{1 \text{ mol O}}{16.00 \text{ g O}} = 6.381 \times 10^{-2} \text{ mol O}$$

$$\frac{6.381 \times 10^{-2} \text{ mol O}}{4.256 \times 10^{-2} \text{ mol S}} = 1.499 \approx \frac{1.5 \text{ mol O}}{\text{mol S}} = \frac{3 \text{ mol O}}{2 \text{ mol S}}$$

Empirical formula: $Na_2S_2O_3$

55. Out of 100 g: $30.4 \text{ g N} \times \frac{1 \text{ mol N}}{14.01 \text{ g N}} = 2.17 \text{ mol N}$

$\%O = 100 - 30.4 = 69.6\%$ O; $69.6 \text{ g O} \times \frac{1 \text{ mol O}}{16.00 \text{ g O}} = 4.35 \text{ mol O}$

$\frac{2.17}{2.17} = 1$ $\frac{4.35}{2.17} = 2$; Empirical formula is NO_2

NO_2 unit has a mass of $14 + 2(16) \sim 46$ g

$\frac{92 \text{ g}}{46 \text{ g}} = 2$; Therefore, the molecular formula is N_2O_4.

57. First, get composition in mass percent.

We assume all of the carbon in 0.213 g CO_2 came from 0.157 g of the compound and that all of the hydrogen in the 0.0310 g H_2O came from the 0.157 g of the compound.

$$0.213 \text{ g } CO_2 \times \frac{12.01 \text{ g C}}{44.01 \text{ g } CO_2} = 0.0581 \text{ g C}$$

$$\%C = \frac{0.0581 \text{ g C}}{0.157 \text{ g compound}} \times 100 = 37.0\% \text{ C}$$

$$0.0310 \text{ g } H_2O \times \frac{2.016 \text{ g H}}{18.02 \text{ g } H_2O} = 3.47 \times 10^{-3} \text{ g H}$$

$$\%H = \frac{3.47 \times 10^{-3} \text{ g}}{0.157 \text{ g}} = 2.21\% \text{ H}$$

We get %N from the second experiment:

$$0.0230 \text{ g } NH_3 \times \frac{14.01 \text{ g N}}{17.03 \text{ g } NH_3} = 1.89 \times 10^{-2} \text{ g N}$$

$$\%N = \frac{1.89 \times 10^{-2} \text{ g}}{0.103 \text{ g}} \times 100 = 18.3\% \text{ N}$$

The mass percent of oxygen is obtained by difference:

$$\%O = 100 - (37.0 + 2.21 + 18.3) = 42.5\%$$

So out of 100 g of compound, there are:

$$37.0 \text{ g C} \times \frac{1 \text{ mol C}}{12.01 \text{ g C}} = 3.08 \text{ mol C}; \quad 2.21 \text{ g H} \times \frac{1 \text{ mol H}}{1.008 \text{ g H}} = 2.19 \text{ mol H}$$

$$18.3 \text{ g N} \times \frac{1 \text{ mol N}}{14.01 \text{ g N}} = 1.31 \text{ mol N}; \quad 42.5 \text{ g O} \times \frac{1 \text{ mol O}}{16.00 \text{ g O}} = 2.66 \text{ mol O}$$

Lastly, and often the hardest part, we need to find simple whole number ratios. We do this by trial and error.

$$\frac{2.19}{1.31} = 1.67 \approx 1\frac{2}{3} \approx \frac{5}{3}$$

So we try $\frac{1.31}{3} = 0.437$ as lowest common denominator.

$$\frac{3.08}{0.437} \approx 7 \quad \frac{2.19}{0.437} \approx 5 \quad \frac{1.31}{0.437} = 3 \quad \frac{2.66}{0.437} \approx 6$$

Empirical formula: $C_7H_5N_3O_6$.

Balancing Chemical Equations

59. a. $Fe_2O_3 + CO \rightarrow Fe_3O_4 + CO_2$ Balance Fe atoms first:

$3\ Fe_2O_3(s) + CO(g) \rightarrow 2\ Fe_3O_4(s) + CO_2(g)$

The equation is balanced (1-C and 10-O atoms on each side).

b. $Fe_2O_4 + CO \rightarrow FeO + CO_2$ Balance Fe atoms first:

$Fe_3O_4(s) + CO(g) \rightarrow 3\ FeO(s) + CO_2(g)$

The equation is balanced (1-C and 5-O atoms on each side).

61. a. $C_{12}H_{22}O_{11}(s) + 12\ O_2(g) \rightarrow 12\ CO_2(g) + 11\ H_2O(g)$

b. $C_6H_6(l) + \frac{15}{2}\ O_2(g) \rightarrow 6\ CO_2(g) + 3\ H_2O(g)$

or $2\ C_6H_6(l) + 15\ O_2(g) \rightarrow 12\ CO_2(g) + 6\ H_2O(g)$

c. $2\ Fe + \frac{3}{2}\ O_2 \rightarrow Fe_2O_3$

or $4\ Fe(s) + 3\ O_2(g) \rightarrow 2\ Fe_2O_3(s)$

d. $C_4H_{10} + \frac{13}{2}\ O_2 \rightarrow 4\ CO_2 + 5\ H_2O$

or $2\ C_4H_{10}(g) + 13\ O_2(g) \rightarrow 8\ CO_2(g) + 10\ H_2O(g)$

e. $2\ FeO(s) + \frac{1}{2}\ O_2(g) \rightarrow Fe_2O_3(s)$

or $4\ FeO(s) + O_2(g) \rightarrow 2\ Fe_2O_3(s)$

63. a. $Cu(s) + 2\ AgNO_3(aq) \rightarrow 2\ Ag(s) + Cu(NO_3)_2(aq)$

b. $Zn(s) + 2\ HCl(aq) \rightarrow ZnCl_2(aq) + H_2(g)$

c. $Au_2S_3(s) + 3\ H_2(g) \rightarrow 2\ Au(s) + 3\ H_2S(g)$

d. $Ca(s) + 2\ H_2O(l) \rightarrow Ca(OH)_2(aq) + H_2(g)$

65. a. $SiO_2(s) + C(s) \rightarrow Si(s) + CO(g)$

Balance oxygen atoms

$SiO_2 + C \rightarrow Si + 2\ CO$

Balance carbon atoms

$SiO_2(s) + 2\ C(s) \rightarrow Si(s) + 2\ CO(g)$

b. $SiCl_4(l) + Mg(s) \rightarrow Si(s) + MgCl_2(s)$

Balance Cl atoms: $SiCl_4 + Mg \rightarrow Si + 2\ MgCl_2$

Balance Mg atoms: $SiCl_4(l) + 2\ Mg(s) \rightarrow Si(s) + 2\ MgCl_2(s)$

c. $Na_2SiF_6(s) + Na(s) \rightarrow Si(s) + NaF(s)$

Balance F atoms: $Na_2SiF_6 + Na \rightarrow Si + 6\ NaF$

Balance Na atoms: $Na_2SiF_6(s) + 4\ Na(s) \rightarrow Si(s) + 6\ NaF(s)$

67. $Pb(NO_3)_2(aq) + H_3AsO_4(aq) \rightarrow PbHAsO_4(s) + 2\ HNO_3(aq)$

Note: The insecticide used is $PbHAsO_4$ and is commonly called lead arsenate. This is not the correct name, however. Correctly, lead arsenate would be $Pb_3(AsO_4)_2$ and $PbHAsO_4$ should be called lead hydrogen arsenate.

Reaction Stoichiometry

69. The balanced reaction is:

$(NH_4)_2Cr_2O_7(s) \rightarrow Cr_2O_3(s) + N_2(g) + 4\ H_2O(g)$

$$10.8\text{ g }(NH_4)_2Cr_2O_7 \times \frac{1\text{ mol }(NH_4)_2Cr_2O_7}{252.08\text{ g}} = 4.28 \times 10^{-2}\text{ mol }(NH_4)_2Cr_2O_7$$

$$4.28 \times 10^{-2}\text{ mol }(NH_4)_2Cr_2O_7 \times \frac{1\text{ mol }Cr_2O_3}{\text{mol }(NH_4)_2Cr_2O_7} \times \frac{152.00\text{ g }Cr_2O_3}{\text{mol }Cr_2O_3} = 6.51\text{ g }Cr_2O_3$$

$$4.28 \times 10^{-2}\text{ mol }(NH_4)_2Cr_2O_7 \times \frac{1\text{ mol }N_2}{\text{mol }(NH_4)_2Cr_2O_7} \times \frac{28.02\text{ g }N_2}{\text{mol }N_2} = 1.20\text{ g }N_2$$

$$4.28 \times 10^{-2}\text{ mol }(NH_4)_2Cr_2O_7 \times \frac{4\text{ mol }H_2O}{\text{mol }(NH_4)_2Cr_2O_7} \times \frac{18.02\text{ g }H_2O}{\text{mol }H_2O} = 3.09\text{ g }H_2O$$

71. $$1.0\text{ kg Al} \times \frac{1000\text{ g Al}}{\text{kg Al}} \times \frac{1\text{ mol Al}}{26.98\text{ g Al}} \times \frac{3\text{ mol }NH_4ClO_4}{3\text{ mol Al}} \times \frac{117.49\text{ g }NH_4ClO_4}{\text{mol }NH_4ClO_4}$$

= 4355 g or 4.4 kg

73. $1.0 \times 10^2 \text{ g } Ca_3(PO_4)_2 \times \frac{1 \text{ mol } Ca_3(PO_4)_2}{310.18 \text{ g } Ca_3(PO_4)_2} \times \frac{3 \text{ mol } H_2SO_4}{\text{mol } Ca_3(PO_4)_2} \times \frac{98.09 \text{ g } H_2SO_4}{\text{mol } H_2SO_4}$

$= 95 \text{ g } H_2SO_4$ are needed

$95 \text{ g } H_2SO_4 \times \frac{100 \text{ g concentrated reagent}}{98 \text{ g } H_2SO_4} = 97$ g of concentrated sulfuric acid

75. a. $1.0 \times 10^2 \text{ mg } NaHCO_3 \times \frac{1 \text{ g}}{1000 \text{ mg}} \times \frac{1 \text{ mol } NaHCO_3}{84.01 \text{ g } NaHCO_3} \times \frac{1 \text{ mol } C_6H_8O_7}{3 \text{ mol } NaHCO_3}$

$\times \frac{192.12 \text{ g } C_6H_8O_7}{\text{mol } C_6H_8O_7} = 0.076$ g or 76 mg $C_6H_8O_7$

b. $0.10 \text{ g } NaHCO_3 \times \frac{1 \text{ mol } NaHCO_3}{84.01 \text{ g } NaHCO_3} \times \frac{3 \text{ mol } CO_2}{3 \text{ mol } NaHCO_3} \times \frac{44.01 \text{ g } CO_2}{\text{mol } CO_2}$

$= 0.052$ g or 52 mg CO_2

Limiting Reactants and Percent Yield

77. a. $Mg(s) + I_2(s) \rightarrow MgI_2(s)$

100 molecules of I_2 reacts completely with 100 atoms of Mg. We have a stoichiometric mixture. Neither is limiting.

b. $150 \text{ atoms Mg} \times \frac{1 \text{ molecule } I_2}{1 \text{ atom Mg}} = 150$ molecules I_2 needed.

We need 150 molecules I_2 to react completely with 150 atoms Mg; we only have 100. I_2 is limiting.

c. $200 \text{ atoms Mg} \times \frac{1 \text{ molecule } I_2}{1 \text{ atom Mg}} = 200$ molecules I_2. Mg is limiting.

d. $0.16 \text{ mol Mg} \times \frac{1 \text{ mol } I_2}{1 \text{ mol Mg}} = 0.16$ mol I_2 needed. Mg is limiting.

e. $0.14 \text{ mol Mg} \times \frac{1 \text{ mol } I_2}{1 \text{ mol Mg}} = 0.14$ mol I_2 needed.

Stoichiometric mixture. Neither is limiting.

f. $0.12 \text{ mol Mg} \times \frac{1 \text{ mol } I_2}{1 \text{ mol Mg}} = 0.12$ mol I_2 needed. I_2 is limiting.

g. $6.078 \text{ g Mg} \times \frac{1 \text{ mol Mg}}{24.31 \text{ g Mg}} \times \frac{1 \text{ mol } I_2}{1 \text{ mol Mg}} \times \frac{253.8 \text{ g } I_2}{\text{mol } I_2} = 63.46 \text{ g } I_2$.

Stoichiometric mixture. Neither is limiting.

h. $1.00 \text{ g Mg} \times \frac{1 \text{ mol Mg}}{24.31 \text{ g Mg}} \times \frac{1 \text{ mol } I_2}{1 \text{ mol Mg}} \times \frac{253.8 \text{ g } I_2}{\text{mol } I_2} = 10.4 \text{ g } I_2$ needed.

10.4 g I_2 needed, we only have 2.00 g. I_2 is limiting.

i. From part h. above, we calculated that 10.4 g I_2 will react completely with 1.00 g Mg. We have 20.00 g I_2. I_2 is in excess. Mg is limiting.

79. $2\,Cu(s) + S(s) \rightarrow Cu_2S(s)$

$1.50 \text{ g Cu} \times \frac{1 \text{ mol Cu}}{63.55 \text{ g Cu}} \times \frac{1 \text{ mol } Cu_2S}{2 \text{ mol Cu}} \times \frac{159.17 \text{ g } Cu_2S}{\text{mol } Cu_2S} = 1.88 \text{ g } Cu_2S$ is theoretical yield.

$\% \text{ yield} = \frac{\text{actual yield}}{\text{theoretical yield}} \times 100 = \frac{1.76 \text{ g}}{1.88 \text{ g}} \times 100 = 93.6\%$

81. a. $2.0 \text{ g Ag} \times \frac{1 \text{ mol Ag}}{107.9 \text{ g Ag}} \times \frac{8 \text{ mol } Ag_2S}{16 \text{ mol Ag}} = 9.3 \times 10^{-3} \text{ mol } Ag_2S$

$2.0 \text{ g } S_8 \times \frac{1 \text{ mol } S_8}{256.56 \text{ g } S_8} \times \frac{8 \text{ mol } Ag_2S}{\text{mol } S_8} = 6.2 \times 10^{-2} \text{ mol } Ag_2S$

When 9.3×10^{-3} mol Ag_2S is formed all of the Ag will be consumed. Ag is limiting.

$9.3 \times 10^{-3} \text{ mol } Ag_2S \times 247.9 \text{ g/mol} = 2.3 \text{ g } Ag_2S$

b. S_8 is left unreacted. 0.3 g S (2.3 g Ag_2S - 2.0 g Ag) is required to make 2.3 g of Ag_2S. Therefore, 1.7 g S_8 remains.

83. a. $C_8H_4O_3 + C_6H_6 \rightarrow C_{14}H_8O_2 + H_2O$

$2.00 \times 10^3 \text{ g } C_8H_4O_3 \times \frac{1 \text{ mol } C_8H_4O_3}{148.11 \text{ g } C_8H_4O_3} \times \frac{1 \text{ mol } C_6H_6}{\text{mol } C_8H_8O_3} \times \frac{78.11 \text{ g } C_6H_6}{\text{mol } C_6H_6} = 1050 \text{ g } C_6H_6$

b. $2.00 \times 10^3 \text{ g } C_8H_4O_3 \times \frac{1 \text{ mol } C_8H_4O_3}{148.11 \text{ g } C_8H_4O_3} \times \frac{1 \text{ mol } C_{14}H_8O_2}{\text{mol } C_8H_4O_3} \times \frac{208.20 \text{ g } C_{14}H_8O_2}{\text{mol } C_{14}H_8O_2}$

$= 2810 \text{ g } C_{14}H_8O_2$

$2.00 \times 10^3 \text{ g } C_8H_4O_3 \times \frac{1 \text{ mol } C_8H_4O_3}{148.11 \text{ g } C_8H_4O_3} \times \frac{1 \text{ mol } H_2O}{\text{mol } C_8H_4O_3} \times \frac{18.02 \text{ g } H_2O}{\text{mol } H_2O} = 243 \text{ g } H_2O$

c. $\%\text{Yield} = \frac{\text{Actual Yield}}{\text{Theoretical Yield}} \times 100 = \frac{1960 \text{ g}}{2810 \text{ g}} \times 100 = 69.8\%$

ADDITIONAL EXERCISES

85. a. $Cu(H_2O)_6Cl_2$; 1(63.55) + 12(1.008 g) + 6(16.00) + 2(35.45) = 242.55 g/mol

b. $NaBrO_3$; 1(22.99) + 1(79.90) + 3(16.00) = 150.89 g/mol

c. $(C_2F_4)_{500}$; 1000(12.01) + 2000(19.00) = 5×10^4 g/mol

d. H_3PO_4; 3(1.008) + 1(30.97) + 4(16.00) = 97.99 g/mol

e. $CaCO_3$; 1(40.08) + 1(12.01) + 3(16.00) = 100.09 g/mol

87. Mass of repeating unit is:

$$8 \text{ mol C}\left(\frac{12 \text{ g}}{\text{mol C}}\right) + 14 \text{ mol H}\left(\frac{1 \text{ g}}{\text{mol H}}\right) + 2 \text{ mol O}\left(\frac{16 \text{ g}}{\text{mol O}}\right) \approx 142 \text{ g}$$

$$\frac{100{,}000}{142} = 704 \text{ or about 700 monomer units}$$

89. $C_8H_{18}(l) + \frac{25}{2}\, O_2(g) \longrightarrow 8\, CO_2(g) + 9\, H_2O(g)$

or $2\, C_8H_{18}(l) + 25\, O_2(g) \longrightarrow 16\, CO_2(g) + 18\, H_2O(g)$

$$1.2 \times 10^{10} \text{ gallon} \times \frac{4 \text{ qt}}{\text{gal}} \times \frac{946 \text{ mL}}{\text{qt}} \times \frac{0.692 \text{ g}}{\text{mL}} = 3.1 \times 10^{13} \text{ g of gasoline}$$

$$3.1 \times 10^{13} \text{ g } C_8H_{18} \times \frac{1 \text{ mol } C_8H_{18}}{114.22 \text{ g } C_8H_{18}} \times \frac{16 \text{ mol } CO_2}{2 \text{ mol } C_8H_{18}} \times \frac{44.01 \text{ g } CO_2}{\text{mol } CO_2} = 9.6 \times 10^{13} \text{ g } CO_2$$

91. $156.8 \text{ mg } CO_2 \times \frac{12.01 \text{ mg C}}{44.01 \text{ mg } CO_2} = 42.79 \text{ mg C}$

$$42.8 \text{ mg } H_2O \times \frac{2.016 \text{ mg H}}{18.02 \text{ mg } H_2O} = 4.79 \text{ mg H}$$

$$\%C = \frac{42.79 \text{ mg}}{47.6 \text{ mg}} \times 100 = 89.9\% \text{ C}; \quad \%H = 100.0 - 89.9 = 10.1\% \text{ H}$$

Out of 100.0 g Cumene, we have

$$89.9 \text{ g C} \times \frac{1 \text{ mol C}}{12.01 \text{ g C}} = 7.49 \text{ mol C}$$

$$10.1 \text{ g H} \times \frac{1 \text{ mol H}}{1.008 \text{ g H}} = 10.0 \text{ mol H}$$

$$\frac{7.49}{10.0} = 0.749 \approx 0.75 \text{ which is a 3:4 ratio}$$

Empirical formula: C_3H_4

Mass of one empirical formula ≈ 3(12) + 4 = 40

So molecular formula is $(C_3H_4)_3$ or C_9H_{12}

93. Out of 100 g of compound there are:

$$83.53 \text{ g Sb} \times \frac{1 \text{ mol Sb}}{121.8 \text{ g Sb}} = 0.6858 \text{ mol Sb}; \qquad 16.47 \text{ g O} \times \frac{1 \text{ mol O}}{16.00 \text{ g O}} = 1.029 \text{ mol O}$$

$\frac{0.6858}{1.029} = 0.6665 \approx \frac{2}{3}$; Empirical formula: Sb_2O_3, Empirical weight = 291.6

Mass of Sb_4O_6 is 583.2, which is in the correct range. Molecular formula: Sb_4O_6.

95. $41.98\text{ mg } CO_2 \times \frac{12.01\text{ mg C}}{44.01\text{ mg }CO_2} = 11.46\text{ mg C}$ $\%C = \frac{11.46\text{ mg}}{19.81\text{ mg}} \times 100 = 57.85\%\text{ C}$

$6.45\text{ mg } H_2O \times \frac{2.016\text{ mg H}}{18.02\text{ mg }H_2O} = 0.722\text{ mg H}$ $\%H = \frac{0.722\text{ mg}}{19.81\text{ mg}} \times 100 = 3.64\%\text{ H}$

$\%O = 100.00 - (57.85 + 3.64) = 38.51\%\text{ O}$

Out of 100.00 g terephthalic acid, there are:

$57.85\text{ g C} \times \frac{1\text{ mol C}}{12.01\text{ g C}} = 4.817\text{ mol C}$

$3.64\text{ g H} \times \frac{1\text{ mol H}}{1.008\text{ g H}} = 3.61\text{ mol H}$

$38.51\text{ g O} \times \frac{1\text{ mol O}}{16.00\text{ g O}} = 2.407\text{ mol O}$

$\frac{4.817}{2.407} = 2,\ \frac{3.61}{2.407} = 1.5,\ \frac{2.407}{2.407} = 1$

C:H:O ratio is 2:1.5:1 or 4:3:2

Empirical formula: $C_4H_3O_2$ Mass of $C_4H_3O_2 \approx 4(12) + 3(1) + 2(16) = 83$

$\frac{166}{83} = 2$; Molecular formula: $C_8H_6O_4$

97. $4\text{ Al} + 3\text{ }O_2 \rightarrow 2\text{ }Al_2O_3$

a. $1.0\text{ mol Al} \times \frac{3\text{ mol }O_2}{4\text{ mol Al}} = 0.75\text{ mol }O_2$; Al is limiting.

b. $2.0\text{ mol Al} \times \frac{3\text{ mol }O_2}{4\text{ mol Al}} = 1.5\text{ mol }O_2$; Al is limiting.

c. $0.50\text{ mol Al} \times \frac{3\text{ mol }O_2}{4\text{ mol Al}} = 0.38\text{ mol }O_2$; Al is limiting.

d. $64.75\text{ g Al} \times \frac{1\text{ mol Al}}{26.98\text{ g Al}} \times \frac{3\text{ mol }O_2}{4\text{ mol Al}} \times \frac{32.00\text{ g }O_2}{\text{mol }O_2} = 57.60\text{ g }O_2$; Al is limiting.

e. $75.89\text{ g Al} \times \frac{1\text{ mol Al}}{26.98\text{ g Al}} \times \frac{3\text{ mol }O_2}{4\text{ mol Al}} \times \frac{32.00\text{ g }O_2}{\text{mol }O_2} = 67.51\text{ g }O_2$; Al is limiting.

f. $51.28\text{ g Al} \times \frac{1\text{ mol Al}}{26.98\text{ g Al}} \times \frac{3\text{ mol }O_2}{4\text{ mol Al}} \times \frac{32.00\text{ g }O_2}{\text{mol }O_2} = 45.62\text{ g }O_2$; Al is limiting.

99. $2.00 \times 10^6\text{ g }CaCO_3 \times \frac{1\text{ mol }CaCO_3}{100.09\text{ g }CaCO_3} \times \frac{1\text{ mol CaO}}{\text{mol }CaCO_3} \times \frac{56.08\text{ g CaO}}{\text{mol CaO}} = 1.12 \times 10^6\text{ g CaO}$

101. Mass of Ni in sample:

$$98.4 \text{ g Ni(CO)}_4 \times \frac{58.69 \text{ g Ni}}{170.73 \text{ g Ni(CO)}_4} = 33.8 \text{ g Ni}$$

$$\%\text{Ni} = \frac{33.8 \text{ g}}{94.2 \text{ g}} = 35.9\% \text{ Ni}$$

103. $C_6H_{10}O_4 + 2\ NH_3 + 4\ H_2 \longrightarrow C_6H_{16}N_2 + 4\ H_2O$

Adip. (Adipic acid) HMD

a. $$1.00 \times 10^3 \text{ g Adip.} \times \frac{1 \text{ mol Adip.}}{146.14 \text{ g Adip}} \times \frac{1 \text{ mol HMD}}{\text{mol Adip.}} \times \frac{116.21 \text{ g HMD}}{\text{mol HMD}} = 795 \text{ g HMD}$$

b. $$\%\text{ Yield} = \frac{765 \text{ g}}{795 \text{ g}} \times 100 = 96.2\%$$

CHALLENGE PROBLEMS

105. Consider the case of aluminum plus oxygen. Aluminum forms Al^{3+} ions; oxygen forms O^{2-} anions. The simplest compound of the two elements is Al_2O_3. Similarly we would expect the formula of any group VI element with Al to be Al_2X_3. Assuming this, out of 100 g of compound there are 18.56 g Al and 81.44 g of the unknown element.

$$18.56 \text{ g Al} \times \frac{1 \text{ mol Al}}{26.98 \text{ g Al}} \times \frac{3 \text{ mol X}}{2 \text{ mol Al}} = 1.032 \text{ mol X}$$

100 g of the compound must contain 1.032 mol of X, if the formula is Al_2X_3.

Therefore, $\dfrac{81.44 \text{ g X}}{1.032 \text{ mol X}} = 78.91 \text{ g X/mol}$

The unknown element is selenium, Se, and the formula is Al_2Se_3.

107. $$5.00 \times 10^6 \text{ g } NH_3 \times \frac{1 \text{ mol } NH_3}{17.03 \text{ g } NH_3} \times \frac{2 \text{ mol HCN}}{2 \text{ mol } NH_3} = 2.94 \times 10^5 \text{ mol HCN}$$

$$5.00 \times 10^6 \text{ g } O_2 \times \frac{1 \text{ mol } O_2}{32.00 \text{ g } O_2} \times \frac{2 \text{ mol HCN}}{3 \text{ mol } O_2} = 1.04 \times 10^5 \text{ mol HCN}$$

$$5.00 \times 10^6 \text{ g } CH_4 \times \frac{1 \text{ mol } CH_4}{16.04 \text{ g } CH_4} \times \frac{2 \text{ mol HCN}}{2 \text{ mol } CH_4} = 3.12 \times 10^5 \text{ mol HCN}$$

O_2 is limiting. Therefore,

$$1.04 \times 10^5 \text{ mol HCN} \times \frac{27.03 \text{ g HCN}}{\text{mol HCN}} = 2.81 \times 10^6 \text{ g HCN}$$

$$1.04 \times 10^5 \text{ mol HCN} \times \frac{6 \text{ mol } H_2O}{2 \text{ mol HCN}} \times \frac{18.02 \text{ g } H_2O}{\text{mol } H_2O} = 5.62 \times 10^6 \text{ g } H_2O$$

109. 10.00 g $XCl_2 \rightarrow$ 12.55 g XCl_4

XCl_4 contains 2.55 g Cl and 10.00 g XCl_2

XCl_2 contains 2.55 g Cl and 7.45 g X

$$2.55 \text{ g Cl} \times \frac{1 \text{ mol Cl}}{35.45 \text{ g Cl}} \times \frac{1 \text{ mol } XCl_2}{2 \text{ mol Cl}} \times \frac{1 \text{ mol X}}{1 \text{ mol } XCl_2} = 3.60 \times 10^{-2} \text{ mol X}$$

$$\frac{7.45 \text{ g X}}{3.60 \times 10^{-2} \text{ mol X}} = \frac{207 \text{ g}}{\text{mol X}};$$ X is Pb.

111. $1.252 \text{ g Cu} \times \frac{1 \text{ mol Cu}}{63.55 \text{ g Cu}} = 1.970 \times 10^{-2} \text{ mol Cu}$

Let x = g Cu_2O and y = g CuO; then x + y = 1.500 and

$$2\left(\frac{x}{143.10}\right) + \frac{y}{79.55} = 1.970 \times 10^{-2} \text{ total mol Cu}$$

$$\begin{array}{r} 1.112\ x + y = 1.567 \\ -x - y = -1.500 \\ \hline 0.112\ x = 0.067 \end{array}$$

x = 0.60 g Cu_2O; 40.% Cu_2O by mass. y = 0.90 g CuO; 60.% CuO by mass.

CHAPTER FOUR: TYPES OF CHEMICAL REACTIONS AND SOLUTION STOICHIOMETRY

QUESTIONS

1. "Slightly soluble" refers to substances that dissolve only to a small extent. A slightly soluble salt may still dissociate completely to ions and, hence, be a strong electrolyte. An example of such a substance is $Mg(OH)_2$. It is a strong electrolyte, but not very soluble. A weak electrolyte is a substance that doesn't dissociate completely to produce ions. A weak electrolyte may be very soluble in water, or it may not be very soluble. Acetic acid is an example of a weak electrolyte that is very soluble in water.

EXERCISES

Aqueous Solutions: Strong and Weak Electrolytes

3. a. $NaBr(s) \rightarrow Na^+(aq) + Br^-(aq)$

   b. $MgCl_2(s) \rightarrow Mg^{2+}(aq) + 2\ Cl^-(aq)$

   c. $Al(NO_3)_3(s) \rightarrow Al^{3+}(aq) + 3\ NO_3^-(aq)$

   d. $(NH_4)_2SO_4(s) \rightarrow 2\ NH_4^+(aq) + SO_4^{2-}(aq)$

   e. $HI\ (g) \rightarrow H^+(aq) + I^-(aq)$

   f. $FeSO_4(s) \rightarrow Fe^{2+}(aq) + SO_4^{2-}(aq)$

   g. $KMnO_4(s) \rightarrow K^+(aq) + MnO_4^-(aq)$

   h. $HClO_4(s) \rightarrow H^+(aq) + ClO_4^-(aq)$

   i. $NH_4C_2H_3O_2(s) \rightarrow NH_4^+(aq) + C_2H_3O_2^-(aq)$

5. $CaCl_2(s) \rightarrow Ca^{2+}(aq) + 2\ Cl^-(aq)$

Solution Concentration: Molarity

7. a. $$1.0\ L \times \frac{0.10\ mol\ NaCl}{L} \times \frac{58.44\ g\ NaCl}{mol} = 5.8\ g\ NaCl$$

   Place 5.8 g NaCl in a 1 L volumetric flask; add water to dissolve the NaCl and fill to the mark.

   b. $$1.0\ L \times \frac{0.10\ mol\ NaCl}{L} \times \frac{1\ L\ stock}{2.5\ mol\ NaCl} = 4.0 \times 10^{-2}\ L$$

   Add 40. mL of 2.5 *M* solution to a 1 L volumetric flask; fill to the mark with water.

   c. $$1.0\ L \times \frac{0.20\ mol\ NaIO_3}{L} \times \frac{197.9\ g\ NaIO_3}{mol\ NaIO_3} = 4.0 \times 10^1\ g\ NaIO_3$$

   As in a, using 40. g $NaIO_3$.

d. $1.0\text{ L} \times \frac{0.010\text{ mol NaIO}_3}{\text{L}} \times \frac{1\text{ L stock}}{0.20\text{ mol NaIO}_3} = 0.050\text{ L}$

As in b, using 50. mL of the 0.20 *M* sodium iodate stock solution.

e. $1.0\text{ L} \times \frac{0.050\text{ mol KHP}}{\text{L}} \times \frac{204.22\text{ g KHP}}{\text{mol KHP}} = 10.2\text{ g KHP} = 10.\text{ g KHP}$

As in a, using 10. g KHP

f. $1.0\text{ L} \times \frac{0.040\text{ mol KHP}}{\text{L}} \times \frac{1\text{ L stock}}{0.50\text{ mol KHP}} = 0.080\text{ L}$

As in b, using 80. mL of the 0.50 *M* KHP stock solution.

9. a. $5.623\text{ g NaHCO}_3 \times \frac{1\text{ mol NaHCO}_3}{84.01\text{ g NaHCO}_3} = 6.693 \times 10^{-2}\text{ mol}$

$M = \frac{6.693 \times 10^{-2}\text{ mol}}{250.0\text{ mL}} \times \frac{1000\text{ mL}}{\text{L}} = 0.2677\ M$

b. $0.1846\text{ g K}_2\text{Cr}_2\text{O}_7 \times \frac{1\text{ mol K}_2\text{Cr}_2\text{O}_7}{294.20\text{ g K}_2\text{Cr}_2\text{O}_7} = 6.275 \times 10^{-4}\text{ mol}$

$M = \frac{6.275 \times 10^{-4}\text{ mol}}{500.0 \times 10^{-3}\text{ L}} = 1.255 \times 10^{-3}\ M$

c. $0.1025\text{ g Cu} \times \frac{1\text{ mol Cu}}{63.55\text{ g Cu}} = 1.613 \times 10^{-3}\text{ mol Cu} = 1.613 \times 10^{-3}\text{ mol Cu}^{2+}$

$M = \frac{1.613 \times 10^{-3}\text{ mol Cu}^{2+}}{200.0\text{ mL}} \times \frac{1000\text{ mL}}{\text{L}} = 8.065 \times 10^{-3}\ M$

11. a. $\text{CaCl}_2(s) \rightarrow \text{Ca}^{2+}(aq) + 2\text{ Cl}^-(aq)$

$M_{\text{Ca}^{2+}} = 0.15\ M;\ M_{\text{Cl}^-} = 2(0.15) = 0.30\ M$

b. $\text{Al(NO}_3)_3(s) \rightarrow \text{Al}^{3+}(aq) + 3\text{ NO}_3^-(aq)$

$M_{\text{Al}^{3+}} = 0.26\ M;\ M_{\text{NO}_3^-} = 3(0.26) = 0.78\ M$

c. $\text{K}_2\text{Cr}_2\text{O}_7(s) \rightarrow 2\text{ K}^+(aq) + \text{Cr}_2\text{O}_7^{2-}(aq)$

$M_{\text{K}^+} = 0.50\ M;\ M_{\text{Cr}_2\text{O}_7^{2-}} = 0.25\ M$

d. $\text{Al}_2(\text{SO}_4)_3(s) \rightarrow 2\text{ Al}^{3+}(aq) + 3\text{ SO}_4^{2-}(aq)$

$M_{\text{Al}^{3+}} = \frac{2.0 \times 10^{-3}\text{ mol Al}_2(\text{SO}_4)_3}{\text{L}} \times \frac{2\text{ mol Al}^{3+}}{\text{mol Al}_2(\text{SO}_4)_3} = 4.0 \times 10^{-3}\ M$

$M_{\text{SO}_4^{2-}} = \frac{2.0 \times 10^{-3}\text{ mol Al}_2(\text{SO}_4)_3}{\text{L}} \times \frac{3\text{ mol SO}_4^{2-}}{\text{mol Al}_2(\text{SO}_4)_3} = 6.0 \times 10^{-3}\ M$

13. $10.8\text{ g }(NH_4)_2SO_4 \times \frac{1\text{ mol}}{132.15\text{ g}} = 8.17 \times 10^{-2}\text{ mol}$

$$\text{Molarity} = \frac{8.17 \times 10^{-2}\text{ mol}}{100.0\text{ mL}} \times \frac{1000\text{ mL}}{\text{L}} = \frac{0.817\text{ mol}}{\text{L}}$$

Moles of $(NH_4)_2SO_4$ in final solution:

$$\frac{0.817\text{ mol}}{\text{L}} \times 10.00 \times 10^{-3}\text{ L} = 8.17 \times 10^{-3}\text{ mol}$$

$$\text{Molarity of final solution} = \frac{8.17 \times 10^{-3}\text{ mol}}{(10.00 + 50.00)\text{ mL}} \times \frac{1000\text{ mL}}{\text{L}} = \frac{0.136\text{ mol}}{\text{L}}$$

$M_{NH_4^+} = 2(0.136) = 0.272\ M$; $M_{SO_4^{2-}} = 0.136\ M$

15. $$\frac{30.\text{ mg }Mg^{2+}}{\text{L}} \times \frac{1\text{ g}}{1000\text{ mg}} \times \frac{1\text{ mol }Mg^{2+}}{24.31\text{ g }Mg^{2+}} = \frac{1.2 \times 10^{-3}\text{ mol}}{\text{L}}$$

$$\frac{75\text{ mg }Ca^{2+}}{\text{L}} \times \frac{1\text{ g}}{1000\text{ mg}} \times \frac{1\text{ mol }Ca^{2+}}{40.08\text{ g }Ca^{2+}} = \frac{1.9 \times 10^{-3}\text{ mol}}{\text{L}}$$

$$\frac{1250\text{ mg }Mg^{2+}}{\text{L}} \times \frac{1\text{ g}}{1000\text{ mg}} \times \frac{1\text{ mol }Mg^{2+}}{24.31\text{ g}} = \frac{0.0514\text{ mol}}{\text{L}}$$

$$\frac{200.\text{ mg }Ca^{2+}}{\text{L}} \times \frac{1\text{ g}}{1000\text{ mg}} \times \frac{1\text{ mol }Ca^{2+}}{40.08\text{ g}} = \frac{4.99 \times 10^{-3}\text{ mol}}{\text{L}}$$

17. $0.5842\text{ g} \times \frac{1\text{ mol}}{90.04\text{ g}} = 6.488 \times 10^{-3}\text{ mol}$

$\frac{6.488 \times 10^{-3}\text{ mol}}{100.0\text{ mL}} \times \frac{1000\text{ mL}}{\text{L}} = 6.488 \times 10^{-2}\ M$ which is the concentration of the initial solution.

Consider, next, the dilution step:

$$\frac{6.488 \times 10^{-2}\text{ mol}}{\text{L}} \times 10.00 \times 10^{-3}\text{ L} = 6.488 \times 10^{-4}\text{ mol}$$

The final solution contains 6.488×10^{-4} mol of oxalic acid in 250.0 mL of solution; or

$$M = \frac{6.488 \times 10^{-4}\text{ mol}}{0.2500\text{ L}} = 2.595 \times 10^{-3}\ M$$

Precipitation Reactions

19. a. $BaCl_2(aq) + Na_2SO_4(aq) \rightarrow BaSO_4(s) + 2\ NaCl(aq)$

$Ba^{2+}(aq) + 2\ Cl^-(aq) + 2\ Na^+(aq) + SO_4^{2-}(aq) \rightarrow BaSO_4(s) + 2\ Na^+(aq) + 2Cl^-(aq)$

$Ba^{2+}(aq) + SO_4^{2-}(aq) \rightarrow BaSO_4(s)$

b. $Pb(NO_3)_2(aq) + 2\ HCl(aq) \rightarrow PbCl_2(s) + 2\ HNO_3(aq)$

$Pb^{2+}(aq) + 2\ NO_3^-(aq) + 2\ H^+(aq) + 2\ Cl^-(aq) \rightarrow PbCl_2(s) + 2\ H^+(aq) + 2\ NO_3^-(aq)$

$Pb^{2+}(aq) + 2Cl^-(aq) \rightarrow PbCl_2(s)$

c. $3\ AgNO_3(aq) + Na_3PO_4(aq) \rightarrow Ag_3PO_4(s) + 3\ NaNO_3(aq)$

$3\ Ag^+(aq) + 3\ NO_3^-(aq) + 3\ Na^+(aq) + PO_4^{3-}(aq) \rightarrow Ag_3PO_4(s) + 3\ Na^+(aq) + 3\ NO_3^-(aq)$

$3\ Ag^+(aq) + PO_4^{3-}(aq) \rightarrow Ag_3PO_4(s)$

d. $3\ NaOH(aq) + Fe(NO_3)_3(aq) \rightarrow Fe(OH)_3(s) + 3\ NaNO_3(aq)$

$3\ Na^+(aq) + 3OH^-(aq) + Fe^{3+}(aq) + 3\ NO_3^-(aq) \rightarrow Fe(OH)_3(s) + 3\ Na^+(aq) + 3\ NO_3^-(aq)$

$Fe^{3+}(aq) + 3\ OH^-(aq) \rightarrow Fe(OH)_3(s)$

21. a. Silver iodide is insoluble. $AgNO_3(aq) + KI(aq) \rightarrow AgI(s) + KNO_3(aq)$

$Ag^+(aq) + I^-(aq) \rightarrow AgI(s)$

b. Copper(II) sulfide is insoluble. $CuSO_4(aq) + Na_2S(aq) \rightarrow CuS(s) + Na_2SO_4(aq)$

$Cu^{2+}(aq) + S^{2-}(aq) \rightarrow CuS(s)$

c. $CoCl_2(aq) + 2\ NaOH(aq) \rightarrow Co(OH)_2(s) + 2\ NaCl(aq)$

$Co^{2+}(aq) + 2\ OH^-(aq) \rightarrow Co(OH)_2(s)$

d. The potential products are $Ni(NO_3)_2$ and HCl. Both are soluble in water. Thus, no reaction occurs.

23. a. $(NH_4)_2SO_4(aq) + Ba(NO_3)_2(aq) \rightarrow 2\ NH_4NO_3(aq) + BaSO_4(s)$

$Ba^{2+}(aq) + SO_4^{2-}(aq) \rightarrow BaSO_4(s)$

b. $Pb(NO_3)_2(aq) + 2\ NaCl(aq) \rightarrow PbCl_2(s) + 2\ NaNO_3(aq)$

$Pb^{2+}(aq) + 2\ Cl^-(aq) \rightarrow PbCl_2(s)$

c. Potassium phosphate and sodium nitrate are both soluble in water. No reaction occurs.

d. No reaction occurs.

e. $CuCl_2(aq) + 2\ NaOH(aq) \rightarrow Cu(OH)_2(s) + 2\ NaCl(aq)$

$Cu^{2+}(aq) + 2\ OH^-(aq) \rightarrow Cu(OH)_2(s)$

25. Three possibilities are:

Addition of H_2SO_4 solution to give a white ppt. of $PbSO_4$.

Addition of HCl solution to give a white ppt. of $PbCl_2$.

Addition of K_2CrO_4 solution to give a bright yellow ppt. of $PbCrO_4$.

27. The reaction is:

$$AgNO_3(aq) + NaBr(aq) \longrightarrow AgBr(s) + NaNO_3(aq)$$

$$100.0 \text{ mL } AgNO_3 \times \frac{1 \text{ L}}{1000 \text{ mL}} \times \frac{0.150 \text{ mol } AgNO_3}{\text{L } AgNO_3} \times \frac{1 \text{ mol AgBr}}{\text{mol } AgNO_3} = 1.50 \times 10^{-2} \text{ mol AgBr}$$

$$20.0 \text{ mL NaBr} \times \frac{1 \text{ L}}{1000 \text{ mL}} \times \frac{1.00 \text{ mol NaBr}}{\text{L NaBr}} \times \frac{1 \text{ mol AgBr}}{\text{mol NaBr}} = 2.00 \times 10^{-2} \text{ mol AgBr}$$

Therefore, $AgNO_3$ is the limiting reagent.

$$1.50 \times 10^{-2} \text{ mol AgBr} \times \frac{187.8 \text{ g AgBr}}{\text{mol AgBr}} = 2.82 \text{ g AgBr}$$

29. The reaction is: $Ni(NO_3)_2(aq) + 2\,NaOH(aq) \longrightarrow Ni(OH)_2(s) + 2\,NaNO_3(aq)$

$$150.0 \text{ mL } Ni(NO_3)_2 \times \frac{1 \text{ L}}{1000 \text{ mL}} \times \frac{0.250 \text{ mol } Ni(NO_3)_2}{\text{L } Ni(NO_3)_2} \times \frac{2 \text{ mol NaOH}}{1 \text{ mol } Ni(NO_3)_2} \times \frac{1 \text{ L NaOH}}{0.100 \text{ mol NaOH}}$$

$= 0.750$ L or 750. mL

Acids and Bases

31. a. $2HClO_4(aq) + Mg(OH_2)(s) \longrightarrow 2\,H_2O(l) + Mg(ClO_4)_2(aq)$

$2\,H^+(aq) + 2ClO_4^-(aq) + Mg(OH)_2(s) \longrightarrow 2\,H_2O(l) + Mg^{2+}(aq) + 2ClO_4^-(aq)$

$2H^+(aq) + Mg(OH)_2(s) \longrightarrow 2H_2O(l) + Mg^{2+}(aq)$

b. $HCN(aq) + NaOH(aq) \longrightarrow H_2O(l) + NaCN(aq)$

$HCN(aq) + Na^+(aq) + OH^-(aq) \longrightarrow H_2O(l) + Na^+(aq) + CN^-(aq)$

$HCN(aq) + OH^-(aq) \longrightarrow H_2O(l) + CN^-(aq)$

c. $HCl(aq) + NaOH(aq) \longrightarrow H_2O(l) + NaCl(aq)$

$H^+(aq) + Cl^-(aq) + Na^+(aq) + OH^-(aq) \longrightarrow H_2O(l) + Na^+(aq) + Cl^-(aq)$

$H^+(aq) + OH^-(aq) \longrightarrow H_2O(l)$

33. a. $KOH(aq) + HNO_3(aq) \rightarrow H_2O(l) + KNO_3(aq)$

$K^+(aq) + OH^-(aq) + H^+(aq) + NO_3^-(aq) \rightarrow H_2O(l) + K^+(aq) + NO_3^-(aq)$

$OH^-(aq) + H^+(aq) \rightarrow H_2O(l)$

b. $Ba(OH)_2(aq) + 2\ HCl(aq) \rightarrow 2\ H_2O(l) + BaCl_2(aq)$

$Ba^{2+}(aq) + 2\ OH^-(aq) + 2\ H^+(aq) + 2\ Cl^-(aq) \rightarrow Ba^{2+}(aq) + 2\ Cl^-(aq) + 2\ H_2O(l)$

$2\ OH^-(aq) + 2\ H^+(aq) \rightarrow 2\ H_2O(l)$ or $OH^-(aq) + H^+(aq) \rightarrow H_2O(l)$

c. $3\ HClO_4(aq) + Fe(OH)_3(s) \rightarrow 3\ H_2O(l) + Fe(ClO_4)_3(aq)$

$3\ H^+(aq) + 3\ ClO_4^-(aq) + Fe(OH)_3(s) \rightarrow 3\ H_2O(l) + Fe^{3+}(aq) + 3ClO_4^-(aq)$

$3\ H^+(aq) + Fe(OH)_3(s) \rightarrow 3\ H_2O(l) + Fe^{3+}(aq)$

35. If we begin with 50.00 mL of 0.200 M NaOH, then

$$50.00 \times 10^{-3}\ L \times \frac{0.200\ mol}{L} = 1.00 \times 10^{-2}\ mol\ NaOH \text{ to be neutralized.}$$

a. $NaOH(aq) + HCl(aq) \rightarrow NaCl(aq) + H_2O(l)$

$$1.00 \times 10^{-2}\ mol\ NaOH \times \frac{1\ mol\ HCl}{mol\ NaOH} \times \frac{1\ L\ soln}{0.100\ mol} = 0.100\ L \text{ or } 100.\ mL$$

b. $3\ NaOH(aq) + H_3PO_4(aq) \rightarrow Na_3PO_4(aq) + 3\ H_2O(l)$

$$1.00 \times 10^{-2}\ mol\ NaOH \times \frac{1\ mol\ H_3PO_4}{3\ mol\ NaOH} \times \frac{1\ L\ soln}{0.200\ mol\ H_3PO_4} = 1.67 \times 10^{-2}\ L \text{ or } 16.7\ mL$$

c. $$50.00\ mL \times \frac{1\ L}{1000\ mL} \times \frac{0.200\ mol\ NaOH}{L} = 1.00 \times 10^{-2}\ mol\ NaOH$$

$HNO_3(aq) + NaOH(aq) \rightarrow H_2O(l) + NaNO_3(aq)$

$$1.00 \times 10^{-2}\ mol\ NaOH \times \frac{1\ mol\ HNO_3}{mol\ NaOH} \times \frac{1\ L}{0.150\ mol\ HNO_3} = 6.67 \times 10^{-2}\ L \text{ or } 66.7\ mL$$

d. $HC_2H_3O_2(aq) + NaOH(aq) \rightarrow H_2O(l) + NaC_2H_3O_2(aq)$

$$1.00 \times 10^{-2}\ mol\ NaOH \times \frac{1\ mol\ HC_2H_3O_2}{mol\ NaOH} \times \frac{1\ L}{0.200\ mol\ HC_2H_3O_2} = 5.00 \times 10^{-2}\ L \text{ or } 50.0\ mL$$

37. $$15\ g\ NaOH \times \frac{1\ mol\ NaOH}{40.00\ g} = 0.38\ mol\ NaOH$$

$$0.15\ L \times \frac{0.25\ mol\ HNO_3}{L} = 0.038\ mol\ HNO_3$$

We have added more moles of NaOH. The reaction is 1:1

$HNO_3(aq) + NaOH(aq) \rightarrow NaNO_3(aq) + H_2O(l)$

We have excess NaOH, the solution will be basic.

In the final solution there will be 0.038 mol NO_3^-, and (0.38 - 0.038) = 0.34 mol of OH^-, and 0.38 mol Na^+.

$$C_{NO_3^-} = \frac{0.038\text{ mol}}{0.15\text{ L}} = \frac{0.25\text{ mol }NO_3^-}{\text{L}};\quad C_{Na^+} = \frac{0.38\text{ mol}}{0.15\text{ L}} = \frac{2.5\text{ mol }Na^+}{\text{L}}$$

$$C_{OH^-} = \frac{0.34\text{ mol}}{0.15\text{ L}} = \frac{2.3\text{ mol }OH^-}{\text{L}}$$

39. $HCl(aq) + NaOH(aq) \rightarrow H_2O(l) + NaCl(aq)$

$$24.16 \times 10^{-3}\text{ L NaOH soln} \times \frac{0.106\text{ mol NaOH}}{\text{L}} \times \frac{1\text{ mol HCl}}{\text{mol NaOH}} = 2.56 \times 10^{-3}\text{ HCl}$$

$$\text{Molarity of HCl} = \frac{2.56 \times 10^{-3}\text{ mol}}{25.00 \times 10^{-3}\text{ L}} = \frac{0.102\text{ mol}}{\text{L}}$$

Oxidation-Reduction Reactions

41. a. K, +1: O, -2: Mn, +7 b. Ni, +4: O, -2

c. $K_4Fe(CN)_6$: Fe, +2

K^+ ions and $Fe(CN)_6^{4-}$ anions; $Fe(CN)_6^{4-}$ composed of Fe^{2+} and CN^-

d. $(NH_4)_2HPO_4$ is made of NH_4^+ cations and HPO_4^{2-} anions. Assign +1 as oxidation number of H and -2 as oxidation number of O. Then we get N, -3: P, +5

e. P, +3: O, -2 h. S, +4: F, -1

f. O, -2: Fe, +8/3 i. C, +2: O, -2

g. O, -2: F, -1: Xe, +6 j. Na, +1: O, -2: C, +3

43. OCl^-: oxidation number of oxygen is (-2).

$-2 + x = -1$: $x = +1$: The oxidation number of Cl in OCl^- is +1.

ClO_2^-: $2(-2) + x = -1$: $x = +3$ ClO_3^-: $3(-2) + x = -1$: $x = +5$ ClO_4^-: $4(-2) + x = -1$: $x = +7$

45.

	Redox?	Oxidizing Agent	Reducing Agent	Substance Oxidized	Substance Reduced
a.	Yes	O_2	CH_4	CH_4 (C)	O_2 (O)
b.	Yes	HCl	Zn	Zn	HCl (H)
c.	No	---	---	---	---
d.	Yes	O_3	NO	NO (N)	O_3 (O)
e.	Yes	H_2O_2	H_2O_2	H_2O_2 (O)	H_2O_2 (O)
f.	Yes	CuCl	CuCl	CuCl (Cu)	CuCl (Cu)

In c, no oxidation numbers change from reactants to products.

47. a. $Zn \rightarrow Zn^{2+} + 2\,e^-$ $\qquad$ $2\,e^- + 2\,HCl \rightarrow H_2 + 2Cl^-$

Adding the two balanced half reactions, $Zn(s) + 2\,HCl(aq) \rightarrow H_2(g) + Zn^{2+}(aq) + 2Cl^-(aq)$

b. $3\,I^- \rightarrow I_3^- + 2e^-$ $\qquad$ $ClO^- \rightarrow Cl^-$

$$2e^- + 2H^+ + ClO^- \rightarrow Cl^- + H_2O$$

Adding the two balanced half reactions so electrons cancel:

$$3\,I^-(aq) + 2\,H^+(aq) + ClO^-(aq) \rightarrow I_3^-(aq) + Cl^-(aq) + H_2O(l)$$

c. $As_2O_3 \rightarrow H_3AsO_4$ $\qquad$ $NO_3^- \rightarrow NO + 2\,H_2O$

$As_2O_3 \rightarrow 2\,H_3AsO_4$ $\qquad$ $4\,H^+ + NO_3^- \rightarrow NO + 2\,H_2O$

left 3 - O $\qquad$ $(3\,e^- + 4\,H^+ + NO_3^- \rightarrow NO + 2\,H_2O) \times 4$

right 6 H and 8 - O

right hand side has 6 extra H and 5 extra O

Balance the oxygen atoms first using H_2O, then balance H using H^+, and finally balance charge using electrons.

$$(5\,H_2O + As_2O_3 \rightarrow 2\,H_3AsO_4 + 4\,H^+ + 4\,e^-) \times 3$$

Common factor is a transfer of 12 e^-. Add half reactions so electrons cancel.

$$12\,e^- + 16\,H^+ + 4\,NO_3^- \rightarrow 4\,NO + 8\,H_2O$$

$$15\,H_2O + 3\,As_2O_3 \rightarrow 6\,H_3AsO_4 + 12\,H^+ + 12\,e^-$$

$$7\,H_2O(l) + 4\,H^+(aq) + 3\,As_2O_3(s) + 4\,NO_3^-(aq) \rightarrow 4\,NO(g) + 6\,H_3AsO_4(aq)$$

d. $(2\ Br^- \rightarrow Br_2 + 2\ e^-) \times 5$ $\quad MnO_4^- \rightarrow Mn^{2+} + 4\ H_2O$

$(5\ e^- + 8\ H^+ + MnO_4^- \rightarrow Mn^{2+} + 4\ H_2O) \times 2$

$10\ Br^- \rightarrow 5\ Br_2 + 10\ e^-$

$10\ e^- + 16\ H^+ + 2\ MnO_4^- \rightarrow 2\ Mn^{2+} + 8\ H_2O$

$16\ H^+(aq) + 2\ MnO_4^-(aq) + 10\ Br^-(aq) \rightarrow 5\ Br_2(l) + 2\ Mn^{2+}(aq) + 8\ H_2O(l)$

e. $CH_3OH + Cr_2O_7^{2-} \rightarrow CH_2O + Cr^{3+}$

$CH_3OH \rightarrow CH_2O$

$(CH_3OH \rightarrow CH_2O + 2\ H^+ + 2\ e^-) \times 3$

$Cr_2O_7^{2-} \rightarrow Cr^{3+}$

$14\ H^+ + Cr_2O_7^{2-} \rightarrow 2\ Cr^{3+} + 7\ H_2O$

$6\ e^- + 14\ H^+ + Cr_2O_7^{2-} \rightarrow 2\ Cr^{3+} + 7\ H_2O$

$3\ CH_3OH \rightarrow 3\ CH_2O + 6\ H^+ + 6\ e^-$

$6\ e^- + 14\ H^+ + Cr_2O_7^{2-} \rightarrow 2\ Cr^{3+} + 7\ H_2O$

$8\ H^+(aq) + 3\ CH_3OH(l) + Cr_2O_7^{2-}(aq) \rightarrow 2\ Cr^{3+}(aq) + 3\ CH_2O(l) + 7\ H_2O(l)$

49. a. $Al \rightarrow Al(OH)_4^-$ $\quad MnO_4^- \rightarrow MnO_2$

$4\ OH^- + Al \rightarrow Al(OH)_4^-$ $\quad 4\ OH^- + 4\ H^+ + MnO_4^- \rightarrow MnO_2 + 2\ H_2O + 4\ OH^-$

$4\ OH^- + Al \rightarrow Al(OH)_4^- + 3\ e^-$ $\quad 2\ H_2O + MnO_4^- \rightarrow MnO_2 + 4\ OH^-$

$3\ e^- + 2\ H_2O + MnO_4^- \rightarrow MnO_2 + 4\ OH^-$

$4\ OH^- + Al \rightarrow Al(OH)_4^- + 3\ e^-$

$3\ e^- + 2\ H_2O + MnO_4^- \rightarrow MnO_2 + 4\ OH^-$

$2\ H_2O(l) + Al(s) + MnO_4^-(aq) \rightarrow Al(OH)_4^-(aq) + MnO_2(s)$

b. $Cl_2 \rightarrow Cl^-$ $\quad Cl_2 \rightarrow ClO^-$

$2\ e^- + Cl_2 \rightarrow 2\ Cl^-$ $\quad 2\ H_2O + Cl_2 \rightarrow 2\ ClO^- + 4\ H^+$

$4\ OH^- + 2\ H_2O + Cl_2 \rightarrow 2ClO^- + 4\ H^+ + 4\ OH^-$

$4\ OH^- + Cl_2 \rightarrow 2\ ClO^- + 2\ H_2O + 2\ e^-$

$$2\,e^- + Cl_2 \rightarrow 2\,Cl^-$$
$$4\,OH^- + Cl_2 \rightarrow 2\,ClO^- + 2\,H_2O + 2\,e^-$$

$$4\,OH^- + 2\,Cl_2 \rightarrow 2\,Cl^- + 2\,ClO^- + 2\,H_2O$$

or $2\,OH^-(aq) + Cl_2(g) \rightarrow Cl^-(aq) + ClO^-(aq) + H_2O(l)$

c.

$$NO_2^- \rightarrow NH_3$$
$$7\,H^+ + NO_2^- \rightarrow NH_3 + 2\,H_2O$$
$$7\,OH^- + 7\,H^+ + NO_2^- \rightarrow NH_3 + 2\,H_2O + 7\,OH^-$$
$$5\,H_2O + NO_2^- \rightarrow NH_3 + 7\,OH^-$$
$$6\,e^- + NO_2^- + 5\,H_2O \rightarrow NH_3 + 7\,OH^-$$

$$Al \rightarrow AlO_2^-$$
$$4\,OH^- + 2\,H_2O + Al \rightarrow AlO_2^- + 4\,H^+ + 4\,OH^-$$
$$4\,OH^- + Al \rightarrow AlO_2^- + 2\,H_2O$$
$$(4\,OH^- + Al \rightarrow AlO_2^- + 2\,H_2O + 3\,e^-) \times 2$$

$$6\,e^- + NO_2^- + 5\,H_2O \rightarrow NH_3 + 7\,OH^-$$
$$8\,OH^- + 2\,Al \rightarrow 2\,AlO_2^- + 4\,H_2O + 6\,e^-$$

$$OH^-(aq) + H_2O(l) + NO_2^-(aq) + 2\,Al(s) \rightarrow NH_3(g) + 2\,AlO_2^-(aq)$$

51. a.

$$\overset{+1\;-1}{NaCl} + \overset{+1\;+6\;-2}{H_2SO_4} + \overset{+4\;-2}{MnO_2} \rightarrow \overset{+1\;+6\;-2}{Na_2SO_4} + \overset{+2\;-1}{MnCl_2} + \overset{0}{Cl_2} + \overset{+1\;-2}{H_2O}$$

$2\,NaCl + H_2SO_4 + MnO_2 \rightarrow Na_2SO_4 + MnCl_2 + Cl_2 + H_2O$

To balance Cl, need two more NaCl for Cl in $MnCl_2$.

$4\,NaCl + H_2SO_4 + MnO_2 \rightarrow Na_2SO_4 + MnCl_2 + Cl_2 + H_2O$

Balance Na and SO_4^{2-}

$4\,NaCl + 2\,H_2SO_4 + MnO_2 \rightarrow 2\,Na_2SO_4 + MnCl_2 + Cl_2 + H_2O$

Left: 4-H and 10-O

Right: 8-O not counting H_2O

Need 2 H_2O on right

$4\,NaCl(aq) + 2\,H_2SO_4(aq) + MnO_2(s) \rightarrow 2\,Na_2SO_4(aq) + MnCl_2(aq) + Cl_2(g) + 2\,H_2O(l)$

53. $H_2C_2O_4 \rightarrow 2\ CO_2 + 2\ H^+$

$(H_2C_2O_4 \rightarrow 2\ CO_2 + 2\ H^+ + 2\ e^-) \times 5$

$(5\ e^- + 8\ H^+ + MnO_4^- \rightarrow Mn^{2+} + 4\ H_2O) \times 2$

$$5\ H_2C_2O_4 \rightarrow 10\ CO_2 + 10\ H^+ + 10\ e^-$$

$$10\ e^- + 16\ H^+ + 2\ MnO_4^- \rightarrow 2\ Mn^{2+} + 8\ H_2O$$

$$6\ H^+ + 5\ H_2C_2O_4 + 2\ MnO_4^- \rightarrow 10\ CO_2 + 2\ Mn^{2+} + 8\ H_2O$$

$$0.1058 \text{ g oxalic acid} \times \frac{1 \text{ mol oxalic acid}}{90.04 \text{ g}} \times \frac{2 \text{ mol } MnO_4^-}{5 \text{ mol oxalic acid}} = 4.700 \times 10^{-4} \text{ mol}$$

$$\text{Molarity} = \frac{4.700 \times 10^{-4} \text{ mol}}{28.97 \text{ mL}} \times \frac{1000 \text{ mL}}{\text{L}} = 1.622 \times 10^{-2}\ M$$

55. a. $Fe^{2+} \rightarrow Fe^{3+} + e^-$

$5\ e^- + 8\ H^+ + MnO_4^- \rightarrow Mn^{2+} + 4\ H_2O$

Balanced equation is:

$$8\ H^+ + MnO_4^- + 5\ Fe^{2+} \rightarrow 5\ Fe^{3+} + Mn^{2+} + 4\ H_2O$$

$$20.62 \times 10^{-3} \text{ L soln} \times \frac{0.0216 \text{ mol } MnO_4^-}{\text{L soln}} \times \frac{5 \text{ mol } Fe^{2+}}{1 \text{ mol } MnO_4^-} = 2.23 \times 10^{-3} \text{ mol } Fe^{2+}$$

$$\text{Molarity} = \frac{2.23 \times 10^{-3} \text{ mol } Fe^{2+}}{50.0 \times 10^{-3} \text{ L}} = 4.46 \times 10^{-2}\ M$$

b. $Fe^{2+} \rightarrow Fe^{3+} + e^-$

$Cr_2O_7^{2-} \rightarrow 2\ Cr^{3+} + 7\ H_2O$

$6\ e^- + 14\ H^+ + Cr_2O_7^{2-} \rightarrow 2\ Cr^{3+} + 7\ H_2O$

The balanced equation is:

$14\ H^+ + Cr_2O_7^{2-} + 6\ Fe^{2+} \rightarrow 6\ Fe^{3+} + 2\ Cr^{3+} + 7\ H_2O$

$$50.00 \times 10^{-3} \text{ L} \times \frac{4.46 \times 10^{-2} \text{ mol } Fe^{2+}}{\text{L}} \times \frac{1 \text{ mol } Cr_2O_7^{2-}}{6 \text{ mol } Fe^{2+}} \times \frac{1 \text{ L}}{0.0150 \text{ mol } Cr_2O_7^{2-}}$$

$$= 2.48 \times 10^{-2} \text{ L or } 24.8 \text{ mL}$$

ADDITIONAL EXERCISES

57. $$4.25 \text{ g Ca} \times \frac{1 \text{ mol Ca}}{40.08 \text{ g Ca}} \times \frac{1 \text{ mol Ca(OH)}_2}{\text{mol Ca}} \times \frac{2 \text{ mol OH}^-}{\text{mol Ca(OH)}_2} = 0.212 \text{ mol OH}^-$$

$$\text{Molarity} = \frac{0.212 \text{ mol}}{225 \times 10^{-3} \text{ L}} = 0.942\ M$$

59. $$Mn + HNO_3 \longrightarrow Mn^{2+} + NO_2$$

$$Mn \longrightarrow Mn^{2+} + 2\,e^-$$

$$HNO_3 \longrightarrow NO_2$$

$$HNO_3 \longrightarrow NO_2 + H_2O$$

$$(e^- + H^+ + HNO_3 \longrightarrow NO_2 + H_2O) \times 2$$

$$Mn \longrightarrow Mn^{2+} + 2\,e^-$$

$$2\,e^- + 2\,H^+ + 2\,HNO_3 \longrightarrow 2\,NO_2 + 2\,H_2O$$

$$2\,H^+(aq) + Mn(s) + 2\,HNO_3(aq) \longrightarrow Mn^{2+}(aq) + 2\,NO_2(g) + 2\,H_2O(l)$$

$$Mn^{2+} + IO_4^- \longrightarrow MnO_4^- + IO_3^-$$

$$(4\,H_2O + Mn^{2+} \longrightarrow MnO_4^- + 8\,H^+ + 5\,e^-) \times 2$$

$$(2\,e^- + 2\,H^+ + IO_4^- \longrightarrow IO_3^- + H_2O) \times 5$$

$$8\,H_2O + 2\,Mn^{2+} \longrightarrow 2\,MnO_4^- + 16\,H^+ + 10\,e^-$$

$$10\,e^- + 10\,H^+ + 5\,IO_4^- \longrightarrow 5\,IO_3^- + 5\,H_2O$$

$$3\,H_2O(l) + 2\,Mn^{2+}(aq) + 5\,IO_4^-(aq) \longrightarrow 2\,MnO_4^-(aq) + 5\,IO_3^-(aq) + 6\,H^+(aq)$$

61. $$1.00 \text{ L} \times \frac{0.200 \text{ mol Na}_2\text{S}_3\text{O}_3}{\text{L}} \times \frac{1 \text{ mol AgBr}}{2 \text{ mol Na}_2\text{S}_2\text{O}_3} \times \frac{187.8 \text{ g AgBr}}{\text{mol AgBr}} = 18.8 \text{ g}$$

63. The molecular weight of $Al(C_9H_6NO)_3$ is

$$1 \text{ mol Al}\left(\frac{26.98 \text{ g}}{\text{mol Al}}\right) + 27 \text{ mol C}\left(\frac{12.01 \text{ g}}{\text{mol C}}\right) + 3 \text{ mol N}\left(\frac{14.01 \text{ g}}{\text{mol N}}\right)$$

$$+ 3 \text{ mol O}\left(\frac{16.00 \text{ g}}{\text{mol O}}\right) + 18 \text{ mol H}\left(\frac{1.008 \text{ g}}{\text{mol H}}\right) = \frac{459.42 \text{ g}}{\text{mol}}$$

$$0.1248 \text{ g Al(C}_9\text{H}_6\text{NO)}_3 \times \frac{26.98 \text{ g Al}}{459.42 \text{ g Al(C}_9\text{H}_6\text{NO)}_3} = 7.329 \times 10^{-3} \text{ g Al}$$

$$\%\text{Al} = \frac{7.329 \times 10^{-3} \text{ g}}{1.8571 \text{ g}} \times 100 = 0.3947\ \%\ \text{Al}$$

65. $CH_3CO_2H(aq) + NaOH(aq) \longrightarrow H_2O(l) + CH_3CO_2Na(aq)$

a. $16.58 \times 10^{-3}\text{ L soln} \times \frac{0.5062\text{ mol NaOH}}{\text{L soln}} \times \frac{1\text{ mol acetic acid}}{\text{mol NaOH}} = 8.393 \times 10^{-3}$ mol acetic acid

$$\text{Concentration of acetic acid} = \frac{8.393 \times 10^{-3}\text{ mol}}{0.01000\text{ L}} = 0.8393\ M$$

b. If we have 1.000 L of solution:

$$\text{Total mass} = 1000.\text{ mL} \times \frac{1.006\text{ g}}{\text{mL}} = 1006\text{ g}$$

$$\text{Mass of acetic acid} = 0.8393\text{ mol} \times \frac{60.05\text{ g}}{\text{mol}} = 50.40\text{ g}$$

$$\%\text{ acetic acid} = \frac{50.40\text{ g}}{1006\text{ g}} \times 100 = 5.010\%$$

67. Since KHP is a monoprotic acid, the reaction is

$$NaOH + HX \longrightarrow NaX + H_2O$$

where HX is short hand for potassium hydrogen phthalate.

$$0.1082\text{ g KHP} \times \frac{1\text{ mol KHP}}{204.22\text{ g KHP}} \times \frac{1\text{ mol NaOH}}{\text{mol KHP}} = 5.298 \times 10^{-4}\text{ mol NaOH}$$

There is 5.298×10^{-4} mol of sodium hydroxide in 20.46 mL of solution. Therefore, the concentration of sodium hydroxide is:

$$\frac{5.298 \times 10^{-4}\text{ mol}}{20.46 \times 10^{-3}\text{ L}} = 2.589 \times 10^{-2}\ M$$

69. a. No, no element shows a change in oxidation number.

b. $1.0\text{ L oxalic acid} \times \frac{0.14\text{ mol oxalic acid}}{\text{L}} \times \frac{1\text{ mol } Fe_2O_3}{6\text{ mol oxalic acid}} \times \frac{159.70\text{ g } Fe_2O_3}{\text{mol } Fe_2O_3}$

$= 3.7\text{ g } Fe_2O_3$

71. Fe^{2+} will react with MnO_4^- (purple) producing Fe^{3+} and Mn^{2+} (almost colorless). There is no reaction between MnO_4^- and Fe^{3+}. Therefore, add a few drops of the potassium permanganate solution. If the purple color is present the solution contains iron(III) sulfate. If the color disappears, iron(II) sulfate is present.

73. $KMnO_4$ added to Sb(III) solution.

$$25.00 \times 10^{-3}\text{ L} \times \frac{0.0233\text{ mol } KMnO_4}{\text{L}} = 5.83 \times 10^{-4}\text{ mol}$$

$KMnO_4$ left unreacted:

$$8\ H^+ + MnO_4^- + 5\ Fe^{2+} \rightarrow Mn^{2+} + 5\ Fe^{3+} + 4\ H_2O$$

$$2.58 \times 10^{-3}\ \text{L}\ Fe^{2+}\ \text{soln} \times \frac{0.0843\ \text{mol}\ Fe^{2+}}{\text{L}} \times \frac{1\ \text{mol}\ MnO_4^-}{5\ \text{mol}\ Fe^{2+}} = 4.35 \times 10^{-5}\ \text{mol}\ MnO_4^-$$

Thus, 4.35×10^{-5} mol MnO_4^- was unreacted.

The amount of $KMnO_4$ that reacted with Sb(III) was:

$$5.83 \times 10^{-4}\ \text{mol} - 0.435 \times 10^{-4}\ \text{mol} = 5.40 \times 10^{-4}\ \text{mol}$$

$MnO_4^- \rightarrow Mn^{2+}$ (change by 5 e^-); Sb(III) $\rightarrow$ Sb(V) (change by 2 e^-)

2 mol MnO_4^- will oxidize 5 mol Sb(III).

$$5.40 \times 10^{-4}\ \text{mol}\ MnO_4^- \times \frac{5\ \text{mol Sb(III)}}{2\ \text{mol}\ MnO_4^-} = 1.35 \times 10^{-3}\ \text{mol Sb}$$

$$1.35 \times 10^{-3}\ \text{mol Sb} \times \frac{1\ \text{mol}\ Sb_2S_3}{2\ \text{mol Sb}} \times \frac{339.8\ \text{g}\ Sb_2S_3}{\text{mol}\ Sb_2S_3} = 0.229\ \text{g}\ Sb_2S_3$$

$$\%Sb_2S_3 = \frac{0.229\ \text{g}\ Sb_2S_3}{0.506\ \text{g ore}} \times 100 = 45.3\%\ Sb_2S_3$$

75. $0.104\ \text{g AgCl} \times \dfrac{35.45\ \text{g}\ Cl^-}{143.4\ \text{g AgCl}} = 2.57 \times 10^{-2}\ \text{g}\ Cl^-$

Chlorisondiamine = $14(12.01) + 18(1.008) + 6(35.45) + 2(14.01) = \dfrac{427.00\ \text{g}}{\text{mol}}$

There are 6(35.45) = 212.70 g chlorine for every mole (427.00 g) of chlorisondiamine.

$$2.57 \times 10^{-2}\ \text{g}\ Cl^- \times \frac{427.00\ \text{g drug}}{212.70\ \text{g}\ Cl^-} = 5.16 \times 10^{-2}\ \text{g drug}$$

$$\%\ \text{drug} = \frac{5.16 \times 10^{-2}\ \text{g}}{1.28\ \text{g}} \times 100 = 4.03\%$$

CHALLENGE PROBLEMS

77. a. HCl(aq) dissociates to $H^+(aq) + Cl^-(aq)$. For simplicity let's use H^+ and Cl^- separately.

$H^+ \rightarrow H_2$ $Fe \rightarrow HFeCl_4$

$(2\ H^+ + 2e^- \rightarrow H_2) \times 3$ $(H^+ + 4\ Cl^- + Fe \rightarrow HFeCl_4 + 3e^-) \times 2$

$$6\ H^+ + 6\ e^- \rightarrow 3\ H_2$$
$$2\ H^+ + 8\ Cl^- + 2\ Fe \rightarrow 2\ HFeCl_4 + 6\ e^-$$

$$8\ H^+ + 8\ Cl^- + 2\ Fe \rightarrow 2\ HFeCl_4 + 3\ H_2$$

or $$8\ HCl(aq) + 2\ Fe(s) \rightarrow 2\ HFeCl_4(aq) + 3\ H_2(g)$$

b.

$$IO_3^- \rightarrow I_3^- \qquad I^- \rightarrow I_3^-$$
$$3\ IO_3^- \rightarrow I_3^- \qquad (3\ I^- \rightarrow I_3^- + 2\ e^-) \times 8$$
$$3\ IO_3^- \rightarrow I_3^- + 9\ H_2O$$
$$16\ e^- + 18\ H^+ + 3\ IO_3^- \rightarrow I_3^- + 9\ H_2O$$

$$16\ e^- + 18\ H^+ + 3\ IO_3^- \rightarrow I_3^- + 9\ H_2O$$
$$24\ I^- \rightarrow 8\ I_3^- + 16\ e^-$$

$$18\ H^+ + 24\ I^- + 3\ IO_3^- \rightarrow 9\ I_3^- + 9\ H_2O$$

or $$6\ H^+(aq) + 8\ I^-(aq) + IO_3^-(aq) \rightarrow 3\ I_3^-(aq) + 3\ H_2O(l)$$

c. $$Ce^{4+} + e^- \rightarrow Ce^{3+}$$

$$Cr(NCS)_6^{4-} \rightarrow Cr^{3+} + NO_3^- + CO_2 + SO_4^{2-}$$

$$54\ H_2O + Cr(NCS)_6^{4-} \rightarrow Cr^{3+} + 6\ NO_3^- + 6\ CO_2 + 6\ SO_4^{2-} + 108\ H^+$$

Charge on left -4

Charge on right (+3) + 6(-1) + 6(-2) + 108(+1) = +93

Add 97 e^- to the right then add two balanced half reactions.

$$54\ H_2O + Cr(NCS)_6^{4-} \rightarrow Cr^{3+} + 6\ NO_3^- + 6\ CO_2 + 6\ SO_4^{2-} + 108\ H^+ + 97\ e^-$$
$$97\ e^- + 97\ Ce^{4+} \rightarrow 97\ Ce^{3+}$$

$$97\ Ce^{4+} + 54\ H_2O + Cr(NCS)_6^{4-} \rightarrow 97\ Ce^{3+} + Cr^{3+} + 6\ NO_3^- + 6\ CO_2 + 6\ SO_4^{2-} + 108\ H^+$$

This is very complicated. A check of the net charge is a good check to see if the equation is balanced. Note: See question for states, i.e. aq, l, or g.

Left: charge = 97 (+4) - 4 = +384

Right: charge = 97 (+3) + 3 + 6(-1) + 6(-2) + 108(+1) = +384

d.

$$Cl_2 \rightarrow Cl^-$$
$$(2\,e^- + Cl_2 \rightarrow 2\,Cl^-) \times 27$$

$$CrI_3 \rightarrow CrO_4^{2-} + IO_4^-$$
$$32\,OH^- + 16\,H_2O + CrI_3 \rightarrow CrO_4^{2-} + 3\,IO_4^- + 32\,H^+ + 32\,OH^-$$

$$32\,OH^- + CrI_3 \rightarrow CrO_4^{2-} + 3\,IO_4^- + 16\,H_2O$$

Net charge on left: -32; Net charge on right: -5

Add 27 e^- to the right.

$$(32\,OH^- + CrI_3 \rightarrow CrO_4^{2-} + 3\,IO_4^- + 16\,H_2O + 27\,e^-) \times 2$$

Common factor is a transfer of 54 e^-

$$54\,e^- + 27\,Cl_2 \rightarrow 54\,Cl^-$$
$$64\,OH^- + 2\,CrI_3 \rightarrow 2\,CrO_4^{2-} + 6\,IO_4^- + 32\,H_2O + 54\,e^-$$

$$64\,OH^-(aq) + 2\,CrI_3(s) + 27\,Cl_2(g) \rightarrow 54\,Cl^-(aq) + 2\,CrO_4^{2-}(aq) + 6\,IO_4^-(aq) + 32\,H_2O(l)$$

e.

$$Ce^{4+} \rightarrow Ce(OH)_3$$
$$(e^- + 3\,OH^- + Ce^{4+} \rightarrow Ce(OH)_3) \times 61$$

$$Fe(CN)_6^{4-} \rightarrow Fe(OH)_3 + CO_3^{2-} + NO_3^-$$
$$3\,OH^- + Fe(CN)_6^{4-} \rightarrow Fe(OH)_3 + 6\,CO_3^{2-} + 6\,NO_3^-$$

36 extra O atoms on right

Add 36 H_2O to left, 72 H^+ to right, then neutralize with 72 OH^- on each side.

$$72\,OH^- + 36\,H_2O + 3\,OH^- + Fe(CN)_6^{4-} \rightarrow Fe(OH)_3 + 6\,CO_3^{2-} + 6\,NO_3^- + 72\,H^+ + 72\,OH^-$$
$$75\,OH^- + Fe(CN)_6^{4-} \rightarrow Fe(OH)_3 + 6\,CO_3^{2-} + 6\,NO_3^- + 36\,H_2O$$

net charge = -79 net charge = -18

Add 61 e^- to the right then add two balanced half reactions.

$$75\,OH^- + Fe(CN)_6^{4-} \rightarrow Fe(OH)_3 + 6\,CO_3^{2-} + 6\,NO_3^- + 36\,H_2O + 61\,e^-$$
$$61\,e^- + 183\,OH^- + 61\,Ce^{4+} \rightarrow 61\,Ce(OH)_3$$

$$258\,OH^-(aq) + Fe(CN)_6^{4-}(aq) + 61\,Ce^{4+}(aq) \rightarrow Fe(OH)_3(s) + 61\,Ce(OH)_3(s) + 6\,CO_3^{2-}(aq) + 6\,NO_3^- + 36\,H_2O(l)$$

f. $Fe(OH)_2 + H_2O_2 \rightarrow Fe(OH)_3$

$Fe(OH)_2 \rightarrow Fe(OH)_3$ $H_2O_2 \rightarrow OH^-$

$(OH^- + Fe(OH)_2 \rightarrow Fe(OH)_3 + e^-) \times 2$ $2\,e^- + H_2O_2 \rightarrow 2\,OH^-$

$$2\,e^- + H_2O_2 \rightarrow 2\,OH^-$$
$$2\,OH^- + 2\,Fe(OH)_2 \rightarrow 2\,Fe(OH)_3 + 2\,e^-$$

$$2\,Fe(OH)_2(s) + H_2O_2(aq) \rightarrow 2\,Fe(OH)_3(s)$$

79. $CaCO_3(s) + H_2SO_4(aq) \rightarrow CaSO_4(aq) + H_2O(l) + CO_2(g)$

81. $35.08 \text{ mL NaOH} \times \frac{1 \text{ L}}{1000 \text{ mL}} \times \frac{2.12 \text{ mol NaOH}}{\text{L NaOH}} \times \frac{1 \text{ mol } H_2SO_4}{2 \text{ mol NaOH}} = 3.72 \times 10^{-2} \text{ mol } H_2SO_4$

$$\text{Molarity} = \frac{3.72 \times 10^{-2} \text{ mol}}{10.00 \text{ mL}} \times \frac{1000 \text{ mL}}{\text{L}} = 3.72\ M$$

83. The pertinent reactions are:

$$2\,NaOH(aq) + H_2SO_4(aq) \rightarrow Na_2SO_4(aq) + 2\,H_2O(l)$$

$$HCl(aq) + NaOH(aq) \rightarrow NaCl(aq) + H_2O(l)$$

Amount of NaOH added:

$$0.0500 \text{ L} \times \frac{0.213 \text{ mol}}{\text{L}} = 1.07 \times 10^{-2} \text{ mol NaOH}$$

Amount of NaOH neutralized by HCl:

$$0.01321 \text{ L HCl} \times \frac{0.103 \text{ mol HCl}}{\text{L HCl}} \times \frac{1 \text{ mol NaOH}}{\text{mol HCl}} = 1.36 \times 10^{-3} \text{ mol NaOH}$$

The difference, 9.3×10^{-3} mol, is the amount of NaOH neutralized by the sulfuric acid.

$$9.3 \times 10^{-3} \text{ mol NaOH} \times \frac{1 \text{ mol } H_2SO_4}{2 \text{ mol NaOH}} = 4.7 \times 10^{-3} \text{ mol } H_2SO_4$$

The concentration of H_2SO_4 is:

$$\frac{4.7 \times 10^{-3} \text{ mol}}{0.100 \text{ L}} = 4.7 \times 10^{-2}\ M$$

85. a. $IO_3^- + I^- \rightarrow I_3^-$ See 4.77b

$IO_3^- \rightarrow I_3^-$ $(3\,I^- \rightarrow I_3^- + 2\,e^-) \times 8$

$16\,e^- + 18\,H^+ + 3\,IO_3^- \rightarrow I_3^- + 9\,H_2O$

$$24\,I^- \rightarrow 8\,I_3^- + 16\,e^-$$

$$16\,e^- + 18\,H^+ + 3\,IO_3^- \rightarrow I_3^- + 9\,H_2O$$

$$18\,H^+ + 24\,I^- + 3\,IO_3^- \rightarrow 9\,I_3^- + 9\,H_2O$$ which simplifies to

$$6\,H^+(aq) + 8\,I^-(aq) + IO_3^-(aq) \rightarrow 3\,I_3^-(aq) + 3\,H_2O(l)$$

b. $S_2O_3^{2-} + I_3^- \rightarrow I^- + S_4O_6^{2-}$

$I_3^- + 2\,e^- \rightarrow 3\,I^-$

$S_2O_3^{2-} \rightarrow S_4O_6^{2-}$

$2\,S_2O_3^{2-} \rightarrow S_4O_6^{2-} + 2\,e^-$

Adding the balanced half reactions gives:

$$2\,S_2O_3^{2-}(aq) + I_3^-(aq) \rightarrow 3\,I^-(aq) + S_4O_6^{2-}(aq)$$

c. $25.00 \times 10^{-3}\ \text{L}\ IO_3^- \times \dfrac{0.0100\ \text{mol}\ IO_3^-}{\text{L}} \times \dfrac{3\ \text{mol}\ I_3^-}{\text{mol}\ IO_3^-} = 7.50 \times 10^{-4}\ \text{mol}\ I_3^-$

$7.50 \times 10^{-4}\ \text{mol}\ I_3^- \times \dfrac{2\ \text{mol}\ S_2O_3^{2-}}{\text{mol}\ I_3^-} = 1.50 \times 10^{-3}\ \text{mol}\ S_2O_3^{2-}$

$M_{S_2O_3^{2-}} = \dfrac{1.50 \times 10^{-3}\ \text{mol}}{32.04 \times 10^{-3}\ \text{L}} = 0.0468\ M$

d. $0.5000\ \text{L} \times \dfrac{0.0100\ \text{mol}\ KIO_3}{\text{L}} \times \dfrac{214.0\ \text{g}\ KIO_3}{\text{mol}\ KIO_3} = 1.07\ \text{g}\ KIO_3$

Place 1.07 g KIO_3 in a 500 mL volumetric flask; add water to dissolve KIO_3; continue adding water to the mark.

87. a. $Fe^{3+}(aq) + 3\,OH^-(aq) \rightarrow Fe(OH)_3(s)$

$Fe(OH)_3$: 55.85 + 3(16.00) + 3(1.008) = 106.87 g/mol

$0.107\ \text{g}\ Fe(OH)_3 \times \dfrac{55.85\ \text{g Fe}}{106.87\ \text{g}\ Fe(OH)_3} = 0.0559\ \text{g Fe}$

b. $Fe(NO_3)_3$: 55.85 + 3(14.01) + 9(16.00) = 241.86 g/mol

$0.559\ \text{g Fe} \times \dfrac{241.86\ \text{g}\ Fe(NO_3)_3}{55.85\ \text{g Fe}} = 0.242\ \text{g}\ Fe(NO_3)_3$

c. $\%Fe(NO_3)_3 = \dfrac{0.242\ \text{g}}{0.456\ \text{g}} \times 100 = 53.1\%$

89. $0.298 \text{ g BaSO}_4 \times \frac{96.07 \text{ g SO}_4^{2-}}{233.4 \text{ g BaSO}_4} = 0.123 \text{ g sulfate}$

$\% \text{ sulfate} = \frac{0.123 \text{ g SO}_4^{2-}}{0.205 \text{ g}} = 60.0\%$

Assume we have 100 g of the mixture of Na_2SO_4 and K_2SO_4. There is:

$$60.0 \text{ g SO}_4^{2-} \times \frac{1 \text{ mol}}{96.07 \text{ g}} = 0.625 \text{ mol SO}_4^{2-}$$

There must be 2 × 0.625 = 1.25 mol of cations.

Let x = number of moles of K^+ and y = number of moles of Na^+

Then, x + y = 1.25

The total mass of Na^+ and K^+ must be 40.0 g.

$$x \text{ mol K}^+ \left(\frac{39.10 \text{ g}}{\text{mol}}\right) + y \text{ mol Na}^+ \left(\frac{22.99 \text{ g}}{\text{mol}}\right) = 40.0 \text{ g}$$

So, we have two equations in two unknowns:

$$x + y = 1.25 \text{ and } 39.10x + 22.99y = 40.0$$

Since x = 1.25 - y, 39.10(1.25 - y) + 22.99y = 40.0

48.9 - 39.10y + 22.99y = 40.0; -16.11y = -8.9

y = 0.55 mol Na^+ and x = 1.25 - 0.55 = 0.70 mol K^+

Therefore:

$$0.70 \text{ mol K}^+ \times \frac{1 \text{ mol K}_2\text{SO}_4}{2 \text{ mol K}^+} = 0.35 \text{ mol K}_2\text{SO}_4$$

$$0.35 \text{ mol K}_2\text{SO}_4 \times \frac{174.27 \text{ g}}{\text{mol}} = 61 \text{ g K}_2\text{SO}_4$$

Since we assumed 100 g then the mixture is 61% K_2SO_4 and 39% Na_2SO_4.

91. a. $\text{ppm Na} = \frac{\mu\text{g Na}}{\text{mL}}$

$$150 \times 10^{-3} \text{ g Na}_2\text{CO}_3 \times \frac{45.98 \text{ g Na}^+}{105.99 \text{ g Na}_2\text{CO}_3} = 6.5 \times 10^{-2} \text{ g Na}^+$$

$$\frac{6.5 \times 10^{-2} \text{ g Na}^+}{1.0 \text{ L}} \times \frac{1 \text{ L}}{1000 \text{ mL}} = \frac{6.5 \times 10^{-5} \text{ g Na}^+}{\text{mL}} = \frac{65 \times 10^{-6} \text{ g Na}^+}{\text{mL}} = 65 \text{ ppm}$$

b. $\text{ppb} = \dfrac{\text{ng dioctylphthalate}}{\text{mL}}$

$$\frac{2.5 \times 10^{-3}\ \text{g}}{500.0\ \text{mL}} = \frac{5.0 \times 10^{-6}\ \text{g}}{\text{mL}} \times \frac{10^9\ \text{ng}}{\text{g}} = \frac{5.0 \times 10^3\ \text{ng}}{\text{mL}} = 5.0 \times 10^3\ \text{ppb}$$

93. We want 100.0 mL of each standard. To make the 100. ppm standard:

$$\frac{100.\ \mu\text{g Cu}}{\text{mL}} \times 100.0\ \text{mL solution} = 1.00 \times 10^4\ \mu\text{g Cu needed}$$

$$1.00 \times 10^4\ \mu\text{g Cu} \times \frac{1\ \text{mL stock}}{1000.0\ \mu\text{g}} = 10.0\ \text{mL of stock solution}$$

Therefore, to make 100.0 mL of 100. ppm solution, transfer 10.0 mL of the 1000.0 ppm stock solution to a 100 mL volumetric flask and dilute to the mark.

Similarly for the:

75.0 ppm standard, dilute 7.50 mL of the 1000.0 ppm stock to 100.0 mL

50.0 ppm standard, dilute 5.00 mL of the 1000.0 ppm stock to 100.0 mL

25.0 ppm standard, dilute 2.50 mL of the 1000.0 ppm stock to 100.0 mL

10.0 ppm standard, dilute 1.00 mL of the 1000.0 ppm stock to 100.0 mL

95. The amount of KHP used is:

$$0.4016\ \text{g} \times \frac{1\ \text{mol}}{204.22\ \text{g}} = 1.967 \times 10^{-3}\ \text{mol KHP}$$

Since one mole of NaOH reacts completely with one mole of KHP, the NaOH solution contains 1.967×10^{-3} mol NaOH.

$$\text{Molarity of NaOH} = \frac{1.967 \times 10^{-3}\ \text{mol}}{25.06 \times 10^{-3}\ \text{L}} = \frac{7.849 \times 10^{-2}\ \text{mol}}{\text{L}}$$

$$\text{Maximum molarity} = \frac{1.967 \times 10^{-3}\ \text{mol}}{25.01 \times 10^{-3}\ \text{L}} = \frac{7.865 \times 10^{-2}\ \text{mol}}{\text{L}}$$

$$\text{Minimum molarity} = \frac{1.967 \times 10^{-3}\ \text{mol}}{25.11 \times 10^{-3}\ \text{L}} = \frac{7.834 \times 10^{-2}\ \text{mol}}{\text{L}}$$

We can express this as 0.07849 $M \pm 0.00016\ M$.

An alternate is to express the molarity as 0.0785 $M \pm 0.0002\ M$.

This second way is consistent with our convention on significant figures. The advantage of the first method is that it shows that we made all of our individual measurements to four significant figures.

97. The first procedure would give the most accurate concentration. The uncertainty in pipetting three times from a 10 mL volumetric pipet would still be less than trying to make a single transfer of 0.050 mL. Each volume measurement with the 10 mL volumetric pipet has an uncertainty of $0.01/10.00 \times 100 = 0.1\%$. The uncertainty in measuring the 0.050 mL aliquot is $0.001/0.050 \times 100 = 2\%$.

CHAPTER FIVE: GASES

QUESTIONS

1. a. Heating the can will increase the pressure of the gas inside the can. $P \propto T$, V & n const. As the pressure increases it may be enough to rupture the can.

   b. As you draw a vacuum in your mouth, atmospheric pressure pushes the liquid up the straw.

   c. The external atmospheric pressure pushes on the can. Since there is no opposing pressure from the air in the inside, the can collapses.

3. PV = nRT = constant at constant n & T. At two sets of conditions, P_1V_1 = const. = P_2V_2.
   $P_1V_1 = P_2V_2$ (Boyle's Law).

   $\frac{V}{T} = \frac{nR}{P}$ = constant at constant n & P. At two sets of conditions, $\frac{V_1}{T_1}$ = const. = $\frac{V_2}{T_2}$

   $\frac{V_1}{T_1} = \frac{V_2}{T_2}$ (Charles's Law)

EXERCISES

Pressure

5. $1 \text{ atm} \times \frac{760 \text{ mm Hg}}{1 \text{ atm}} \times \frac{1 \text{ cm}}{10 \text{ mm}} \times \frac{1 \text{ in}}{2.54 \text{ cm}} = 29.92$ in of Hg

7. $6.5 \text{ cm} \times \frac{10 \text{ mm}}{\text{cm}} = 65$ mm Hg or 65 torr

   $65 \text{ torr} \times \frac{1 \text{ atm}}{760 \text{ torr}} = 8.6 \times 10^{-2}$ atm

   $8.6 \times 10^{-2} \text{ atm} \times \frac{1.01 \times 10^5 \text{ Pa}}{\text{atm}} = 8.7 \times 10^3$ Pa

9. If the levels of Hg in each arm of the manometer are equal, then the pressure in the flask is equal to atmospheric pressure. When they are unequal, the difference in height in mm will be equal to the difference in pressure in mm Hg (torr) between the flask and the atmosphere. Which level is higher will tell us whether the pressure in the flask is less than or greater than atmospheric pressure.

   a. P flask < P atm; P flask = 760. - 140. = 620. torr

   $620. \text{ torr} \times \frac{1 \text{ atm}}{760 \text{ torr}} = 0.816$ atm

   $0.816 \text{ atm} \times \frac{1.013 \times 10^5 \text{ Pa}}{\text{atm}} = 8.27 \times 10^4$ Pa

b. P flask > P atm; P flask = 760. + 175 torr = 935 torr

$$935 \text{ torr} \times \frac{1 \text{ atm}}{760 \text{ torr}} = 1.23 \text{ atm}$$

$$1.23 \text{ atm} \times \frac{1.013 \times 10^5 \text{ Pa}}{\text{atm}} = 1.25 \times 10^5 \text{ Pa}$$

c. P flask = 635 - 140. = 495 torr.; P flask = 635 + 175 = 810. torr

Gas Laws

11. a. PV = nRT

$$V = \frac{nRT}{P} = \frac{(2.00 \text{ mol})\left(\frac{0.08206 \text{ L atm}}{\text{mol K}}\right)(155 + 273) \text{ K}}{5.00 \text{ atm}} = 14.0 \text{ L}$$

b. PV = nRT

$$n = \frac{PV}{RT} = \frac{0.300 \text{ atm} \times 2.00 \text{ L}}{\frac{0.08206 \text{ L atm}}{\text{mol K}} \times 155 \text{ K}} = 4.72 \times 10^{-2} \text{ mol}$$

c. PV = nRT

$$T = \frac{PV}{nR} = \frac{4.47 \text{ atm} \times 25.0 \text{ L}}{2.01 \text{ mol} \times \frac{0.08206 \text{ L atm}}{\text{mol K}}} = 678 \text{ K}$$

d. PV = nRT

$$P = \frac{nRT}{V} = \frac{(10.5 \text{ mol})\left(\frac{0.08206 \text{ L atm}}{\text{mol K}}\right)(273 + 75) \text{ K}}{2.25 \text{ L}} = 133 \text{ atm}$$

13. $R = 0.08206 \frac{\text{L atm}}{\text{mol K}} \times 1.01325 \times 10^5 \frac{\text{Pa}}{\text{atm}} = 8.315 \times 10^3 \frac{\text{L Pa}}{\text{mol K}}$

15. Use the relationship $P_1V_1 = P_2V_2$

For H_2: $P_2 = \frac{P_1 V_1}{V_2} = 475 \text{ torr} \times \frac{1.00 \text{ L}}{1.50 \text{ L}} = 317 \text{ torr}$

For N_2: $P_2 = 45 \text{ kPa} \times \frac{0.50 \text{ L}}{1.50 \text{ L}} = 15 \text{ kPa}$

$$15 \text{ kPa} \times \frac{1000 \text{ Pa}}{\text{kPa}} \times \frac{760 \text{ torr}}{1.013 \times 10^5 \text{ Pa}} = 110 \text{ torr}$$

$P_{total} = P_{H_2} + P_{N_2} = 317 + 110 = 430 \text{ torr}$

17. PV = nRT, Assume n is constant.

$\frac{PV}{T} = nR = \text{constant}; \quad \frac{P_1 V_1}{T_1} = \frac{P_2 V_2}{T_2}$

$$\frac{V_2}{V_1} = \frac{T_2 P_1}{T_1 P_2} = \frac{(273 + 15)\ K \times 720.\ torr}{(273 + 25)\ K \times 605\ torr} = 1.15$$

$V_2 = 1.15\ V_1$ or the volume has increased by 15%

19. $PV = nRT;\ n = \frac{PV}{RT} = \dfrac{145\ atm \times 75 \times 10^{-3}\ L}{\frac{0.08206\ L\ atm}{mol\ K} \times 295\ K} = 0.45\ mol\ O_2$

21. $1.00\ g\ H_2 \times \frac{1\ mol\ H_2}{2.016\ g\ H_2} = 0.496\ mol\ H_2$

$1.00\ g\ He \times \frac{1\ mol\ He}{4.003\ g\ He} = 0.250\ mol\ He$ $\quad P_{H_2} = \chi_{H_2} P_{tot};\quad \chi_{H_2} = \frac{n_{H_2}}{n_{tot}}$

$$P_{H_2} = \left(\frac{0.496}{0.250 + 0.496}\right)(0.480\ atm) = 0.319\ atm$$

$P_{H_2} + P_{He} = P_{tot} = 0.480\ atm$

$P_{He} = 0.480 - 0.319 = 0.161\ atm$

23. PV = nRT, P and n constant

$\frac{V}{T} = \frac{nR}{P} = \text{constant},\ \frac{V_1}{T_1} = \frac{V_2}{T_2},\ V_2 = \frac{V_1 T_2}{T_1};\ T_1 = 273 + 20. = 293\ K$

$V_2 = 700.\ mL \times \frac{100.\ K}{293\ K} = 239\ mL$

25. a. $PV = nRT;\ 175\ g\ Ar \times \frac{1\ mol\ Ar}{39.95\ g\ Ar} = 4.38\ mol\ Ar$

$$T = \frac{PV}{nR} = \frac{2.50\ L \times 10.0\ atm}{4.38\ mol \times \frac{0.08206\ L\ atm}{mol\ K}} = 69.6\ K$$

b. $PV = nRT;\ P = \frac{nRT}{V} = \dfrac{4.38\ mol \times \frac{0.08206\ L\ atm}{mol\ K} \times 225\ K}{2.50\ L} = 32.3\ atm$

27. PV = nRT, V and n constant; $\frac{P}{T} = \frac{nR}{V} = \text{constant},\ \frac{P_1}{T_1} = \frac{P_2}{T_2},$

$$P_2 = \frac{P_1 T_2}{T_1} = 13.7\ MPa \times \frac{(273 + 450)\ K}{(273 + 23)\ K} = 33\ MPa$$

29. $P_{total} = 1.00\ atm = 760.\ torr$

$760. = P_{N_2} + P_{H_2O} = P_{N_2} + 17.5;\ P_{N_2} = 743\ torr$

$$PV = nRT;\ n = \frac{PV}{RT} = \frac{(743\ \text{torr})(2.50 \times 10^2\ \text{mL})}{\left(\frac{0.08206\ \text{L atm}}{\text{mol K}}\right)(293\ \text{K})} \times \frac{1\ \text{atm}}{760\ \text{torr}} \times \frac{1\ \text{L}}{1000\ \text{mL}}$$

$n = 1.02 \times 10^{-2}$ mol N_2

$$1.02 \times 10^{-2}\ \text{mol}\ N_2 \times \frac{28.02\ \text{g}\ N_2}{\text{mol}\ N_2} = 0.286\ \text{g}\ N_2$$

Gas Density, Molar Mass, and Reaction Stoichiometry

31. $PV = nRT;\ \frac{n}{V} = \frac{P}{RT}$

$\frac{nM}{V} = \frac{PM}{RT}$ and $\frac{nM}{V} = d$, M = molar mass, so $d = \frac{PM}{RT}$

For $SiCl_4$, $M = 28.09 + 4(35.45) = \frac{169.89\ \text{g}}{\text{mol}}$

$$d = \frac{758\ \text{torr} \times \frac{169.89\ \text{g}}{\text{mol}}}{\frac{0.08206\ \text{L atm}}{\text{mol K}} \times 358\ \text{K}} \times \frac{1\ \text{atm}}{760\ \text{torr}} = 5.77\ \text{g/L for}\ SiCl_4$$

For $SiHCl_3$, $M = 28.09 + 1.008 + 3(35.45) = \frac{135.45\ \text{g}}{\text{mol}}$

$$d = \frac{PM}{RT} = \frac{758\ \text{torr} \times \frac{135.45\ \text{g}}{\text{mol}}}{\frac{0.08206\ \text{L atm}}{\text{mol K}} \times 358\ \text{K}} \times \frac{1\ \text{atm}}{760\ \text{torr}} = 4.60\ \text{g/L for}\ SiHCl_3$$

33. Out of 100 g of compound, there are:

$$87.4\ \text{g N} \times \frac{1\ \text{mol N}}{14.01\ \text{g N}} = 6.24\ \text{mol N};\ \frac{6.24}{6.24} = 1$$

$$12.6\ \text{g H} \times \frac{1\ \text{mol H}}{1.008\ \text{g H}} = 12.5\ \text{mol H};\ \frac{12.5}{6.24} = 2$$

Empirical formula, NH_2

$PV = nRT \quad n = \frac{g}{M}$ where g = mass in g, M = molar mass in $\frac{g}{mol}$

$PV = \frac{g}{M}RT,\ PM = \frac{g}{V}RT = dRT$

$$M = \frac{dRT}{P} = \frac{\frac{0.977\ \text{g}}{\text{L}} \times \frac{0.08206\ \text{L atm}}{\text{mol K}} \times 373\ \text{K}}{710.\ \text{torr}} \times \frac{760\ \text{torr}}{\text{atm}} = 32.0\ \text{g/mol}$$

Mass of NH_2 = 16.0 g. Therefore, molecular formula is N_2H_4.

35. $27.37 \times 10^9 \text{ lb} \times \frac{1 \text{ kg}}{2.2046 \text{ lb}} \times \frac{1000 \text{ g}}{\text{kg}} \times \frac{1 \text{ mol } NH_3}{17.03 \text{ g}} = 7.290 \times 10^{11} \text{ mol of } NH_3$

$7.290 \times 10^{11} \text{ mol } NH_3 \times \frac{1 \text{ mol } N_2}{2 \text{ mol } NH_3} = 3.645 \times 10^{11} \text{ mol } N_2$

$$PV = nRT;\ V = \frac{nRT}{P} = \frac{3.645 \times 10^{11} \text{ mol} \times \frac{0.08206 \text{ L atm}}{\text{mol K}} \times 273.2 \text{ K}}{1.000 \text{ atm}}$$

$V = 8.172 \times 10^{12} \text{ L of } N_2 \text{ at STP}$

$7.290 \times 10^{11} \text{ mol } NH_3 \times \frac{3 \text{ mol } H_2}{2 \text{ mol } NH_3} = 1.094 \times 10^{12} \text{ mol } H_2$

N_2 and H_2 are measured at the same pressure and temperature.

PV = nRT, P and T constant

$\frac{V}{n} = \frac{RT}{P} = \text{constant},\ \frac{V_1}{n_1} = \frac{V_2}{n_2}$

$$V_{H_2} = \frac{V_{N_2} \times n_{H_2}}{n_{N_2}} = 8.172 \times 10^{12} \text{ L} \times \frac{1.094 \times 10^{12} \text{ mol}}{3.645 \times 10^{11} \text{ mol}}$$

$V_{H_2} = 2.453 \times 10^{13} \text{ L of } H_2 \text{ at STP}$

There are several other ways to do this problem.

37. The unbalanced reaction is: $C_8H_{18} + O_2 \rightarrow CO_2 + H_2O$

The balanced equation is: $2\ C_8H_{18} + 25\ O_2 \rightarrow 16\ CO_2 + 18\ H_2O$

$125 \text{ g } C_8H_{18} \times \frac{1 \text{ mol } C_8H_{18}}{114.22 \text{ g } C_8H_{18}} \times \frac{25 \text{ mol } O_2}{2 \text{ mol } C_8H_{18}} = 13.7 \text{ mol } O_2$

$$PV = nRT,\ V = \frac{nRT}{P} = \frac{13.7 \text{ mol} \times \frac{0.08206 \text{ L atm}}{\text{mol K}} \times 273 \text{ K}}{1.00 \text{ atm}} = 307 \text{ L of } O_2$$

Or we can make use of the fact that at STP one mole of an ideal gas occupies a volume of 22.42 L. This can be calculated using the ideal gas law.

So: $13.7 \text{ mol } O_2 \times \frac{22.42 \text{ L}}{\text{mol}} = 307 \text{ L}$

39. a. $CH_4(g) + NH_3(g) + O_2(g) \rightarrow HCN(g) + H_2O(g)$

$CH_4 + NH_3 + O_2 \rightarrow HCN + 3\ H_2O;\ CH_4 + NH_3 + \frac{3}{2} O_2 \rightarrow HCN + 3\ H_2O$

$2\ CH_4(g) + 2\ NH_3(g) + 3\ O_2(g) \rightarrow 2\ HCN(g) + 6\ H_2O(g)$

b. $PV = nRT$; $n \propto V$ since P & T are constant.

$$20.0 \text{ L } CH_4 \times \frac{2 \text{ L HCN}}{2 \text{ L } CH_4} = 20.0 \text{ L HCN};\ 20.0 \text{ L } NH_3 \times \frac{2 \text{ L HCN}}{2 \text{ L } NH_3} = 20.0 \text{ L HCN}$$

$$20.0 \text{ L } O_2 \times \frac{2 \text{ L HCN}}{3 \text{ L } O_2} = 13.3 \text{ L HCN};$$ O_2 is limiting, 13.3 L of HCN is produced.

41. $P_{total} = P_{N_2} + P_{H_2O}$

$$P_{N_2} = 726 \text{ torr} - 23.8 \text{ torr} = 702 \text{ torr} \times \frac{1 \text{ atm}}{760 \text{ torr}} = 0.924 \text{ atm}$$

$$PV = nRT;\quad n = \frac{PV}{RT} = \frac{0.924 \text{ atm} \times 31.8 \times 10^{-3} \text{ L}}{\frac{0.08206 \text{ L atm}}{\text{mol K}} \times 298 \text{ K}} = 1.20 \times 10^{-3} \text{ mol } N_2$$

$$\text{Mass of N in compound} = 1.20 \times 10^{-3} \text{ mol} \times \frac{28.02 \text{ g } N_2}{\text{mol}} = 3.36 \times 10^{-2} \text{ g}$$

$$\%N = \frac{3.36 \times 10^{-2} \text{ g}}{0.253 \text{ g}} \times 100 = 13.3\% \text{ N}$$

43. $$3.7 \text{ g } KClO_3 \times \frac{1 \text{ mol } KClO_3}{122.55 \text{ g } KClO_3} \times \frac{3 \text{ mol } O_2}{2 \text{ mol } KClO_3} = 4.5 \times 10^{-2} \text{ mol } O_2$$

$P_{total} = P_{O_2} + P_{H_2O}$

$$P_{O_2} = P_{total} - P_{H_2O} = 735 - 26.7 = 708 \text{ torr} \times \frac{1 \text{ atm}}{760 \text{ torr}} = 0.932 \text{ atm}$$

$$V = \frac{nRT}{P} = \frac{4.5 \times 10^{-2} \text{ mol} \times \frac{0.08206 \text{ L atm}}{\text{mol K}} \times 300. \text{ K}}{0.932 \text{ atm}} = 1.2 \text{ L} = 1200 \text{ mL}$$

Kinetic Molecular Theory and Real Gases

45. $(KE)_{avg} = 3/2\ RT$

$$\text{at 273 K: } (KE)_{avg} = \frac{3}{2} \times \frac{8.3145 \text{ J}}{\text{mol K}} \times 273 \text{ K} = 3.40 \times 10^3 \text{ J/mol}$$

$$\text{at 546 K: } (KE)_{avg} = \frac{3}{2} \times \frac{8.3145 \text{ J}}{\text{mol K}} \times 546 \text{ K} = 6.81 \times 10^3 \text{ J/mol}$$

47. $u_{rms} = \left(\frac{3\,RT}{M}\right)^{1/2}$, where $R = \frac{8.3145 \text{ J}}{\text{mol K}}$ and M = molar mass in kg; 1.604×10^{-2} kg/mol for CH_4

$$\text{For } CH_4 \text{ at 273 K, } u_{rms} = \left(\frac{\frac{3 \times 8.3145 \text{ J}}{\text{mol K}} \times 273 \text{ K}}{1.604 \times 10^{-2} \text{ kg/mol}}\right)^{1/2} = 652 \text{ m/s}$$

Similarly u_{rms} for CH_4 at 546 K is 921 m/s.

49. No, the number calculated in 5.45 is the average energy. There is a distribution of energies.

51. a. They will all have the same average kinetic energy since they are all at the same temperature.

b. Flask C, H_2 has the smallest molar mass. Lightest molecules are fastest.

53. $\frac{R_1}{R_2} = \left(\frac{M_2}{M_1}\right)^{1/2}$; $\frac{31.50}{30.50} = \left(\frac{32.00}{M}\right)^{1/2} = 1.033$, M = molar mass of the unknown;

$\frac{32.00}{M} = 1.067$, so M = 29.99

Of the choices, the gas would be NO, nitric oxide.

55. a. PV = nRT

$$P = \frac{nRT}{V} = \frac{0.5000 \text{ mol}\left(\frac{0.08206 \text{ L atm}}{\text{mol K}}\right)(25.0 + 273.2)\text{ K}}{1.000 \text{ L}} = 12.24 \text{ atm}$$

b. $\left[P + a\left(\frac{n}{V}\right)^2\right](V - nb) = nRT$; For N_2: $a = 1.39$ atm L^2/mol^2 and $b = 0.0391$ L/mol

$$\left[P + 1.39\left(\frac{0.5000}{1.000}\right)^2 \text{atm}\right](1.000 \text{ L} - 0.5000 \times 0.0391 \text{ L}) = 12.24 \text{ L atm}$$

(P + 0.348 atm)(0.9805 L) = 12.24 L atm (carry extra significant figure)

$$P = \frac{12.24 \text{ L atm}}{0.9805 \text{ L}} - 0.348 \text{ atm} = 12.48 - 0.35 = 12.13 \text{ atm}$$

c. The ideal gas is high by 0.11 atm or $\frac{0.11}{12.13} \times 100 = 0.91\%$ if we carry extra digits through the entire calculation.

Atmospheric Chemistry

57. $\chi_{NO} = 5 \times 10^{-7}$ from Table 5.4.

$P_{NO} = \chi_{NO} P_{total} = 5 \times 10^{-7} \times 1.0 \text{ atm} = 5 \times 10^{-7}$ atm

$$PV = nRT;\ \frac{n}{V} = \frac{P}{RT} = \frac{5 \times 10^{-7} \text{ atm}}{\left(\frac{0.08206 \text{ L atm}}{\text{mol K}}\right)(273 \text{ K})} = 2 \times 10^{-8} \text{ mol/L}$$

$$\frac{2 \times 10^{-8} \text{ mol}}{\text{L}} \times \frac{1 \text{ L}}{1000 \text{ cm}^3} \times \frac{6.022 \times 10^{23} \text{ molecules}}{\text{mol}} = 1 \times 10^{13} \text{ molecules/cm}^3$$

59. At 100. km, T ≈ -90° C and P ≈ $10^{-5.5} \approx 3 \times 10^{-6}$ atm

$$PV = nRT;\ \frac{PV}{T} = nR = \text{Const.};\ \frac{P_1V_1}{T_1} = \frac{P_2V_2}{T_2}$$

$$V_2 = \frac{V_1P_1T_2}{T_1P_2} = \frac{10.0\ \text{L} \times 3 \times 10^{-6}\ \text{atm} \times 273\ \text{K}}{183\ \text{K} \times 1.0\ \text{atm}} = 4 \times 10^{-5}\ \text{L} = 0.04\ \text{mL}$$

61. $N_2(g) + O_2(g) \rightarrow 2\ NO(g)$ (automobile combustion or caused by lightning)

$2\ NO(g) + O_2(g) \rightarrow 2\ NO_2(g)$ (reaction with atmospheric O_2)

$2\ NO_2(g) + H_2O(l) \rightarrow HNO_3(aq) + HNO_2(aq)$ (reaction with atmospheric H_2O)

$S(s) + O_2(g) \rightarrow SO_2(g)$ (combustion of coal)

$2\ SO_2(g) + O_2(g) \rightarrow 2\ SO_3(g)$ (reaction with atmospheric O_2)

$H_2O(l) + SO_3(g) \rightarrow H_2SO_4(aq)$ (reaction with atmospheric H_2O)

ADDITIONAL EXERCISES

63. a. PV = nRT, n and T Const

PV = Constant

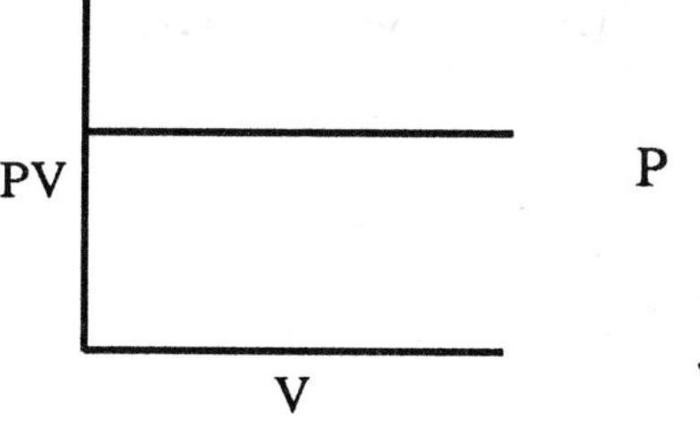

b. PV = nRT

$$P = \left(\frac{nR}{V}\right)T = \text{Const} \times T$$

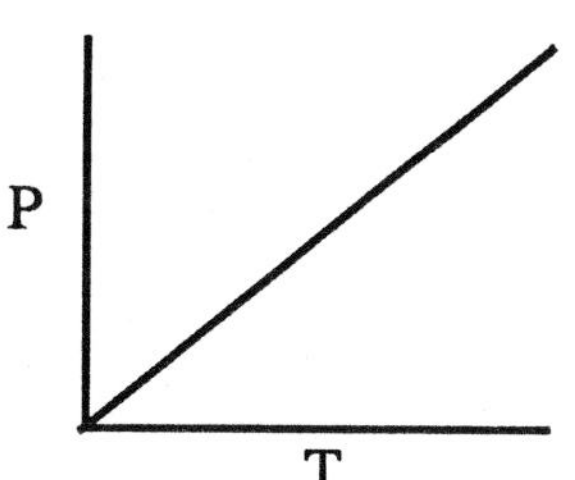

c. PV = nRT

$$T = \left(\frac{P}{nR}\right)V = \text{Const} \times V$$

T

V

d. PV = nRT

$$P = \frac{nRT}{V} = \frac{\text{Const}}{V}$$

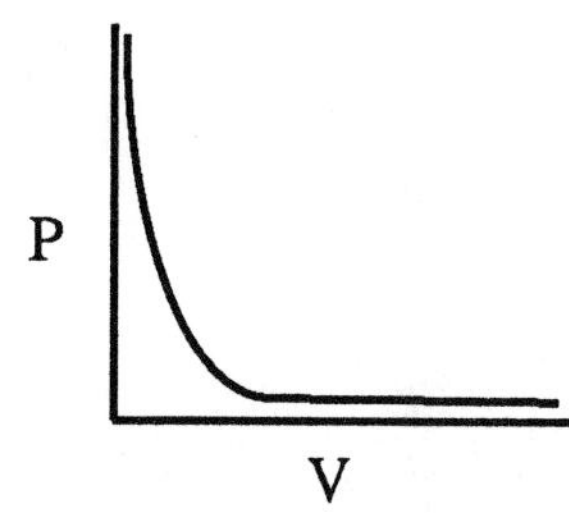

e. $$P = \frac{nRT}{V} = \frac{\text{Const}}{V} = \text{Const}\left(\frac{1}{V}\right)$$

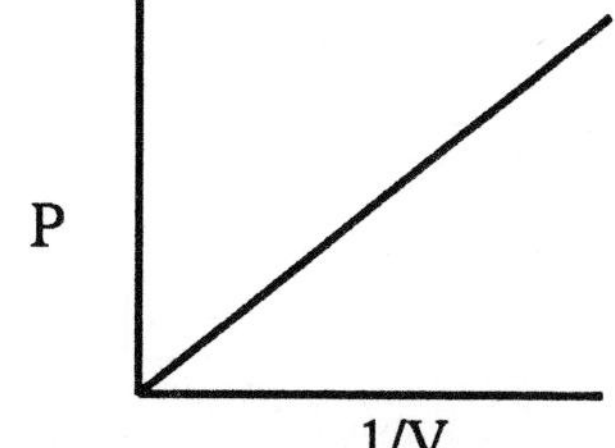

f. PV = nRT

$$\frac{PV}{T} = nR = \text{Const}$$

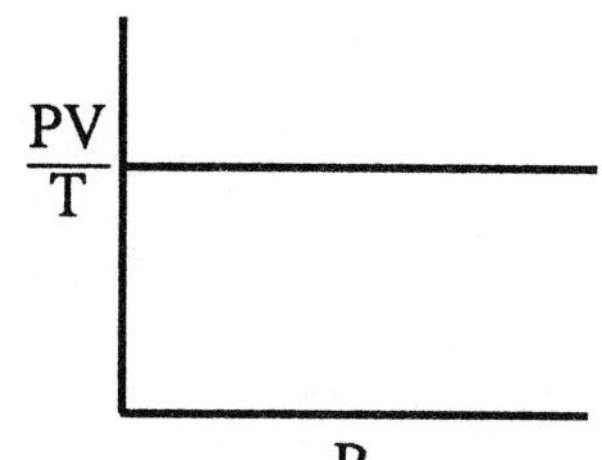

65. a. PV = nRT, n and V are constant.

$$\frac{P}{T} = \frac{nR}{V} \text{ or } \frac{P_1}{T_1} = \frac{P_2}{T_2};\ P_2 = \frac{P_1 T_2}{T_1} = 40.0 \text{ atm} \times \frac{318 \text{ K}}{273 \text{ K}} = 46.6 \text{ atm}$$

b. $$\frac{P_1}{T_1} = \frac{P_2}{T_2};\ T_2 = \frac{T_1 P_2}{P_1} = 273 \text{ K} \times \frac{150. \text{ atm}}{40.0 \text{ atm}} = 1.02 \times 10^3 \text{ K}$$

c. $$T_2 = \frac{T_1 P_2}{P_1} = 273 \text{ K} \times \frac{25.0 \text{ atm}}{40.0 \text{ atm}} = 171 \text{ K}$$

67. $P_{He} + P_{H_2O} = 1.0$ atm = 760 torr, $P_{H_2O} = 23.8$ torr; $P_{He} = 736$ torr = 740 torr (2 sig. figs.)

$P_{tot}V_{tot} = n_{tot}RT$, $V_{tot} = V_{He} + V_{H_2O} = V_{wet}$

$$P_{He} = \chi_{He}P_{tot} = \frac{n_{He}}{n_{tot}}(760) = 740;\ n_{He} = 0.586 \text{ g} \times \frac{1 \text{ mol}}{4.003 \text{ g}} = 0.146 \text{ mol}$$

$$740 = \frac{0.146 \times 760}{n_{tot}},\ n_{tot} = 0.15 \text{ mol};\ V_{wet} = \frac{0.15 \times 0.08206 \times 298}{1.0} = 3.7 \text{ L}$$

69. $Br_2 + 3\ F_2 \rightarrow 2$ X; Two moles of X must contain two moles of Br and 6 moles of F; X must have the formula, BrF_3.

71. $$PV = nRT;\ \frac{nT}{P} = \frac{V}{R} = \text{Const.};\ \frac{n_1 T_1}{P_1} = \frac{n_2 T_2}{P_2};$$

$$\text{mol} \times \frac{\text{g}}{\text{mol}} = \text{g};\ \frac{n_1 M T_1}{P_1} = \frac{n_2 M T_2}{P_2};\ \frac{g_1 T_1}{P_1} = \frac{g_2 T_2}{P_2}$$

$$g_2 = \frac{g_1 T_1 P_2}{T_2 P_1} = \frac{500. \text{ g} \times 291 \text{ K} \times 980. \text{ psi}}{299 \text{ K} \times 2050. \text{ psi}} = 233 \text{ g}$$

73. If Be^{3+}, formula is $Be(C_5H_7O_2)_3$ and M ≈ 13.5 + 15(12) + 21(1) + 6(16) = 311 g/mol

If Be^{2+}, formula is $Be(C_5H_7O_2)_2$ and M ≈ 9.0 + 10(12) + 14(1) + 4(16) = 207 g/mol

Data Set I:

$$M = \frac{gRT}{PV} = \frac{0.2022 \text{ g} \times \frac{0.08206 \text{ L atm}}{\text{mol K}} \times 286 \text{ K} \times 760 \frac{\text{torr}}{\text{atm}}}{22.6 \times 10^{-3} \text{ L} \times 765.2 \text{ torr}} = 209 \text{ g/mol}$$

Data Set II:

$$M = \frac{gRT}{PV} = \frac{0.2224 \text{ g} \times \frac{0.08206 \text{ L atm}}{\text{mol K}} \times 290. \text{ K} \times 760 \frac{\text{torr}}{\text{atm}}}{26.0 \times 10^{-3} \text{ L} \times 764.6 \text{ torr}} = 202 \text{ g/mol}$$

These results are close to the expected value of 207 g/mol for $Be(C_5H_7O_2)_2$. Thus, we conclude from these data that beryllium is a divalent element with an atomic weight of 9.0 g/mol.

75. Out of 100.0 g compounds there are:

$$58.51 \text{ g C} \times \frac{1 \text{ mol C}}{12.01 \text{ g C}} = 4.872 \text{ mol C}; \; \frac{4.872}{2.435} = 2$$

$$7.37 \text{ g H} \times \frac{1 \text{ mol H}}{1.008 \text{ g H}} = 7.31 \text{ mol H}; \; \frac{7.31}{2.435} = 3$$

$$34.12 \text{ g N} \times \frac{1 \text{ mol N}}{14.01 \text{ g N}} = 2.435 \text{ mol N}$$

Empirical formula: C_2H_3N

$$\frac{\text{Rate (1)}}{\text{Rate (2)}} = \left(\frac{M_2}{M_1}\right)^{1/2}; \text{ Let Gas (1) = He;} \qquad 3.20 = \left(\frac{M_2}{4.003}\right)^{1/2}$$

$M_2 = 41.0$, Mass of C_2H_3N: 2(12.0) + 3(1.0) + 1(14.0) = 41.0

So molecular formula is also C_2H_3N

77. $PV = nRT = \text{Const.}$; $P_1V_1 = P_2V_2$

Let Set (1) correspond to He from the tank that can be used to fill balloons.

P_1 = 199 atm (200. - 1) = 199. We must leave 1.0 atm of He in the tank.

V_1 = 15.0 L

Set (2) will correspond to the filled balloon. P_2 = 1.00 atm

V_2 = N(2.00 L) where N is the number of filled balloons.

(199)(15.0) = (1.00)(2.00) N; N = 1492.5; We can't fill 0.5 of a balloon.

So N = 1492 balloons or to 3 significant figures 1490 balloons.

79. The van der Waals' constant, *b*, is a measure of the size of the molecule. Thus, C_3H_8 should have the largest value of *b*.

CHALLENGE PROBLEMS

81. $M = \frac{dRT}{P}$, P & M constant.

$dT = \frac{PM}{R} = \text{const}$ or $d_1T_1 = d_2T_2$, where T is the absolute temperature.

$T = °C + x$; $1.2930(0.0 + x) = 0.9460(100.0 + x)$

$1.2930x = 0.9460x + 94.60$; $0.3470x = 94.60$; $x = 272.6$

From these data absolute zero would be -272.6 °C. Actual value: -273.15 °C.

83. $\frac{\text{Rate (1)}}{\text{Rate (2)}} = \left(\frac{M_2}{M_1}\right)^{1/2}$; Let Gas (1) = He, Gas (2) = Cl_2

$$\frac{\frac{1.0\ L}{4.5\ min}}{\frac{1.0\ L}{t}} = \left(\frac{70.90}{4.003}\right)^{1/2}; \quad \frac{t}{4.5\ min} = 4.21; \ t = 19\ min$$

85. $\left(P + \frac{an^2}{V^2}\right)(V\text{-}nb) = nRT$; $PV + \frac{an^2 V}{V^2} - nbP - \frac{an^3 b}{V^2} = nRT$; $PV + \frac{an^2}{V^2} - nbP - \frac{an^3 b}{V^2} = nRT$

At low P and high T, the molar volume of a gas will be relatively large. $\frac{an^2}{V}$ and $\frac{an^3 b}{V^2}$ become negligible because V is large.

Since nb is the actual volume of the gas molecules themselves: nb << V and -nbP is negligible compared to PV. Thus PV = nRT.

87. PV = nRT, P constant; $\frac{nT}{V} = \frac{P}{R}$, $\frac{n_1 T_1}{V_1} = \frac{n_2 T_2}{V_2}$

$$\frac{n_2}{n_1} = \frac{T_1 V_2}{T_2 V_1} = \frac{294\ K}{335\ K} \times \frac{4.20 \times 10^3\ m^3}{4.00 \times 10^3\ m^3} = 0.921$$

89. 10.10 atm - 7.62 atm = 2.48 atm is a pressure of the amount of F_2 reacted.

PV = nRT, V and T are constant. $\frac{P}{n}$ = constant, $\frac{P_1}{n_1} = \frac{P_2}{n_2}$ or $\frac{P_1}{P_2} = \frac{n_1}{n_2}$

$$\frac{\text{moles } F_2 \text{ reacted}}{\text{moles Xe reacted}} = \frac{2.48\ atm}{1.24\ atm} = 2; \ \text{So Xe} + 2\ F_2 \longrightarrow XeF_4$$

91. PV = nRT, V and T are constant. $\frac{P_1}{n_1} = \frac{P_2}{n_2}$, $\frac{P_2}{P_1} = \frac{n_2}{n_1}$

Let's calculate the partial pressure of C_3H_3N that can be produced from each of the starting materials.

$$P = 0.50\ MPa \times \frac{2\ mol\ C_3H_3N}{2\ mol\ C_3H_6} = 0.50\ MPa \text{ if } C_3H_6 \text{ is limiting.}$$

$$P = 0.80\ MPa \times \frac{2\ mol\ C_3H_3N}{2\ mol\ NH_3} = 0.80\ MPa \text{ if } NH_3 \text{ is limiting.}$$

$$P = 1.50\ MPa \times \frac{2\ mol\ C_3H_3N}{3\ mol\ O_2} = 1.00\ MPa \text{ if } O_2 \text{ is limiting.}$$

Thus, C_3H_6 is limiting. The partial pressure of C_3H_3N after the reaction is:

$$0.50 \times 10^6\ Pa \times \frac{1\ atm}{1.013 \times 10^5\ Pa} = 4.9\ atm$$

$$n = \frac{PV}{RT} = \frac{4.9 \text{ atm} \times 150. \text{ L}}{\frac{0.08206 \text{ L atm}}{\text{mol K}} \times 298 \text{ K}} = 30. \text{ mol } C_3H_3N$$

$$30. \text{ mol} \times \frac{53.06 \text{ g}}{\text{mol}} = 1.6 \times 10^3 \text{ g}$$

93. a. If we have 10^6 L of air, there are 3.0×10^2 L of CO.

$$P_{CO} = \chi_{CO}P_{tot}; \quad \chi_{CO} = \frac{V_{CO}}{V_{tot}} \text{ since } V \propto n.$$

$$P_{CO} = \frac{3.0 \times 10^2}{1.0 \times 10^6} \times 628 = 0.19 \text{ torr}$$

b. $n_{CO} = \frac{P_{CO}V}{RT}$ Assume 1.0 m^3 air, 1.0 m^3 = 1.0×10^3 L

$$n_{CO} = \frac{\frac{0.19}{760} \text{ atm} \times 1.0 \times 10^3 \text{ L}}{\frac{0.08206 \text{ L atm}}{\text{mol K}} \times 273 \text{ K}} = 1.1 \times 10^{-2} \text{ mol}$$

$$1.1 \times 10^{-2} \text{ mol} \times \frac{6.022 \times 10^{23} \text{ molecules}}{\text{mol}} = 6.6 \times 10^{21} \text{ molecules in the 1.0 m}^3 \text{ of air}$$

c. $$\frac{6.6 \times 10^{21} \text{ molecules}}{m^3} \times \left(\frac{1 \text{ m}}{100 \text{ cm}}\right)^3 = \frac{6.6 \times 10^{15} \text{ molecules}}{\text{cm}^3}$$

95. For benzene: $101 \times 10^{-9} \text{ g} \times \frac{1 \text{ mol}}{78.11 \text{ g}} = 1.29 \times 10^{-9}$ mol

$$V = \frac{nRT}{P} = \frac{1.29 \times 10^{-9} \text{ mol} \times \frac{0.08206 \text{ L atm}}{\text{mol K}} \times 296 \text{ K}}{748 \text{ torr}} \times \frac{760 \text{ torr}}{\text{atm}} = 3.18 \times 10^{-8} \text{ L}$$

$$\text{Mixing ratio} = \frac{3.18 \times 10^{-8} \text{ L}}{2.0 \text{ L}} \times 10^6 = 1.6 \times 10^{-2} \text{ ppmv}$$

or better yet: Use ppbv, defined similarly to ppmv in problem 5.93.

$$\text{ppbv} = \frac{\text{vol. of } x}{\text{total vol.}} \times 10^9$$

In this case Mixing ratio = 16 ppbv

$$\frac{1.29 \times 10^{-9} \text{ mol benzene}}{2.0 \text{ L}} \times \frac{1 \text{ L}}{1000 \text{ cm}^3} \times \frac{6.022 \times 10^{23} \text{ molecules}}{\text{mol}}$$

$$= 3.9 \times 10^{11} \text{ molecules benzene/cm}^3$$

For toluene: 274×10^{-9} g $C_7H_8 \times \dfrac{1 \text{ mol}}{92.13 \text{ g}} = 2.97 \times 10^{-9}$ mol toluene

$$V_{tol} = \frac{nRT}{P} = \frac{2.97 \times 10^{-9} \text{ mol} \times \dfrac{0.08206 \text{ L atm}}{\text{mol K}} \times 296 \text{ K}}{748 \text{ torr}} \times \frac{760 \text{ torr}}{\text{atm}} = 7.33 \times 10^{-8} \text{ L}$$

$$\text{Mixing ratio} = \frac{7.33 \times 10^{-8} \text{ L}}{2.0 \text{ L}} \times 10^9 = 37 \text{ ppbv}$$

$$\frac{2.97 \times 10^{-9} \text{ mol toluene}}{2.0 \text{ L}} \times \frac{1 \text{ L}}{1000 \text{ cm}^3} \times \frac{6.022 \times 10^{23} \text{ molecules}}{\text{mol}}$$

$$= 8.9 \times 10^{11} \text{ molecules toluene/cm}^3$$

CHAPTER SIX: THERMOCHEMISTRY

QUESTIONS

1. A solution calorimeter is at constant (atmospheric) pressure. The heat at constant P is ΔH. A bomb calorimeter is at constant volume. The heat at constant volume is ΔE.

3. A state function is a function whose change depends only on the initial and final states and not on how one got from the initial to the final state. If H and E were not state functions, the law of conservation of energy (first law) would not be true.

5. We can only measure differences in enthalpy. Thus, in order for different workers to compare measured values of ΔH to each other, a common reference (or zero) point must be chosen. The definition of ΔH_f° establishes the pure elements in their standard states as that common reference point.

EXERCISES

Potential and Kinetic Energy

7. $KE = \frac{1}{2}mv^2$, convert mass and velocity to SI units. $1\ J = \frac{1\ kg\ m^2}{s^2}$

$$5.25\ oz \times \frac{1\ lb}{16\ oz} \times \frac{1\ kg}{2.205\ lb} = 0.149\ kg = mass$$

$$Velocity = \frac{1.0 \times 10^2\ mi}{hr} \times \frac{1\ hr}{60\ min} \times \frac{1\ min}{60\ s} \times \frac{1760\ yd}{mi} \times \frac{1\ m}{1.094\ yd} = \frac{45\ m}{s}$$

$$KE = \frac{1}{2}mv^2 = \frac{1}{2}\ (0.149\ kg)\left(\frac{45\ m}{s}\right)^2 = 150\ J$$

9. $KE = \frac{1}{2}mv^2 = \frac{1}{2}\ (2.0\ kg)\left(\frac{1.0\ m}{s}\right)^2 = 1.0\ J$

$$KE = \frac{1}{2}mv^2 = \frac{1}{2}\ (1.0\ kg)\left(\frac{2.0\ m}{s}\right)^2 = 2.0\ J$$

The 1.0 kg object with a velocity of 2.0 m/s has the greater kinetic energy.

Heat and Work

11. a. $\Delta E = q + w = 51\ kJ + (-15\ kJ) = 36\ kJ$

    b. $\Delta E = 100.\ kJ + (-65\ kJ) = 35\ kJ$ c. $\Delta E = -65 - 20. = -85\ kJ$

    d. When the system delivers work to the surroundings, $w < 0$. This is the case in all these examples, a, b and c.

13. $\Delta E = q + w = +45 - 29 = +16\ kJ$

15. $w = -P\Delta V = -P(V_f - V_i) = -2.0\ atm\ (5.0 \times 10^{-3}\ L - 5.0\ L)$

$$w = -2.0\ atm\ (-5.0\ L) = 10.\ L\ atm$$

We can also calculate the work in Joules.

$$1 \text{ atm} = 1.013 \times 10^5 \text{ Pa} = 1.013 \times 10^5 \frac{\text{N}}{\text{m}^2}$$

$$1 \text{ m}^3 = (10 \text{ dm})^3 = 1000 \text{ dm}^3 = 1000 \text{ L}$$

$$1 \text{ L atm} = 10^{-3} \text{ m}^3 \times 1.013 \times 10^5 \frac{\text{N}}{\text{m}^2} = 101.3 \text{ N m} = 101.3 \text{ J}$$

$$w = 10. \text{ L atm} \times \frac{101.3 \text{ J}}{\text{L atm}} = 1013 \text{ J} = 1.0 \times 10^3 \text{ J}$$

17. $q = \frac{20.8 \text{ J}}{°\text{C mol}} \times 39.1 \text{ mol} \times (38.0 - 0.0) \text{ °C}$

$q = 30{,}900 \text{ J} = 30.9 \text{ kJ}$

$w = -P\Delta V = -1.00 \text{ atm}(998 \text{ L} - 875 \text{ L}) = -123 \text{ L atm} \times \frac{101.3 \text{ J}}{\text{L atm}}$

$w = -12{,}500 \text{ J} = -12.5 \text{ kJ}$

$\Delta E = q + w = 30.9 \text{ kJ} + (-12.5 \text{ kJ}) = 18.4 \text{ kJ}$

Properties of Enthalpy

19. a. endothermic b. exothermic c. exothermic d. endothermic

21. $S(s) + O_2(g) \longrightarrow SO_2(g) \quad \Delta H° = \frac{-296 \text{ kJ}}{\text{mol}}$

a. $275 \text{ g S} \times \frac{1 \text{ mol S}}{32.07 \text{ g S}} \times \frac{296 \text{ kJ}}{\text{mol S}} = 2.54 \times 10^3 \text{ kJ}$

b. $25 \text{ mol S} \times \frac{296 \text{ kJ}}{\text{mol S}} = 7.4 \times 10^3 \text{ kJ}$

c. $150. \text{ g } SO_2 \times \frac{1 \text{ mol } SO_2}{64.07 \text{ g } SO_2} \times \frac{296 \text{ kJ}}{\text{mol } SO_2} = 693 \text{ kJ}$

Calorimetry and Heat Capacity

23. a. $\Delta H = q = 850. \text{ g} \times \frac{0.900 \text{ J}}{\text{g °C}} \times (94.6 - 22.8) \text{ °C} = 5.49 \times 10^4 \text{ J or } 54.9 \text{ kJ}$

b. $\frac{0.900 \text{ J}}{\text{g °C}} \times \frac{26.98 \text{ g}}{\text{mol}} = \frac{24.3 \text{ J}}{\text{mol °C}}$

25. Specific heat capacity (C): J/g•°C

$C = \frac{78.2 \text{ J}}{45.6 \text{ g} \times 13.3 \text{ °C}} = \frac{0.129 \text{ J}}{\text{g °C}}$

Molar heat capacity $= \frac{0.129 \text{ J}}{\text{g °C}} \times \frac{207.2 \text{ g}}{\text{mol}} = \frac{26.7 \text{ J}}{\text{mol °C}}$

27. Heat gained by water = Heat lost by nickel

Heat gain = $\frac{4.18 \text{ J}}{\text{g °C}} \times 150. \text{ g} \times (25.0 \text{ °C} - 23.5 \text{ °C}) = 940 \text{ J}$

Note: A temperature change of one Kelvin is the same as a temperature change of one degree Celsius.

Heat loss $= C \times m \times \Delta T$

$940 \text{ J} = C \times 28.2 \text{ g} \times (99.8 - 25.0) \text{ °C}$

$C = \frac{940 \text{ J}}{28.2 \text{ g} \times 74.8 \text{ °C}} = \frac{0.45 \text{ J}}{\text{g °C}}$

29. Heat lost by solution = Heat gained by KBr

Heat lost by solution = $\frac{4.18 \text{ J}}{\text{g °C}} \times 136 \text{ g} \times 3.1 \text{ °C} = 1800 \text{ J}$

Heat gained by KBr = $\frac{1800 \text{ J}}{10.5 \text{ g}} = \frac{170 \text{ J}}{\text{g}}$

$$\frac{170 \text{ J}}{\text{g KBr}} \times \frac{119.00 \text{ g KBr}}{\text{mol KBr}} \times \frac{1 \text{ kJ}}{1000 \text{ J}} = \frac{20.2 \text{ kJ}}{\text{mol}} = \frac{2.0 \times 10^1 \text{ kJ}}{\text{mol}} = \Delta H$$

31. Heat lost by camphor = Heat gained by calorimeter

Heat lost by combustion of camphor = $\frac{5903.6 \text{ kJ}}{\text{mol}} \times \frac{1 \text{ mol}}{152.23 \text{ g}} \times 0.1204 \text{ g} = 4.669 \text{ kJ}$

Heat gained by calorimeter = $C \times \Delta T$; $C \times (2.28 \text{ °C}) = 4.669 \text{ kJ}$; $C = 2.05 \text{ kJ/°C}$

33. $50.0 \times 10^{-3} \text{ L} \times \frac{0.100 \text{ mol}}{\text{L}} = 5.00 \times 10^{-3}$ mol of both $AgNO_3$ and HCl is reacted.

Thus, 5.00×10^{-3} mol of AgCl will be produced.

Heat lost by chemicals = Heat gain by solution

Heat Gain = $\frac{4.18 \text{ J}}{\text{g °C}} \times 100. \text{ g} \times 0.8\text{°C} = 300 \text{ J}$

Heat Loss = 300 J: this is the heat evolved (exothermic reaction) when 5.00×10^{-3} mol of AgCl is produced.

Thus, $\Delta H = \frac{-300 \text{ J}}{5.00 \times 10^{-3} \text{ mol}} \times \frac{1 \text{ kJ}}{1000 \text{ J}} = -60 \text{ kJ/mol}$

Since, we only know the temperature difference to one significant figure, $\Delta H = -60$ kJ/mol (without intermediate rounding we get $\Delta H = -66.9$ kJ/mol ≈ -70 kJ/mol).

Hess's Law

35. $S + 3/2\ O_2 \rightarrow SO_3$ $\Delta H° = -395.2$ kJ

$SO_3 \rightarrow SO_2 + 1/2\ O_2$ $\Delta H° = +198.2$ kJ/2 = +99.1 kJ ÷(-2)

$S(s) + O_2(g) \rightarrow SO_2(g)$ $\Delta H° = -296.1$ kJ

37. $4\ HNO_3 \rightarrow 2\ N_2O_5 + 2\ H_2O$ $\Delta H° = 2(+76.6$ kJ) ×(-2)

$2\ N_2 + 6\ O_2 + 2\ H_2 \rightarrow 4\ HNO_3$ $\Delta H° = 4(-174.1$ kJ) ×4

$2\ H_2O \rightarrow 2\ H_2 + O_2$ $\Delta H° = 2(+285.8$ kJ) ×(-2)

$2\ N_2(g) + 5\ O_2\ (g) \rightarrow 2\ N_2O_5(g)$ $\Delta H° = 28.4$ kJ

39. $NO + O_3 \rightarrow NO_2 + O_2$ $\Delta H° = -199$ kJ

$3/2\ O_2 \rightarrow O_3$ $\Delta H° = +427$ kJ/2 ÷(-2)

$O \rightarrow 1/2\ O_2$ $\Delta H° = -495$ kJ/2 ÷(-2)

$NO(g) + O(g) \rightarrow NO_2(g)$ $\Delta H° = -233$ kJ

41. $1/2\ O_2 + 1/2\ H_2 \rightarrow OH$ $\Delta H° = 77.9$ kJ/2 ÷2

$O \rightarrow 1/2\ O_2$ $\Delta H° = -495$ kJ/2 ÷(-2)

$H \rightarrow 1/2\ H_2$ $\Delta H° = -435.9$ kJ/2 ÷(-2)

$O(g) + H(g) \rightarrow OH(g)$ $\Delta H° = -427$ kJ

Standard Enthalpies of Formation

43. The enthalpy change for the formation of one mole of a compound at 25°C from its elements, with all substances in their standard states.

$Na(s) + 1/2\ Cl_2(g) \rightarrow NaCl(s)$

$H_2(g) + 1/2\ O_2(g) \rightarrow H_2O(l)$

$6\ C(graphite) + 6\ H_2(g) + 3\ O_2(g) \rightarrow C_6H_{12}O_6(s)$

$Pb(s) + S(s) + 2\ O_2(g) \rightarrow PbSO_4(s)$

45. In general:

$$\Delta H° = \Sigma\Delta H_f° \text{ (products)} - \Sigma\Delta H_f° \text{ (reactants)}$$

a. $2\ NH_3(g) + 3\ O_2(g) + 2\ CH_4(g) \longrightarrow 2\ HCN(g) + 6\ H_2O(g)$

$\Delta H° = 2 \text{ mol HCN} \times \Delta H_f° (HCN) + 6 \text{ mol } H_2O(g) \times \Delta H_f° (H_2O,g)$

$- [2 \text{ mol } NH_3 \times \Delta H_f° (NH_3) + 2 \text{ mol } CH_4 \times \Delta H_f° (CH_4)]$

$\Delta H° = 2(135.1) + 6(-242) - [2(-46) + 2(-75)] = -940.\ kJ$

b. $\Delta H° = 3 \text{ mol } CaSO_4 \text{ (-1433 kJ/mol)} + 2 \text{ mol } H_3PO_4(l) \text{ (-1267 kJ/mol)}$

$-[1 \text{ mol } Ca_3(PO_4)_2 \text{ (-4126 kJ/mol)} + 3 \text{ mol } H_2SO_4(l) \text{ (-814 kJ/mol)}]$

$\Delta H° = -6833\ kJ - (-6568\ kJ) = -265\ kJ$

c. $\Delta H° = 1 \text{ mol } NH_4Cl \times \Delta H_f° (NH_4Cl)$

$- [1 \text{ mol } NH_3 \times \Delta H_f° (NH_3) + 1 \text{ mol HCl} \times \Delta H_f° (HCl)]$

$\Delta H° = 1 \text{ mol (-314 kJ/mol)} - [1 \text{ mol (-46 kJ/mol)} + 1 \text{ mol (-92 kJ/mol)}]$

$\Delta H° = -314\ kJ + 138\ kJ = -176\ kJ$

47. a. $4\ NH_3(g) + 5\ O_2(g) \longrightarrow 4\ NO(g) + 6\ H_2O(g)$

$\Delta H° = 4 \text{ mol (90 kJ/mol)} + 6 \text{ mol (-242 kJ/mol)} - 4 \text{ mol (-46 kJ/mol)} = -908\ kJ$

$2\ NO(g) + O_2(g) \longrightarrow 2\ NO_2(g)$

$\Delta H° = 2 \text{ mol (34 kJ/mol)} - 2 \text{ mol (90 kJ/mol)} = -112\ kJ$

$3\ NO_2(g) + H_2O(l) \longrightarrow 2\ HNO_3(aq) + NO(g)$

$\Delta H° = 2 \text{ mol (-207 kJ/mol)} + 1 \text{ mol (90 kJ/mol)}$

$-[3 \text{ mol (34 kJ/mol)} + 1 \text{ mol (-286 kJ/mol)}] = -140.\ kJ$

b. $12\ NH_3(g) + 15\ O_2(g) \longrightarrow 12\ NO(g) + 18\ H_2O(g)$

$12\ NO(g) + 6\ O_2(g) \longrightarrow 12\ NO_2(g)$

$12\ NO_2(g) + 4\ H_2O(l) \longrightarrow 8\ HNO_3(aq) + 4\ NO(g)$

$4\ H_2O(g) \longrightarrow 4\ H_2O(l)$

$12\ NH_3(g) + 21\ O_2(g) \longrightarrow 8\ HNO_3(aq) + 4\ NO(g) + 14\ H_2O(g)$

The overall reaction is exothermic, since each step is exothermic.

49. $3\ Al(s) + 3\ NH_4ClO_4(s) \longrightarrow Al_2O_3(s) + AlCl_3(s) + 3\ NO(g) + 6\ H_2O(g)$

$\Delta H° = 6\text{ mol }(-242\text{ kJ/mol}) + 3\text{ mol }(90\text{ kJ/mol}) + 1\text{ mol }(-704\text{ kJ/mol})$

$+ 1\text{ mol }(-1676\text{ kJ/mol}) - [3\text{ mol }(-295\text{ kJ/mol})] = -2677\text{ kJ}$

51. $C(s) + H_2O(g) \longrightarrow H_2(g) + CO(g)$

$\Delta H° = -110.5\text{ kJ} - (-242\text{ kJ}) = +132\text{ kJ}$

53. $2\ ClF_3(g) + 2\ NH_3(g) \longrightarrow N_2(g) + 6\ HF(g) + Cl_2(g)$

$\Delta H° = 6\ \Delta H_f^\circ\ (HF) - [2\ \Delta H_f^\circ\ (ClF_3) + 2\ \Delta H_f^\circ\ (NH_3)]$

$-1196\text{ kJ} = 6\text{ mol }(-271\text{ kJ/mol}) - 2\ \Delta H_f^\circ - 2\text{ mol }(-46\text{ kJ/mol})$

$-1196\text{ kJ} = -1626\text{ kJ} - 2\ \Delta H_f^\circ + 92\text{ kJ}$

$$\Delta H_f^\circ = \frac{(-1626 + 92 + 1196)\text{ kJ}}{2\text{ mol}} = \frac{-169\text{ kJ}}{\text{mol}}$$

Energy Consumption and Sources

55. $4.19 \times 10^6\text{ kJ} \times \frac{1\text{ mol } CH_4}{891\text{ kJ}} \times \frac{22.42\text{ L } CH_4 @\text{ STP}}{\text{mol } CH_4} = 1.05 \times 10^5\text{ L}$

57. $CO(g) + 2\ H_2(g) \longrightarrow CH_3OH(l)$

$\Delta H° = (-239\text{ kJ}) - (-110.5\text{ kJ}) = -129\text{ kJ}$

ADDITIONAL EXERCISES

59. a. $1.00\text{ g } CH_4 \times \frac{1\text{ mol } CH_4}{16.04\text{ g } CH_4} \times \frac{(-891\text{ kJ})}{\text{mol } CH_4} = -55.5\text{ kJ}$

b. $PV = nRT$

$$n = \frac{PV}{RT} = \frac{\frac{740.}{760}\text{ atm} \times 1.00 \times 10^3\text{ L}}{\frac{0.08206\text{ L atm}}{\text{mol K}} \times 298\text{ K}} = 39.8\text{ mol}$$

$$\frac{-891\text{ kJ}}{\text{mol}} \times 39.8\text{ mol} = -3.55 \times 10^4\text{ kJ}$$

61. The specific heat capacities are: 0.89 J/g•°C (Al) and 0.45 J/g•°C (Fe)
Al would be the better choice. It has a higher heat capacity and a lower density than Fe. Using Al, the same amount of heat could be dissipated by a smaller mass, keeping the mass of the amplifier down.

63. $V = 10.0\text{ m} \times 4.0\text{ m} \times 3.0\text{ m} = 1.2 \times 10^2\text{ m}^3 \times (100\text{ cm/m})^3 = 1.2 \times 10^8\text{ cm}^3$

Mass of water = 1.2×10^8 g since the density of water is 1.0 g/cm^3

$$\text{Heat} = \frac{4.18\text{ J}}{\text{g °C}} \times 1.2 \times 10^8\text{ g} \times 9.8\text{°C} = 4.9 \times 10^9\text{ J or } 4.9 \times 10^6\text{ kJ}$$

65. $2\ K(s) + 2\ H_2O(l) \longrightarrow 2\ KOH(aq) + H_2(g)$

$\Delta H° = 2(-481\ kJ) - 2(-286\ kJ) = -390.\ kJ$

$5.00\ g\ K \times \frac{1\ mol\ K}{39.10\ g\ K} \times \frac{390.\ kJ}{2\ mol\ K} = 24.9$ kJ of heat released upon reaction of 5.00 g of potassium.

$24{,}900\ J = \frac{4.18\ J}{g\ °C} \times 1.00 \times 10^3\ g \times \Delta T$

$\Delta T = \frac{24{,}900}{4.18 \times 1.00 \times 10^3} = 5.96°C$

Final Temperature = 24.0 + 5.96 = 30.0°C

67. $C_2H_2(g) + 5/2\ O_2(g) \longrightarrow 2\ CO_2(g) + H_2O(g)$

$\Delta H° = 2(-393.5\ kJ) + (-242\ kJ) - (227\ kJ) = -1256\ kJ$

If one mole of acetylene is burned with a stoichiometric amount of oxygen, the products will be 2 mol CO_2 and 1 mol H_2O. The heat capacity of this mixture is:

$C = 2.00\ mol\ CO_2 \times \frac{37.1\ J}{mol\ °C} + 1.00\ mol\ H_2O \times \frac{33.6\ J}{mol\ °C}$

$C = 108\ J/°C = 0.108\ kJ/°C$

$1256\ kJ = C \times \Delta T$

$\Delta T = \frac{1256\ kJ}{\frac{0.108\ kJ}{°C}} = 11{,}600°C$

or final temperature is $11{,}600 + 25 \approx 12{,}000°C$

Note: The actual temperature of an oxyacetylene torch is about 3000°C. The assumption of no heat loss is not a very good one.

69. We need 1.3×10^8 J of energy or 1.3×10^5 kJ.

For:

$C_3H_8(g) + 5\ O_2(g) \longrightarrow 3\ CO_2(g) + 4\ H_2O(l)$

$\Delta H° = 4(-286\ kJ) + 3(-393.5\ kJ) - (-104\ kJ) = -2221\ kJ$

So the mass of C_3H_8 is:

$1.3 \times 10^5\ kJ \times \frac{1\ mol\ C_3H_8}{2221\ kJ} \times \frac{44.09\ g\ C_3H_8}{mol} = 2600$ g of C_3H_8

71. a. $C_2H_4(g) + O_3(g) \longrightarrow CH_3CHO(g) + O_2(g)$

$\Delta H° = -166\ kJ - [143\ kJ + 52\ kJ] = -361\ kJ$

b. $O_3 + NO \rightarrow NO_2 + O_2$

$\Delta H° = 1 \text{ mol}(34 \text{ kJ/mol}) - [1 \text{ mol } (90 \text{ kJ/mol}) + 1 \text{ mol } (143 \text{ kJ/mol})] = -199 \text{ kJ}$

c. $SO_3(g) + H_2O(l) \rightarrow H_2SO_4(aq)$

$\Delta H° = 1 \text{ mol } (-909 \text{ kJ/mol}) - [1 \text{ mol } (-396 \text{ kJ/mol}) + 1 \text{ mol } (-286 \text{ kJ/mol})] = -227 \text{ kJ}$

d. $2\ NO(g) + O_2(g) \rightarrow 2\ NO_2(g)$

$\Delta H° = 2(34) - 2(90) = 68 \text{ kJ} - 180. \text{ kJ} = -112 \text{ kJ}$

CHALLENGE PROBLEMS

73. One should try to cool the reaction mixture or provide some means of removing heat. The reaction is very exothermic. The $H_2SO_4(aq)$ will get very hot and possibly boil.

75. To avoid fractions, let's first calculate ΔH for the reaction:

$6\ FeO(s) + 6\ CO(g) \rightarrow 6\ Fe(s) + 6\ CO_2(g)$

$6\ FeO + 2\ CO_2 \rightarrow 2\ Fe_3O_4 + 2\ CO$	$\Delta H° = 2(-18 \text{ kJ})$	×(-2)
$2\ Fe_3O_4 + CO_2 \rightarrow 3\ Fe_2O_3 + CO$	$\Delta H° = +39 \text{ kJ}$	(-)
$3\ Fe_2O_3 + 9\ CO \rightarrow 6\ Fe + 9\ CO_2$	$\Delta H° = 3(-23 \text{ kJ})$	×3
$6\ FeO(s) + 6\ CO(g) \rightarrow 6\ Fe(s) + 6\ CO_2(g)$	$\Delta H° = -66 \text{ kJ}$	

So for

$FeO(s) + CO(g) \rightarrow Fe(s) + CO_2(g)$ $\qquad \Delta H° = \frac{-66 \text{ kJ}}{6} = -11 \text{ kJ}$

77. $400 \text{ kcal} \times \frac{4.18 \text{ kJ}}{\text{kcal}} = 1.67 \times 10^3 \text{ kJ} \approx 2 \times 10^3 \text{ kJ}$

$$E = mgz = \left(180 \text{ lb} \times \frac{1 \text{ kg}}{2.205 \text{ lb}}\right)\left(\frac{9.8 \text{ m}}{s^2}\right)\left(8 \text{ in} \times \frac{2.54 \text{ cm}}{\text{in}} \times \frac{1 \text{ m}}{100 \text{ cm}}\right) = 160 \text{ J} \approx 200 \text{ J}$$

200 J of energy are needed to climb one step.

$2 \times 10^6 \text{ J} \times \frac{1 \text{ step}}{200 \text{ J}} = 1 \times 10^4 \text{ steps}$

CHAPTER SEVEN: ATOMIC STRUCTURE AND PERIODICITY

QUESTIONS

1. Planck found that heated bodies only give off certain frequencies of light. Einstein's studies of the photoelectric effect.

3. Very small particles (small mass) at large velocities.

5. a. A discrete bundle of light energy.

   b. A number describing a discrete state of an electron.

   c. The lowest energy state of the electron(s) in an atom or ion.

   d. An allowed energy state that is higher in energy than the ground state.

7. The 2p orbitals differ from each other in the direction in which they point in space.

9. A nodal surface in an atomic orbital is a surface in which the probability of finding an electron is zero.

11. The electrostatic energy of repulsion, from Coulomb's Law will be of the form e^2/r where r is the distance between the two electrons. From the Heisenberg Uncertainty Principle, we cannot know precisely the position of each electron. Thus, we cannot precisely know the distance between the electrons nor the value of the electrostatic repulsions.

13. There is a higher probability of finding the 4s electron very close to the nucleus than for the 3d electron.

15. If one more electron is added to a half-filled subshell, electron-electron repulsions will increase.

17. Each element has a characteristic spectrum. Thus, the presence of the characteristic spectral lines of an element confirms its presence in any particular sample.

19. The electron is no longer part of that atom. The proton and electron are completely separated.

EXERCISES

Light and Matter

21. $99.5 \text{ MHz} = 99.5 \times 10^6 \text{ Hz} = 99.5 \times 10^6 \text{ s}^{-1}$

$$\lambda = \frac{c}{\nu} = \frac{2.998 \times 10^8 \text{ m/s}}{99.5 \times 10^6 \text{ s}^{-1}} = 3.01 \text{ m}$$

23. $$\nu = \frac{c}{\lambda} = \frac{3.00 \times 10^8 \text{ m/s}}{1.0 \times 10^{-2} \text{ m}} = 3.0 \times 10^{10} \text{ s}^{-1}$$

$$E = h\nu = 6.63 \times 10^{-34} \text{ J s} \times 3.0 \times 10^{10} \text{ s}^{-1} = 2.0 \times 10^{-23} \text{ J}$$

$$2.0 \times 10^{-23} \text{ J/photon} \times 6.02 \times 10^{23} \text{ photons/mol} = 12 \text{ J/mol}$$

25. $\Delta E = 7.21 \times 10^{-19}\ J = E_{photon}$

$E_{photon} = h\nu$ and $\lambda\nu = c$. So, $\nu = \frac{c}{\lambda}$ and $E = \frac{hc}{\lambda}$

$$\lambda = \frac{hc}{E_{photon}} = \frac{(6.626 \times 10^{-34}\ J\ s)(2.998 \times 10^{8}\ m/s)}{7.21 \times 10^{-19}\ J}$$

$\lambda = 2.76 \times 10^{-7}\ m = 276 \times 10^{-9}\ m = 276\ nm$

27. $\frac{492\ kJ}{mol} \times \frac{1\ mol}{6.022 \times 10^{23}} = 8.17 \times 10^{-22}\ kJ = 8.17 \times 10^{-19}\ J$

$E = \frac{hc}{\lambda}$

$$\lambda = \frac{hc}{E} = \frac{(6.626 \times 10^{-34}\ J\ s)(2.998 \times 10^{8}\ m/s)}{(8.17 \times 10^{-19}\ J)} = 2.43 \times 10^{-7}\ m = 243\ nm$$

29. $\frac{890.1\ kJ}{mol} \times \frac{1\ mol}{6.022 \times 10^{23}\ atoms} = \frac{1.478 \times 10^{-21}\ kJ}{atom} = \frac{1.478 \times 10^{-18}\ J}{atom}$

$E = \frac{hc}{\lambda}$

$$\lambda = \frac{hc}{E} = \frac{(6.626 \times 10^{-34}\ J\ s)(2.9979 \times 10^{8}\ m/s)}{1.478 \times 10^{-18}\ J} = 1.344 \times 10^{-7}\ m = 134.4\ nm$$

No, it will take light with a wavelength of 134.4 nm or less to ionize gold. A photon of light with a wavelength of 225 nm is a longer wavelength and, thus, less energy than 134.4 nm light.

Hydrogen Atom: Bohr Model

31. For the H-atom: $E_n = \frac{-R_H}{n^2}$, $R_H = 2.178 \times 10^{-18}\ J$

For a spectral transition $\Delta E = E_f - E_i$

$$\Delta E = \left(\frac{-R_H}{n_f^2}\right) - \left(\frac{-R_H}{n_i^2}\right) = R_H\left(\frac{1}{n_i^2} - \frac{1}{n_f^2}\right)$$

Where n_i and n_f are the quantum numbers of the initial and final states, respectively. If we follow this convention a positive value of ΔE will correspond to absorption of light and a negative value of ΔE will correspond to emission of light. However, we will know whether we are observing emission or absorption of light. For the remainder, all examples will be worked so that the value of ΔE is always positive.

a. $\Delta E = 2.178 \times 10^{-18}\ J\left(\frac{1}{2^2} - \frac{1}{3^2}\right) = 2.178 \times 10^{-18}\ J\left(\frac{1}{4} - \frac{1}{9}\right)$

$= 2.178 \times 10^{-18}\ J\ (0.2500 - 0.1111) = 3.025 \times 10^{-19}\ J$

The photon of light must have precisely this energy.

$$\Delta E = E_{photon} = h\nu = \frac{hc}{\lambda} \quad \text{or} \quad \lambda = \frac{hc}{\Delta E} = \frac{(6.626 \times 10^{-34}\text{ J s})(2.9979 \times 10^{8}\text{ m/s})}{3.025 \times 10^{-19}\text{ J}}$$

$$= 6.567 \times 10^{-7}\text{ m or } 656.7\text{ nm}$$

b. $\Delta E = 2.178 \times 10^{-18}\text{ J}\left(\frac{1}{2^2} - \frac{1}{4^2}\right) = 4.084 \times 10^{-19}\text{ J}$

$$\lambda = \frac{hc}{\Delta E} = \frac{(6.626 \times 10^{-34}\text{ J s})(2.9979 \times 10^{8}\text{ m/s})}{4.084 \times 10^{-19}\text{ J}} = 4.864 \times 10^{-7}\text{ m or } 486.4\text{ nm}$$

c. $\Delta E = 2.178 \times 10^{-18}\text{ J}\left(\frac{1}{1^2} - \frac{1}{2^2}\right) = 1.634 \times 10^{-18}\text{ J}$

$$\lambda = \frac{(6.626 \times 10^{-34}\text{ J s})(2.9979 \times 10^{8}\text{ m/s})}{1.634 \times 10^{-18}\text{ J}} = 1.216 \times 10^{-7}\text{ m or } 121.6\text{ nm}$$

33. Ionization from n = 1 corresponds to the transition n = 1 ⟶ n = ∞ and $E_\infty = 0$

$$\Delta E = E_\infty - E_1 = -E_1 = 2.178 \times 10^{-18}\left(\frac{1}{1^2}\right) = 2.178 \times 10^{-18}\text{ J}$$

$$\lambda = \frac{hc}{\Delta E} = \frac{(6.626 \times 10^{-34}\text{ J s})(2.9979 \times 10^{8}\text{ m/s})}{2.178 \times 10^{-18}\text{ J}} = 9.120 \times 10^{-8}\text{ m or } 91.20\text{ nm}$$

To ionize from n = 2, $\Delta E = E_\infty - E_2 = 0 - E_2$

$$\Delta E = 2.178 \times 10^{-18}\left(\frac{1}{2^2}\right) = 5.445 \times 10^{-19}\text{ J}$$

$$\lambda = \frac{(6.626 \times 10^{-34}\text{ J s})(2.9979 \times 10^{8}\text{ m/s})}{5.445 \times 10^{-19}\text{ J}} = 3.648 \times 10^{-7}\text{ m or } 364.8\text{ nm}$$

35.

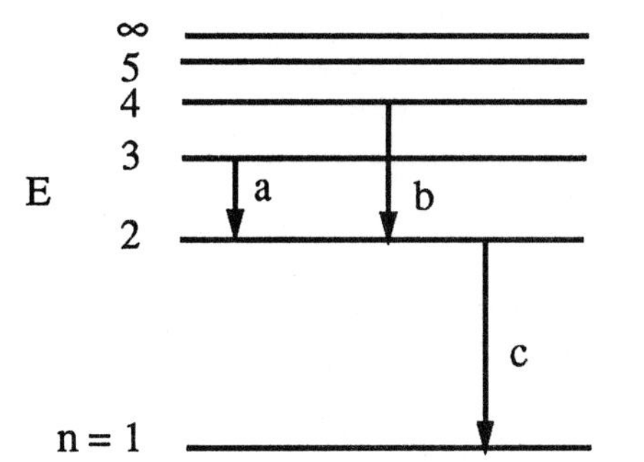

a. 3 ⟶ 2

b. 4 ⟶ 2

c. 2 ⟶ 1

37. The longest wavelength light emitted will correspond to the transition with the lowest energy change. This is the transition from n = 6 to n = 5.

$$\Delta E = -2.178 \times 10^{-18}\text{ J}\left(\frac{1}{5^2} - \frac{1}{6^2}\right) = -2.662 \times 10^{-20}\text{ J}$$

$$\lambda = \frac{hc}{\Delta E} = \frac{(6.626 \times 10^{-34}\text{ J s})(2.9979 \times 10^{8}\text{ m/s})}{2.662 \times 10^{-20}\text{ J}} = 7.462 \times 10^{-6}\text{ m} = 7462\text{ nm}$$

The shortest wavelength emitted will correspond to the largest ΔE; this is n = 6 $\longrightarrow$ n = 1.

$$\Delta E = -2.178 \times 10^{-18}\left(\frac{1}{1^2} - \frac{1}{6^2}\right) = -2.118 \times 10^{-18}\text{ J}$$

$$\lambda = \frac{hc}{\Delta E} = \frac{(6.626 \times 10^{-34}\text{ J s})(2.9979 \times 10^{8}\text{ m/s})}{2.118 \times 10^{-18}\text{ J}} = 9.379 \times 10^{-8}\text{ m} = 93.79\text{ nm}$$

Wave Mechanics, Quantum Numbers, and Orbitals

39. $\lambda = \frac{h}{mv}$

a. 5.0% of speed of light = $0.050 \times 3.00 \times 10^8$ m/s = 1.5×10^7 m/s

$$\lambda = \frac{6.63 \times 10^{-34}\text{ J s}}{1.67 \times 10^{-27}\text{ kg} \times (1.5 \times 10^{7}\text{ m/s})} = 2.6 \times 10^{-14}\text{ m} = 2.6 \times 10^{-5}\text{ nm}$$

Note: for units to come out the mass must be in kg since $J = \frac{kg\ m^2}{s^2}$

b. $$\lambda = \frac{6.63 \times 10^{-34}\text{ J s}}{9.11 \times 10^{-31}\text{ kg} \times (0.15 \times 3.00 \times 10^{8}\text{ m/s})} = 1.6 \times 10^{-11}\text{ m} = 1.6 \times 10^{-2}\text{ nm}$$

c. $$m = 5.2\text{ oz} \times \frac{1\text{ lb}}{16\text{ oz}} \times \frac{1\text{ kg}}{2.205\text{ lb}} = 0.15\text{ kg}$$

$$v = \frac{100.8\text{ mi}}{\text{hr}} \times \frac{1\text{ hr}}{3600\text{ s}} \times \frac{1760\text{ yd}}{\text{mi}} \times \frac{0.9144\text{ m}}{\text{yd}} = 45.06\text{ m/s}$$

$$\lambda = \frac{h}{mv} = \frac{6.63 \times 10^{-34}\text{ J s}}{0.15\text{ kg} \times 45.06\text{ m/s}} = 9.8 \times 10^{-35}\text{ m} = 9.8 \times 10^{-26}\text{ nm}$$

This number is so small it is essentially zero. We cannot detect a wavelength this small. The meaning of this number is that we do not have to worry about wave properties of large objects at low velocities.

41. $\lambda = \frac{h}{mv}$, $v = \frac{h}{\lambda m}$

for $\lambda = 1.0 \times 10^2$ nm = 1.0×10^{-7} m

$$v = \frac{6.63 \times 10^{-34}\ \text{J s}}{(9.11 \times 10^{-31}\ \text{kg})(1.0 \times 10^{-7}\ \text{m})} = 7.3 \times 10^{3}\ \text{m/s}$$

for $\lambda = 1.0$ nm

$$v = \frac{h}{m\lambda} = \frac{6.63 \times 10^{-34}\ \text{J s}}{(9.11 \times 10^{-31}\ \text{kg})(1.0 \times 10^{-9}\ \text{m})} = 7.3 \times 10^{5}\ \text{m/s}$$

43. $\Delta p = m\Delta v$

$$\Delta p = m\Delta v = 9.11 \times 10^{-31}\ \text{kg} \times 0.100\ \text{m/s} = \frac{9.11 \times 10^{-32}\ \text{kg m}}{\text{s}}$$

$$\Delta p \Delta x \geq \frac{h}{4\pi};\ \Delta x = \frac{h}{4\pi\Delta p} = \frac{6.626 \times 10^{-34}\ \text{J s}}{4 \times 3.1416 \times 9.11 \times 10^{-32}\ \text{kg m/s}}$$

$\Delta x = 5.79 \times 10^{-4}$ m = 0.579 mm = 5.79×10^{5} nm

Diameter of H atom is roughly 1.0×10^{-8} cm. The uncertainty in position is much larger than the size of the atom.

Polyelectronic Atoms

45. n = 1, 2, 3, ...

l = 0, 1, 2, ... (n - 1)

$m_l = -l \ldots -2, -1, 0, 1, 2, \ldots +l$

47. b. l must be smaller than n.

d. For $l = 0$, $m_l = 0$ only allowed value.

49. $E_n = \frac{-R_H Z^2}{n^2}$, $R_H = 2.178 \times 10^{-18}$ J

Normal ionization energy is for transition n = 1 $\rightarrow$ n = ∞, E = 0. Thus, the I.E. is given by the energy of the state n = 1. Ionization energies are positive.

a. $\text{IE} = 2.178 \times 10^{-18}\ \text{J} \left(\frac{1^2}{1^2}\right) = 2.178 \times 10^{-18}\ \text{J}$

$$\text{IE} = \frac{2.178 \times 10^{-18}\ \text{J}}{\text{atom}} \times \frac{1\ \text{kJ}}{1000\ \text{J}} \times \frac{6.022 \times 10^{23}\ \text{atoms}}{\text{mol}} = 1311.6\ \text{kJ/mol}$$

IE = 1312 kJ/mol

For brevity, the I.E. of heavier one electron species can be given as:

$$\text{IE} = \frac{1311.6\ \text{kJ}}{\text{mol}} \left(\frac{Z^2}{n^2}\right)$$ (We will carry an extra significant figure.)

We get this by combining $IE(J) = 2.178 \times 10^{-18}\ J \left(\frac{Z^2}{n^2}\right)$

and $\frac{2.178 \times 10^{-18}\ J}{atom} \times \frac{6.022 \times 10^{23}\ atoms}{mol} \times \frac{1\ kJ}{1000\ J} = \frac{1311.6\ kJ}{mol}$

b. He^+, Z = 2 IE = 5246 kJ/mol

c. Li^{2+}, Z = 3 IE = 1311.6 kJ/mol × 3^2 = 1.180 × 10^4 kJ/mol

The size of the 1s orbitals would be proportional to 1/Z, that is as Z increases, the electrons are more strongly attracted to the nucleus and will be drawn in closer. Thus the relative sizes would be:

H : He^+ : Li^{2+} $1 : \frac{1}{2} : \frac{1}{3}$ 1 : 0.50 : 0.33

51. 5p: three $3d_{z^2}$: one 4d: five

n = 5: l = 0 (1 orbital), l = 1 (3 orbitals), l = 2 (5 orbitals)

l = 3 (7 orbitals), l = 4 (9 orbitals)

Total for n = 5 is 25 orbitals.

n = 4: l = 0 (1), l = 1 (3), l = 2 (5), l = 3 (7)

Total for n = 4 is 16 orbitals.

53. a. For n = 4, l can be 0, 1, 2, or 3. Thus we have s(2 e^-), p(6 e^-), d(10 e^-) and f(14 e^-) orbitals present. Total number of electrons to fill these orbitals is 32.

b. n = 5, m_l = +1 For n = 5, l = 0, 1, 2, 3, 4.

l = 1, 2, 3, 4 can have m_l = +1. Four distinct orbitals, thus 8 electrons.

c. n = 5, l = 0, 1, 2, 3, 4

number of orbitals 1, 3, 5, 7, 9 for each value of l respectively.

There are 25 orbitals with n = 5. They can hold 50 electrons, 25 electrons can have m_s = +1/2.

d. n = 3, l = 2 defines a set of 3d orbitals.

5 orbitals, 10 electrons

e. n = 2, l = 1 defines a set of 2p orbitals.

3 orbitals, 6 electrons

55. Sc : $1s^22s^22p^63s^23p^64s^23d^1$

Fe : $1s^22s^22p^63s^23p^64s^23d^6$

S : $1s^22s^22p^63s^23p^4$

P : $1s^22s^22p^63s^23p^3$

Cs : $1s^22s^22p^63s^23p^64s^23d^{10}4p^65s^24d^{10}5p^66s^1$

Eu : $1s^22s^22p^63s^23p^64s^23d^{10}4p^65s^24d^{10}5p^66s^24f^65d^1$*

Pt : $1s^22s^22p^63s^23p^64s^23d^{10}4p^65s^24d^{10}5p^66s^24f^{14}5d^8$*

Xe : $1s^22s^22p^63s^23p^64s^23d^{10}4p^65s^24d^{10}5p^6$

Br : $1s^22s^22p^63s^23p^64s^23d^{10}4p^5$

Se : $1s^22s^22p^63s^23p^64s^23d^{10}4p^4$

*Note: These electron configurations were written down using only the periodic table. Actual electron configurations are:

Eu: $[Xe]6s^24f^7$ and Pt: $[Xe]6s^14f^{14}5d^9$

57. Exceptions: Cr, Cu, Nb, Mo, Tc, Ru, Rh, Pd, Ag, Pt, Au

Tc, Ru, Rh, Pd and Pt do not correspond to the supposed extra stability of half-filled and filled subshells.

59. a. The smallest halogen is fluorine: $1s^22s^22p^5$

b. K: $1s^22s^22p^63s^23p^64s^1$

c. Be: $1s^22s^2$ Mg: $1s^22s^22p^63s^2$ Ca: $1s^22s^22p^63s^23p^64s^2$

d. In: $[Kr]5s^24d^{10}5p^1$

e. C: $1s^22s^22p^2$ Si: $1s^22s^22p^63s^23p^2$

f. This will be element #118: $[Rn]7s^25f^{14}6d^{10}7p^6$

61. B : $1s^22s^22p^1$

	n	l	m_l	m_s
1s	1	0	0	+1/2
1s	1	0	0	-1/2
2s	2	0	0	+1/2
2s	2	0	0	-1/2
2p*	2	1	-1	+1/2

C : $1s^22s^22p^2$

	n	l	m_l	m_s	
1s	1	0	0	+1/2	
1s	1	0	0	-1/2	
2s	2	0	0	+1/2	
2s	2	0	0	-1/2	
2p*	2	1	-1	+1/2	or
2p	2	1	0	+1/2	-1/2

*This is only one of several possibilities for 2p electrons. For example, the 2p electron in B could have m_l = -1, 0, +1, and m_s = +1/2 or -1/2, a total of six possibilities.

N : $1s^22s^22p^3$

	n	l	m_l	m_s
1s	1	0	0	+1/2
1s	1	0	0	-1/2
2s	2	0	0	+1/2
2s	2	0	0	-1/2
2p	2	1	-1	+1/2
2p	2	1	0	+1/2
2p	2	1	+1	+1/2

(or all 2p electrons could have m_s = -1/2.)

O : $1s^22s^22p^4$

	n	l	m_l	m_s	
1s	1	0	0	+1/2	
1s	1	0	0	-1/2	
2s	2	0	0	+1/2	
2s	2	0	0	-1/2	
2p*	2	1	-1	+1/2	
2p	2	1	-1	-1/2	
2p	2	1	0	+1/2 }	or
2p	2	1	+1	+1/2 }	-1/2

* only one of several possibilities for $2p^4$

63. O: $1s^22s^22p_x^22p_y^2$ ↑↓ ↑↓ __

There are no unpaired electrons in this oxygen atom. This configuration would be an excited state and in going to the ground state ↑↓ ↑ ↑ energy would be released.

65. We get the number of unpaired electrons by looking at the incompletely filled subshells.

Sc:	$[Ar]4s^23d^1$:	$3d^1$:	↑ __ __ __ __	one unpaired e^-
Ti:	$[Ar]4s^23d^2$	$3d^2$:	↑ ↑ __ __ __	two unpaired e^-
Al:	$[Ne]3s^23p^1$	$3p^1$:	↑ __ __	one unpaired e^-
Sn:	$[Kr]5s^24d^{10}5p^2$	$5p^2$	↑ ↑ __	two unpaired e^-
Te:	$[Kr]5s^24d^{10}5p^4$	$5p^4$	↑↓ ↑ ↑	two unpaired e^-
Br:	$[Ar]4s^23d^{10}4p^5$	$4p^5$	↑↓ ↑↓ ↑	one unpaired e^-

The Periodic Table and Periodic Properties

67. a. Be < Mg < Ca b. Xe < I < Te c. Ge < Ga < In

69. a. Ca < Mg < Be b. Te < I < Xe c. In < Ga < Ge

71. $IE \propto \frac{Z_{eff}^2}{n^2}$ or $Z_{eff} \propto n\,(IE)^{1/2}$

The effect of n is greater than the effect of the ionization energy.

a. Be < Mg < Ca b. Te < I < Xe c. Ga < Ge < In

73. Ge: $[Ar]3d^{10}4s^24p^2$; As: $[Ar]3d^{10}4s^24p^3$; Se: $[Ar]3d^{10}4s^24p^4$

There are extra electron-electron repulsions in Se because two electrons are in the same 4p orbital, resulting in a lower ionization energy.

75. a. Li b. P c. O^+

d. Ar < Cl < S and Kr > Ar. Since variation in size down a family is greater than the variation across a period, we would predict Cl to be the smallest of the three.

e. Cu

77. See graphs. mp ≈ 22°C, IE ≈ 340 kJ/mol

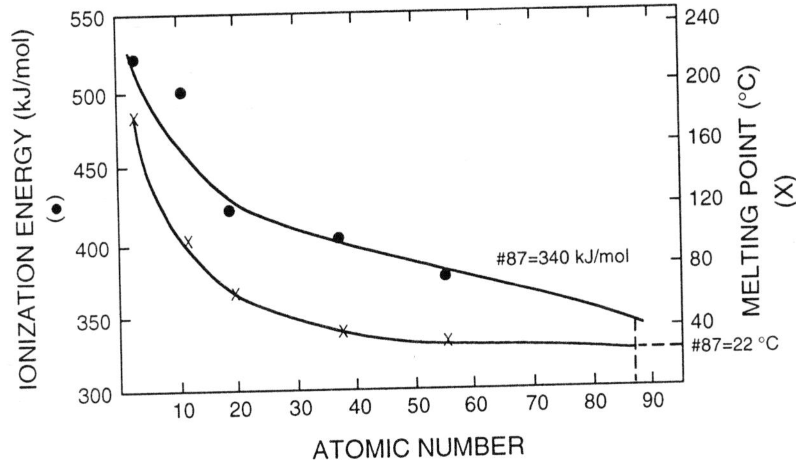

79. a O < S, S more exothermic

b. I < Br < F < Cl, Cl most exothermic

81. a. more favorable EA: Li and S

b. higher IE: Li and S

c. larger radius: K and Sc

83. Yes, since it is an exothermic EA there are conditions in which we might expect Na^- to exist. Such compounds were first synthesized by James L. Dye at Michigan State University.

85. $P(g) \longrightarrow P^+(g) + e^-$

87. a. The electron affinity of Mg^{2+} is ΔH for

$$Mg^{2+}(g) + e^- \longrightarrow Mg^+(g)$$

This is just the reverse of the second ionization energy, or

$$EA(Mg^{2+}) = -IE_2(Mg) = -1445 \text{ kJ/mol}$$

b. EA of Al^+ is ΔH for: $Al^+(g) + e^- \longrightarrow Al(g)$

$EA(Al^+) = -IE_1(Al) = -580$ kJ/mol

The Alkali Metals

89. Li^+ ions will be the smallest of the alkali metal cations and will be most strongly attracted to the water molecules.

91. It should be potassium peroxide, K_2O_2. K^{2+} ions are not stable; the second ionization energy of K is very large compared to the first.

93. $\nu = \frac{c}{\lambda} = \frac{2.9979 \times 10^8 \text{ m/s}}{455.5 \times 10^{-9} \text{ m}} = 6.582 \times 10^{14} \text{ s}^{-1}$

$E = h\nu = (6.626 \times 10^{-34} \text{ J s})(6.582 \times 10^{14} \text{ s}^{-1}) = 4.361 \times 10^{-19} \text{ J}$

95. a. Li_3N lithium nitride b. NaBr sodium bromide c. K_2S potassium sulfide

97. a. $4\,Li(s) + O_2(g) \longrightarrow 2\,Li_2O(s)$ b. $2\,K(s) + S(s) \longrightarrow K_2S(s)$

ADDITIONAL EXERCISES

99. It should be element #119 with ground state electron configuration: [Rn] $7s^25f^{14}6d^{10}7p^68s^1$

101. $E = \frac{310.\text{ kJ}}{\text{mol}} \times \frac{1 \text{ mol}}{6.022 \times 10^{23}} = 5.15 \times 10^{-22} \text{ kJ} = 5.15 \times 10^{-19} \text{ J}$

$E = \frac{hc}{\lambda};\ \lambda = \frac{hc}{E} = \frac{6.626 \times 10^{-34} \text{ J s} \times 2.998 \times 10^8 \text{ m/s}}{5.15 \times 10^{-19}} = 3.86 \times 10^{-7} \text{ m} = 386 \text{ nm}$

103. a. n = 3 We can have 3s, 3p, and 3d orbitals. Nine orbitals can hold 18 electrons.

b. n = 2, l = 0 specifies a 2s orbital. 2 electrons

c. n = 2, l = 2 is not possible. No electrons can have this combination of quantum numbers.

d. These four quantum numbers completely specify a single electron.

105. No, for n = 2, the allowed values of l are 0 and 1; for n = 3 the allowed values of l are 0, 1, and 2.

107. He: $1s^2$; Ne: $1s^22s^22p^6$; Ar: $1s^22s^22p^63s^23p^6$

Each peak in the diagram corresponds to a subshell with different values of n. Corresponding subshells are closer to the nucleus for heavier elements because of the increased nuclear charge.

109. b and f are the only possible sets of quantum numbers.

a. For l = 0, m_l can only be 0.

c. m_s can only be +1/2 or -1/2.

d. For n = 1, l can only be 0.

e. For l = 2, m_l cannot be -3; lowest allowed value is -2.

111. Sb: $1s^22s^22p^63s^23p^64s^23d^{10}4p^65s^24d^{10}5p^3$

a. l = 1 designates p orbitals. There are 21 electrons in p orbitals (l = 1).

b. $m_l = 0$ All s electrons, 2 out of each set of 2p, 3p, 4p electrons, 2 out of each set of 3d and 4d electrons, and one of the 5p electrons have m_l = 0. 10 + 6 + 4 + 1 = 21 electrons with m_l = 0.

c. $m_l = 1$:

2 out of each set of 2p, 3p amd 4 p	(6)
2 out of each set of 3d and 4d	(4)
1 of the 5p	(1)

11 electrons have m_l = 1.

113. Yes, the electron configuration is $1s^22s^22p^63s^23p^2$. There should be another big jump when the thirteenth electron is removed, i.e. when the 1s electrons begin to be removed.

115. Expected order:

Li < Be < B < C < N < O < F < Ne

B and O are out of order. The IE of O (and also F and Ne) is lower because of the extra electron-electron repulsions present when two electrons are paired in the same orbital. B is out of order because of the different penetrating abilities of the 2p electron in B compared to the 2s electrons in Be.

117. Electron-electron repulsions become more important when we try to add electrons to an atom. From the standpoint of electron-electron repulsions, large atoms would have more favorable (more exothermic) electron affinities. Considering only electron-nucleus attractions, smaller atoms would be expected to have more favorable (more exothermic) EA's. These trends are exactly the opposite of each other. Thus, the overall variation is not as great as ionization energy in which attractions to the nucleus dominate.

119.

$1s^22s^22p^63s^23p^6$	:	Ar	:	1.5205 MJ/mol	:	0.94 Å
$1s^22s^22p^63s^2$	:	Mg	:	0.7377 MJ/mol	:	1.60 Å
$1s^22s^22p^63s^23p^64s^1$	:	K	:	0.4189 MJ/mol	:	1.97 Å

Size: Ar < Mg < K; IE: K < Mg < Ar

121. a.

$$Cu^+(g) + e^- \rightarrow Cu(g) \qquad -I_1 = -746 \text{ kJ}$$

$$Cu^+(g) \rightarrow Cu^{2+}(g) + e^- \qquad I_2 = 1958 \text{ kJ}$$

$$2\,Cu^+(g) \rightarrow Cu(g) + Cu^{2+}(g) \qquad \Delta H = 1212 \text{ kJ}$$

b. $Na^-(g) \rightarrow Na(g) + e^-$ $\quad -EA = 52$ kJ

$Na^+(g) + e^- \rightarrow Na(g)$ $\quad -I_1 = -495$ kJ

$Na^-(g) + Na^+(g) \rightarrow 2\ Na(g)$ $\quad \Delta H = -443$ kJ

c. $Mg^{2+}(g) + e^- \rightarrow Mg^+(g)$ $\quad -I_2 = -1445$ kJ

$K(g) \rightarrow K^+(g) + e^-$ $\quad I_1 = 419$ kJ

$Mg^{2+}(g) + K(g) \rightarrow Mg^+(g) + K^+(g)$ $\quad \Delta H = -1026$ kJ

d. $Na(g) \rightarrow Na^+(g) + e^-$ $\quad I_1 = 495$ kJ

$Cl(g) + e^- \rightarrow Cl^-(g)$ $\quad EA = -349$ kJ

$Na(g) + Cl(g) \rightarrow Na^+(g) + Cl^-(g)$ $\quad \Delta H = 146$ kJ

e. $Mg(g) \rightarrow Mg^+(g) + e^-$ $\quad \Delta H = I_1 = 735$ kJ

$F(g) + e^- \rightarrow F^-(g)$ $\quad \Delta H = EA = -328$ kJ

$Mg(g) + F(g) \rightarrow Mg^+ + F^-(g)$ $\quad \Delta H = 407$ kJ

f. $Mg^+(g) \rightarrow Mg^{2+}(g) + e^-$ $\quad \Delta H = I_2 = 1445$ kJ

$F(g) + e^- \rightarrow F^-(g)$ $\quad \Delta H = EA = -328$ kJ

$Mg^+(g) + F(g) \rightarrow Mg^{2+} + F^-(g)$ $\quad \Delta H = 1117$ kJ

g. From parts e and f we get:

$Mg(g) + F(g) \rightarrow Mg^+(g) + F^-(g)$ $\quad \Delta H = 407$ kJ

$Mg^+(g) + F(g) \rightarrow Mg^{2+} + F^-(g)$ $\quad \Delta H = 1117$ kJ

$Mg(g) + 2\ F(g) \rightarrow Mg^{2+}(g) + 2\ F^-(g)$ $\quad \Delta H = 1524$ kJ

CHALLENGE PROBLEMS

123. Size also decreases going across a period. Sc & Ti and Y & Zr are adjacent elements. There are 14 elements (the Lanthanides) between La and Hf, making Hf considerably smaller.

125. a. Each orbital could hold 3 electrons. c. 15

b. First period: 3. Second period: 12. d. 21

127. None of the noble gases and no subatomic particles had been discovered when Mendeleev published his periodic table. Thus, there was not an element out of place in terms of reactivity. There was no reason to predict an entire family of elements. Mendeleev ordered his table by mass, he had no way of knowing there were gaps in atomic numbers (they hadn't been invented yet).

CHAPTER EIGHT: BONDING - GENERAL CONCEPTS

QUESTIONS

1. a. electronegativity: The ability of an atom in a molecule to attract electrons to itself.

   electron affinity: The energy change for the reaction:

   $$M(g) + e^- \longrightarrow M^-(g)$$

   EA deals with isolated atoms in the gas phase.

   b. Covalent bond: sharing of electron pair(s); polar covalent bond: unequal sharing of electron pair(s).

   c. Ionic bond: electrons are no longer shared. Completely unequal sharing (i.e. transfer) of electrons from one atom to another.

3. No, we would expect the more highly charged ions to have a greater attraction for electrons. Thus, the electronegativity of an element does depend on its oxidation state.

5. The two requirements for a polar molecular are:

   1. polar bonds.

   2. a structure such that the bond dipoles do not cancel.

   In addition some molecules that have no polar bonds, but contain unsymmetrical lone pairs are polar. For example, PH_3 and AsH_3 are slightly polar.

EXERCISES

Chemical Bonds and Electronegativity

7. Using the periodic table we expect the general trend for electronegativity to be: 1) increase as we go from left to right across a period and 2) decrease as we go down a group.

   a. C < N < O b. Se < S < Cl c. Sn < Ge < Si d. Tl < Ge < S

9. The greater the difference in electronegativity between two atoms will result in a more polar bond.

   a. Ge—F b. P—Cl c. S—F d. Ti—Cl

11. The general trends in electronegativity used on Exercises 8.7 and 8.9 are only rules of thumb. In this exercise we use experimental values of electronegativities and can begin to see several exceptions.

   Order of EN from Figure 8.3.

   a. C (2.5) < N (3.0) < O (3.5) same as predicted.

   b. Se (2.4) < S (2.5) < Cl (3.0) same

   c. Si = Ge = Sn (1.8) different

   d. Tl (1.8) = Ge (1.8) < S (2.5) different

Most polar bonds using actual EN values

a. Si — F and Ge — F equal polarity (Ge — F predicted)

b. P — Cl (same as predicted)

c. S — F (same as predicted)

d. Ti—Cl (correctly predicted) Si — Cl and Ge — Cl equal polarity

Ionic Compounds

13. a. $Cu > Cu^+ > Cu^{2+}$ b. $Pt^{2+} > Pd^{2+} > Ni^{2+}$ c. $Se^{2-} > S^{2-} > O^{2-}$

d. $La^{3+} > Eu^{3+} > Gd^{3+} > Yb^{3+}$ e. $Te^{2-} > I^- > Xe > Cs^+ > Ba^{2+} > La^{3+}$

15. a. Mg^{2+}: $1s^22s^22p^6$; Sn^{2+}: $[Kr]\ 5s^24d^{10}$

K^+: $1s^22s^22p^63s^23p^6$; Al^{3+}: $1s^22s^22p^6$

Tl^+: $[Xe]6s^24f^{14}5d^{10}$; As^{3+}: $[Ar]4s^23d^{10}$

b. N^{3-}, O^{2-} and F^-: $1s^22s^22p^6$; Te^{2-}: $[Kr]5s^24d^{10}5p^6$

c. Be^{2+}: $1s^2$; Rb^+: $[Ar]4s^23d^{10}4p^6$; Ba^{2+}: $[Kr]5s^24d^{10}5p^6$

Se^{2-}: $[Ar]4s^23d^{10}4p^6$; I^-: $[Kr]5s^24d^{10}5p^6$

17. a. Sc^{3+} b. Te^{2-} c. Ce^{4+} and Ti^{4+} d. Ba^{2+}

19. Se^{2-}, Br^-, Rb^+, Sr^{2+}, Y^{3+}, Zr^{4+} are all isoelectronic with Kr.

21. Lattice energy is proportional to $\frac{q_1 q_2}{r}$

In general, charge effects are much greater than size effects.

a. NaCl, Na^+ smaller than K^+.

b. LiF, F^- smaller than Cl^-.

c. MgO, O^{2-} greater charge than OH^-.

d. $Fe(OH)_3$, Fe^{3+} greater charge than Fe^{2+}.

e. Na_2O, greater negative charge on O.

f. MgO, smaller ions.

23.

Reaction	ΔH	
$Na(s) \rightarrow Na(g)$	$\Delta H = 108$ kJ	(sublimation)
$Na(g) \rightarrow Na^+(g) + e^-$	$\Delta H = 495$ kJ	(ionization energy)
$1/2\ Cl_2(g) \rightarrow Cl(g)$	$\Delta H = 239/2$ kJ	(bond energy)
$Cl(g) + e^- \rightarrow Cl^-(g)$	$\Delta H = -348$ kJ	(electron affinity)
$Na^+(g) + Cl^-(g) \rightarrow NaCl(s)$	$\Delta H = -786$ kJ	(lattice energy)
$Na(s) + 1/2\ Cl_2(g) \rightarrow NaCl(s)$	$\Delta H_f^\circ = -412$ kJ/mol	

25. a. From the data given, it costs less energy to produce $Mg^{+}(g) + O^{-}(g)$ than to produce $Mg^{2+}(g) + O^{2-}(g)$. However, the lattice energy for $Mg^{2+}O^{2-}$ will be much larger than for $Mg^{+}O^{-}$.

b. Mg^{+} and O^{-} both have unpaired electrons. In Mg^{2+} and O^{2-} there are no unpaired electrons. Hence, $Mg^{+}O^{-}$ would be paramagnetic; Mg^{2+} and O^{2-} would be diamagnetic. Paramagnetism can be detected by measuring the mass of a sample in the presence and absence of a magnetic field. The apparent mass of a paramagnetic substance will be larger in a magnetic field because of a force between the unpaired electrons and the field.

27. Ca^{2+} has greater charge than Na^{+} and Se^{2-} is smaller than Te^{2-}.

The effect of charge on the lattice energy is greater than the effect of size. We expect the trend from most exothermic to least exothermic to be:

CaSe	>	CaTe	>	Na_2Se	>	Na_2Te	
(-2862)		(-2721)		(-2130)		(-2095)	This is what we observe.

29. a. Li^{+} and N^{3-} are expected ions. The formula of the compound would be Li_3N (lithium nitride).

b. Ga^{3+} and O^{2-}, Ga_2O_3, gallium(III) oxide

c. Rb^{+} and Cl^{-}, RbCl, rubidium chloride

d. Ba^{2+} and S^{2-}, BaS, barium sulfide

Bond Energies (Assume all bond energies are ±1 kJ)

31.

$$H_3C{-}N{\equiv}C \longrightarrow H_3C{-}C{\equiv}N$$

Break C—N (305 kJ/mol)
Make C—C (347 kJ/mol)
$\Delta H° = 305 - 347 = -42$ kJ

33.

$$4\ (CH_3)HN{-}NH_2 + 5\ O_2N{-}NO_2 \longrightarrow 12\ H{-}O{-}H + 9\ N{\equiv}N + 4\ O{=}C{=}O$$

Bonds broken:

9 N—N	(160 kJ/mol)
4 N—C	(305 kJ/mol)
12 C—H	(413 kJ/mol)
12 N—H	(391 kJ/mol)
10 N=O	(607 kJ/mol)
10 N—O	(201 kJ/mol)

Bonds made:

24 O—H	(467 kJ/mol)
9 N≡N	(941 kJ/mol)
8 C=O	(799 kJ/mol)

(Assume all ±1 kJ)

$\Delta H = \Sigma$ D(broken) - Σ D (made)

$= 9(160) + 4(305) + 12(413) + 12(391) + 10(607) + 10(201)$

$- [24(467) + 9(941) + 8(799)]$

$= 20{,}388$ kJ - $26{,}069$ kJ = -5681 kJ

From standard enthalpies of formation we got ΔH = -4594 kJ. The bond energy gives us an estimate that is about 24% high. We could do better. In 6.71 we used ΔH_f° for N_2O_4(l) and $CH_3N_2H_3$(l). It takes heat to convert liquids to the gas. ΔH for the gas phase reaction using bond energies will be more exothermic than for the liquids.

35.

$$H_2C{=}CH_2 + O{=}O{-}O \longrightarrow H{-}CH_2{-}C({=}O){-}H + O{=}O$$

Bonds broken:

1 C=C	(614 kJ/mol)
1 O—O	(146 kJ/mol)

Bonds made:

1 C—C	(347 kJ/mol)
1 C=O	(799 kJ/mol)

$\Delta H = 614 + 146 - (347 + 799) = -386$ kJ

From Ex. 6.71, $\Delta H^\circ = -361$ kJ

37. a. Using SF_4 data: $SF_4(g) \longrightarrow S(g) + 4\,F(g)$

$\Delta H^\circ = 4\,D_{SF} = 278.8 + 4(79.0) - (-775) = 1370.$ kJ

$$D_{SF} = \frac{1370.\text{ kJ}}{4\text{ mol SF bonds}} = 342.5\text{ kJ/mol}$$

Using SF_6 data: $SF_6(g) \longrightarrow S(g) + 6\,F(g)$

$\Delta H^\circ = 6\,D_{SF} = 278.8 + 6(79.0) - (-1209) = 1962$ kJ

$$D_{SF} = \frac{1962\text{ kJ}}{6} = 327.0\text{ kJ/mol}$$

b. The S—F bond energy in the table is 327 kJ/mol. The value in the table was based on the S—F bond in SF_6.

c. S(g) and F(g) are not the most stable form of the element at 25°C. The most stable forms are S_8(s) and F_2(g); $\Delta H_f^\circ = 0$ for these two species.

Lewis Structures and Resonance

39. a. HCN has 1 + 4 + 5 = 10 valence electrons.

H—C—N   H—C≡N:

uses 4: (6 e^- left, 3 pairs)

b. PH_3 has 5 + 3(1) = 8 valence electrons

uses 6: (2 e^- left, 1 pair)

c. $CHCl_3$ has 4 + 1 + 3(7) = 26 valence electrons.

uses 8 e^-: (18 e^- left, 9 pairs)

d. NH_4^+ has 5 + 4(1) - 1 = 8 valence electrons

e. BF_4^- has 3 + 4(7) + 1 = 32 electrons.

f. SeF_2 has 6 + 2(7) = 20 valence electrons

41. a. NO_2^- has 5 + 2(6) + 1 = 18 valence electrons.

O—N—O

skeletal structure

complete octet of more electronegative oxygen (16 e^- used)

We have 1 more pair of e^-; putting this pair on the nitrogen we get:

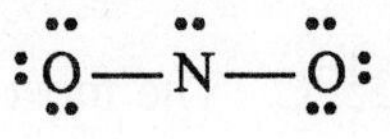

:Ö—N̈—Ö: This accounts for all 18 electrons, but N does not obey the octet rule

To get an octet about the nitrogen and not use any more electrons, one of the unshared pairs on an oxygen must be shared between N and O instead.

[:O=N̈—Ö:]⁻ (0 0 -1) ⟷ [:Ö—N̈=O:]⁻ (-1 0 0)

Since, there is no prior reason to have the double bond to either oxygen atom, we can draw two resonance structures. Formal charges are shown.

HNO_2 has 1 + 5 + 2(6) = 18 valence electrons.

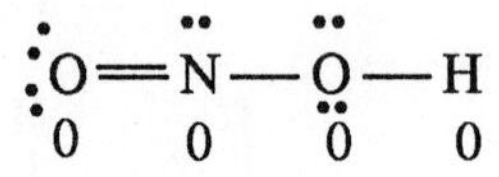

:O=N̈—Ö—H (0 0 0 0)

NO_3^- has 5 + 3(6) + 1 = 24 valence electrons.

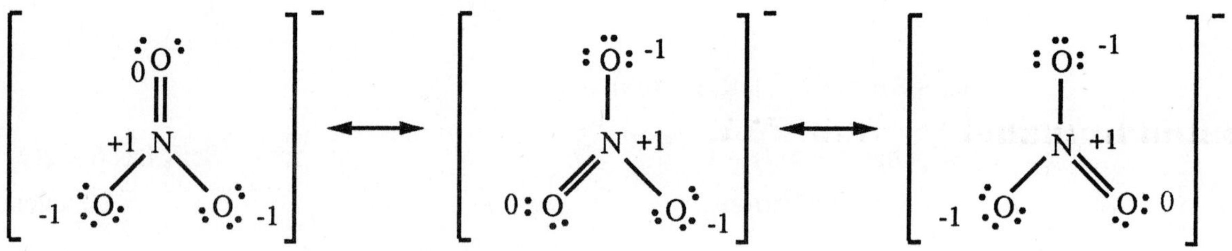

HNO_3 has 1 + 5 + 3(6) = 24 valence electrons.

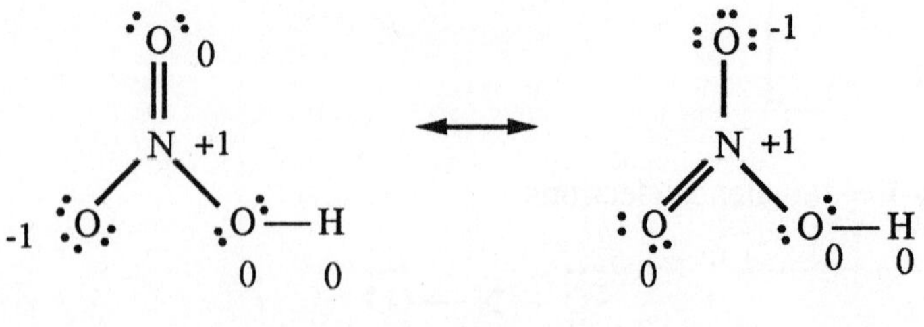

b. SO_4^{2-} see Exercise 8.40a.

HSO_4^- has 1 + 6 + 4(6) + 1 = 32 valence electrons.

```
        ••
   -1 :O:
        |  +2          ] -
   ••   |     ••
-1 :O — S — O — H
   ••   |     ••
        |     0    0
       :O:
        ••
        -1
```

H_2SO_4 has 2 + 6 + 4(6) = 32 valence electrons.

```
        ••
       :O: -1
        |  +2
   ••   |     ••
-1 :O — S — O — H
   ••   |     ••
        |     0    0
       :O: 0
        |
        H
        0
```

c. CN^-, 4 + 5 + 1 = 10 e^- 	HCN, 1 + 4 + 5 = 10 e^-

```
[:C≡N:]-          H—C≡N:
  -1  0           0  0  0
```

d. OCN^- has 6 + 4 + 5 + 1 = 16 valence electrons.

```
O—C—N              :Ö—C—N̈:          16 e- but no octet about C.
```

skeletal structure 	complete octets of O and N

We need to convert 2 nonbonding pairs to bonding pairs.

```
[:Ö—C≡N:]-   ←→   [:O=C=N:]-   ←→   [:O≡C—N̈:]-
  -1  0  0           0  0  -1          +1  0  -2
```

Third structure not likely because of formal charge.

SCN^- has 6 + 4 + 5 + 1 = 16 valence electrons.

$[:\ddot{S}-C\equiv N:]^-$ (-1, 0, 0) ⟷ $[\dot{S}=C=\dot{N}]^-$ (0, 0, -1) ⟷ $[:S\equiv C-\ddot{N}:]^-$ (+1, 0, -2)

First two structures, and especially the second, best from formal charge.

N_3^- has 3(5) + 1 = 16 valence electrons.

$[:\ddot{N}-N\equiv N:]^-$ (-2, +1, 0) ⟷ $[N=N=N]^-$ (-1, +1, -1) ⟷ $[:N\equiv N-\ddot{N}:]^-$ (0, +1, -2)

Second structure best from formal charge.

e. NCCN has 2(5) + 2(4) = 18 valence electrons.

$:N\equiv C-C\equiv N:$ (0, 0, 0, 0)

43. PAN ($H_3C_2NO_5$) has 3 + 2(4) + 5 + 5(6) = 46 valence electrons.

$H_3C-C(-\ddot{O}:)-\ddot{O}-\ddot{O}-N(-\ddot{O}:)(-\ddot{O}:)$ skeletal structure with complete octets about oxygen atoms (46 electrons used).

The skeletal structure has used all 46 electrons, but there are only six electrons around one of the carbon atoms and the nitrogen atom. Two unshared pairs must become shared, that is we form two double bonds.

$H_3C-C(=O)-\ddot{O}-\ddot{O}-N(=O)(-\ddot{O}:)$ ⟷ $H_3C-C(=O)-\ddot{O}-\ddot{O}-N(-\ddot{O}:)(=O)$

⟷ $H_3C-C(-\ddot{O}:^{-1})=\ddot{O}^{+1}-\ddot{O}^{+1}=N^{+1}(-\ddot{O}:^{-1})(-\ddot{O}:^{-1})$ (last form not important)

45.

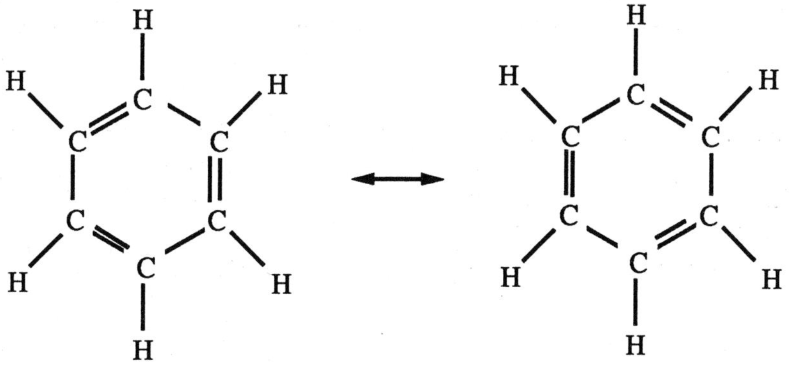

H H
| |
H C H H C H
C C C C
| || || |
C C C C
H C H H C H
| |
H H

(resonance structures ⟷)

Benzene has 6(4) + 6(1) = 30 valence electrons.

47. PF_5, 5 + 5(7) = 40 e^-

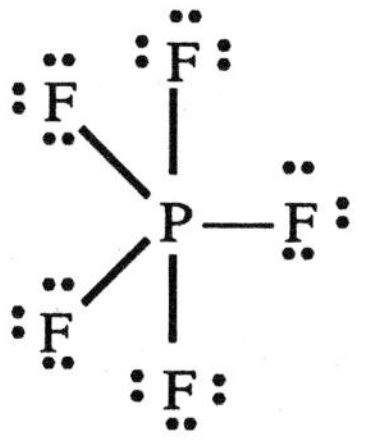

:F: :F: P—F: :F: :F:

$Be(CH_3)_2$, 2 + 2[4 + 3(1)] = 16 e^-

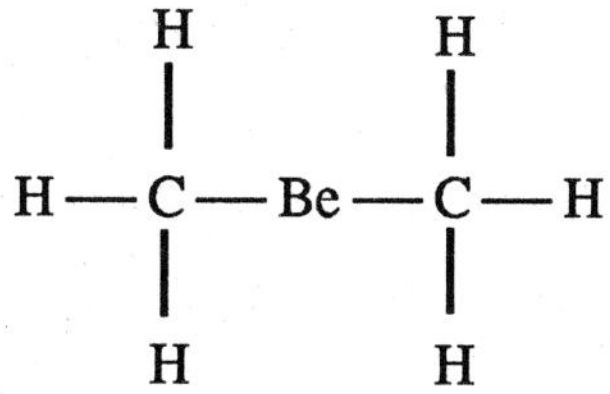

H—C—Be—C—H (each C bonded to two further H)

BCl_3, 3 + 3(7) = 24 e^-

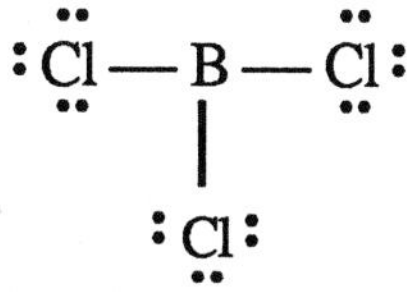

:Cl—B—Cl: with :Cl: below B

$XeOF_4$, 8 + 6 + 4(7) = 42 e^-

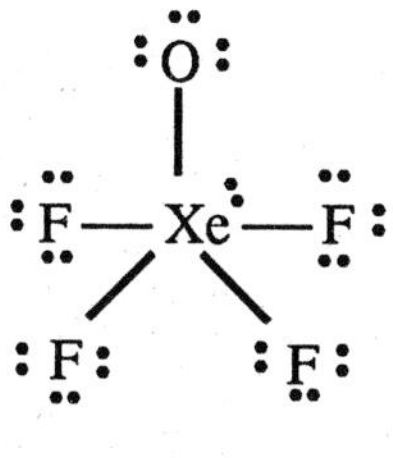

:O: above Xe; :F—Xe—F: ; :F: :F:

XeF_6, 8 + 6(7) = 50 e^-

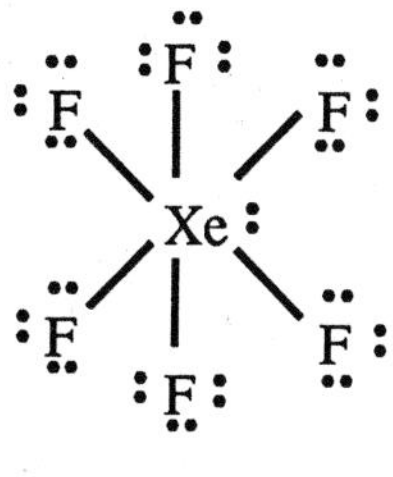

Xe: surrounded by six :F:

49.

CO :C≡O: triple bond between C and O

CO_2 :O=C=O: double bond between C and O

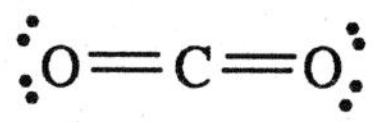

CO_3^{2-}

average of 1 1/3 bond between C and O

CH_3OH

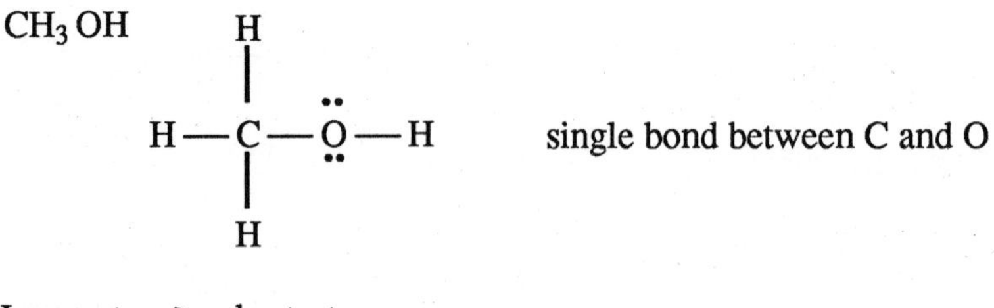

Longest → shortest

$CH_3OH > CO_3^{2-} > CO_2 > CO$

Weakest → strongest

$CH_3OH < CO_3^{2-} < CO_2 < CO$

Formal Charge

51. The Lewis structure

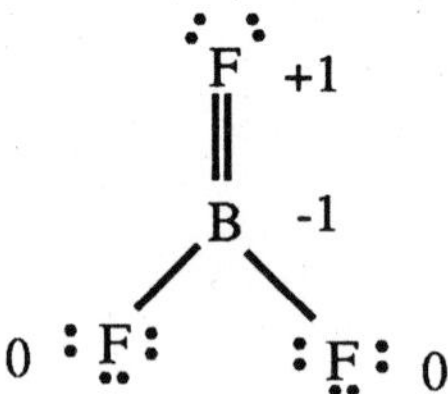

obeys the octet rule, but has a +1 formal charge on the most electronegative element there is, fluorine, and a negative formal charge on a much less electronegative element (boron). This is just the opposite of what we expect; negative formal charge on F and positive formal charge on B. Therefore, BF_3 does not follow the octet rule and there are only six electrons around the boron atom.

53. See Exercise 8.40a for Lewis structures of a, b, c, and d.

a. $POCl_3$: P, FC = 5 - 1/2(8) = +1

b. SO_4^{2-}: S, FC = 6 - 1/2(8) = +2

c. ClO_4^-: Cl, FC = 7 - 1/2(8) = +3

d. PO_4^{3-}: P, FC = 5 - 1/2(8) = +1

e. SO_2Cl_2, 6 + 2(6) + 2(7) = 32 e^-

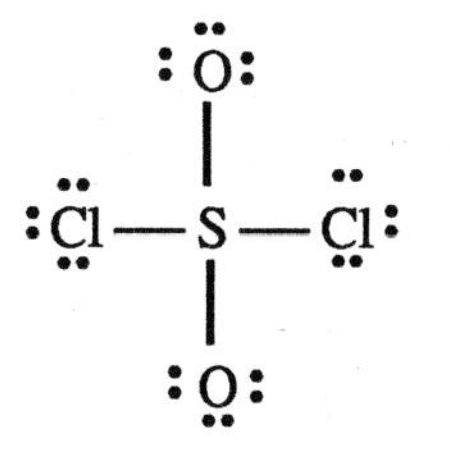

S, FC = 6 - 1/2(8) = +2

f. XeO_4, 8 + 4(6) = 32 e^-

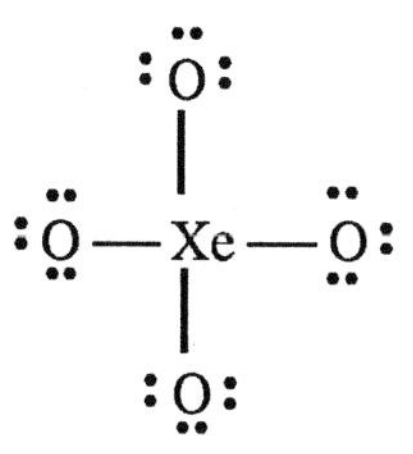

Xe, FC = 8 - 1/2(8) = +4

g. ClO_3^-, 7 + 3(6) + 1 = 26 e^-

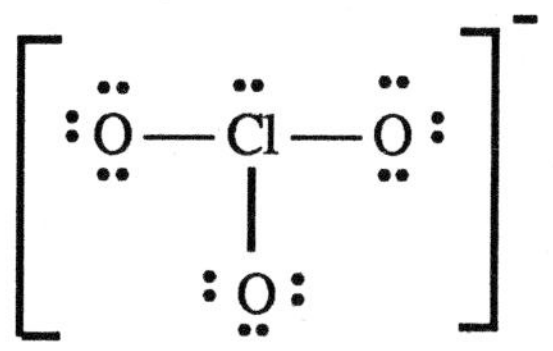

Cl, FC = 7 - 2 - 1/2(6) = +2

h. NO_4^{3-}, 5 + 4(6) + 3 = 32 e^-

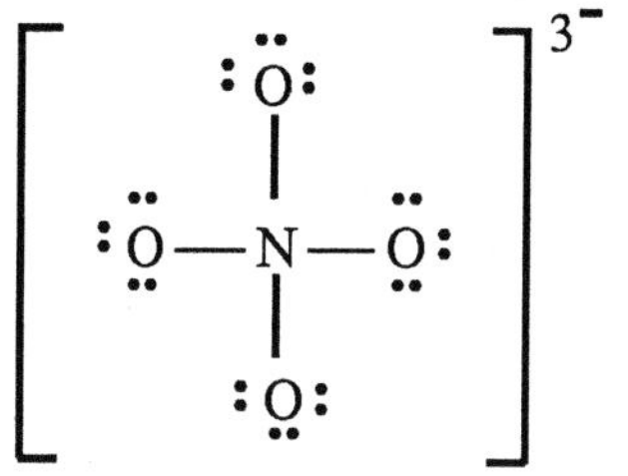

N, FC = 5 - 1/2(8) = +1

Molecular Geometry and Polarity

55. 8.39 a. linear, 180°

b. trigonal pyramid, < 109.5°

c. tetrahedral, ≈ 109.5°

d. tetrahedral, 109.5°

e. tetrahedral, 109.5°

f. V-shaped or bent, < 109.5°

8.41 a. NO_2^-, bent, < 120°

HNO_2, bent about O and N, HON angle < 109.5°, ONO angle < 120°

NO_3^-, trigonal planar, 120°

HNO_3, trigonal planar about N, 120°

bent about N—O—H, angle < 109.5°

b. tetrahedral about S in all three, angle ≈ 109.5°

bent about S—O—H, angle < 109.5°

c. HCN is linear, 180°

d. all are linear, 180°

e. linear about both carbons, 180°

57. a. BF_3, 3 + 3(7) = 24 e^-

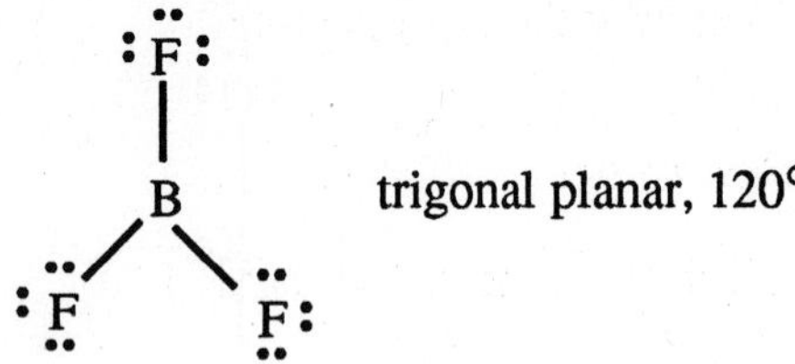

trigonal planar, 120°

b. BH_2^-, 3 + 2(1) + 1 = 6 e^-

$$\left[\begin{array}{c} \ddot{B} \\ H \quad\quad H \end{array}\right]^-$$

bent, < 120°

c. $COCl_2$, 4 + 6 + 2(7) = 24 e^-

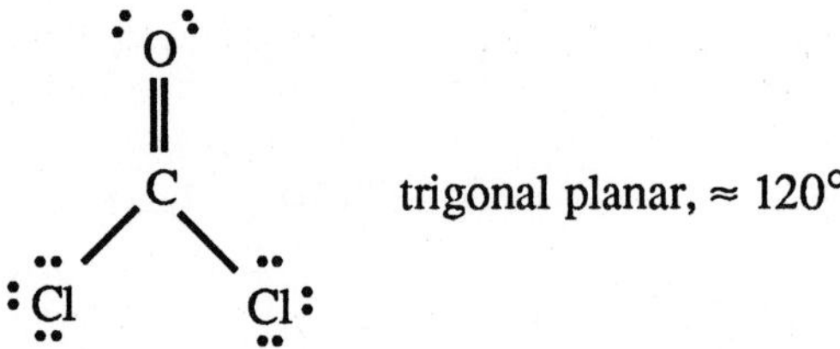

trigonal planar, ≈ 120°

Note: All of these structures have three effective pairs of electrons (a double bond counts as one effective pair for geometry) about the central atom. All of the structures are based on a trigonal planar geometry, but only a and c are described as having a trigonal planar structure.

59. a. I_3^- has 3(7) + 1 = 22 valence electrons.

$:\ddot{I}-\ddot{I}-\ddot{I}:$ Complete octets, but only 20 electrons used.

Expand octet of Central I.

$$\left[:\ddot{I}-\ddot{I}-\ddot{I}:\right]^-$$

In general we add extra pairs of electrons to the central atom. We can see a more specific reason when we look at how I_3^- forms in solution:

$$I_2(aq) + I^-(aq) \rightarrow I_3^-(aq)$$

$$:\ddot{I}-\ddot{I}: + \left[:\ddot{I}:\right]^- \rightarrow \left[:\ddot{I}-\ddot{I}-\ddot{I}:\right]^-$$

There are 5 pairs of electrons about the central ion. The structure will be based on a trigonal bipyramid geometry. The lone pairs require more room and will occupy the equatorial positions; the bonding pairs will occupy the axial positions.

axial: apex of the trigonal pyramid

equatorial: around the middle, the corners of the trigonal pyramid

Thus, I_3^- is linear with a 180° bond angle.

b. ClF_3 has 7 + 3(7) = 28 valence electrons.

T-shaped, FClF angles are ≈ 90°. Since the lone pair will take up more space, the FClF bond angles will probably be slightly less than 90°.

c. IF_4^+ has 7 + 4(7) - 1 = 34 valence electrons.

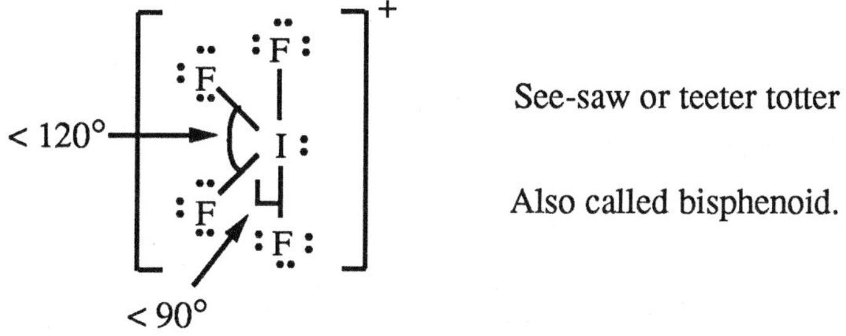

See-saw or teeter totter

Also called bisphenoid.

d. SF_5^+ has 6 + 5(7) - 1 = 40 valence electrons.

120°
90°

trigonal bipyramid

Note: All of the species in this exercise have 5 pairs of electrons around the central atom. All of the structures are based on a trigonal bipyramid, but only in SF_5^+ are all of the pairs bonding pairs. Thus, SF_5^+ is the only one we describe as being trigonal bipyramidal. Still, we had to begin with the trigonal bipyramid geometry to get to the structures of the others.

61. SbF_5 has 5 + 5(7) = 40 valence electrons.

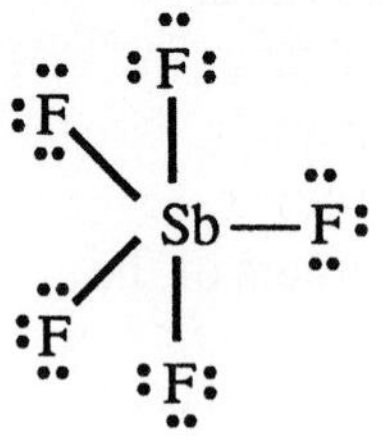

trigonal bipyramid

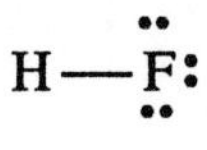

linear

SbF_6^- has 5 + 6(7) + 1 = 48 valence electrons.

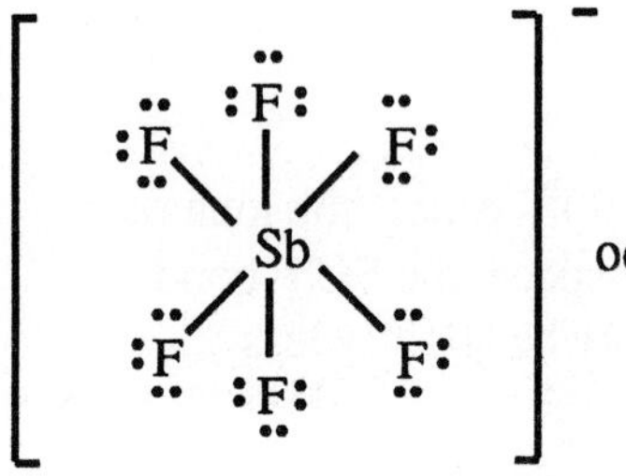

octahedral

H_2F^+ has 2 + 7 - 1 = 8 valence electrons.

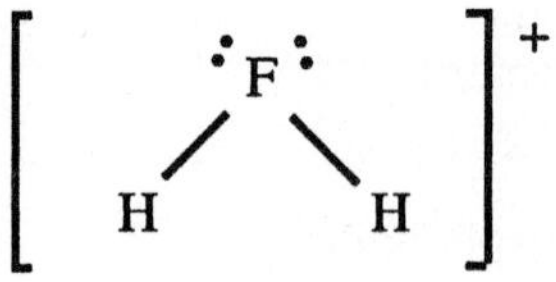

bent or V-shaped

63. a. OCl_2 has 6 + 2(7) = 20 valence electrons.

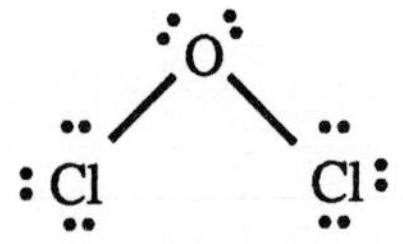

bent, polar

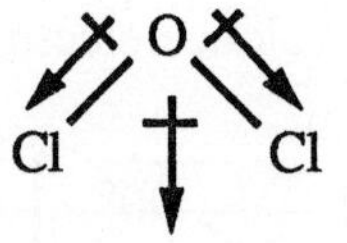

Molecule is polar because the O — Cl bond dipoles don't cancel. The resultant dipole moment is shown in the drawing.

Br_3^- has 3(7) + 1 = 22 valence electrons.

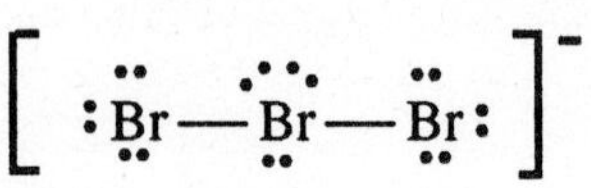

linear, non-polar

All bonds are non-polar and molecule is non-polar.

BeH_2 has 2 + 2(1) = 4 valence electrons.

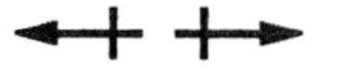

H—Be—H linear, non-polar Be—H bond dipoles are equal and point in opposite directions. They cancel each other. BeH_2 is non-polar.

BH_2^- has 3 + 2 + 1 = 6 valence electrons.

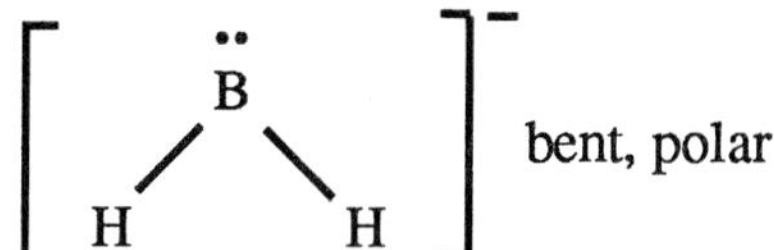

bent, polar Bond dipoles do not cancel. BH_2^- is polar.

Note: All four species contain three atoms. They have different structures because the number of lone pairs of electrons around the central atom are different in each case.

b. BCl_3 has 3 + 3(7) = 24 valence electrons.

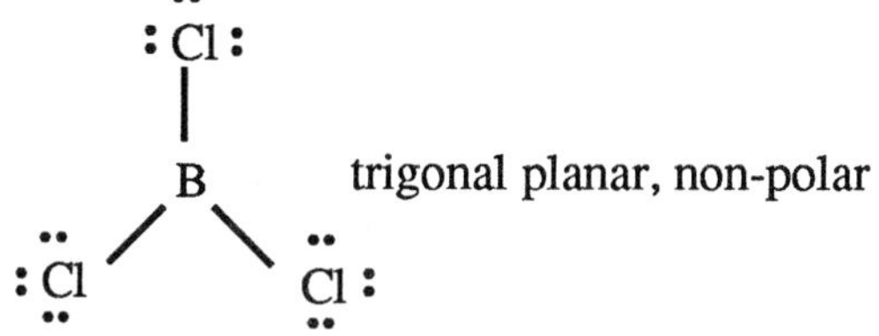

Bond dipoles cancel.

NF_3 has 5 + 3(7) = 26 valence electrons.

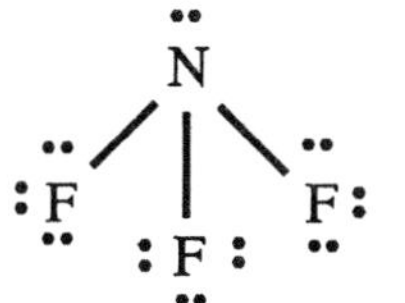

trigonal pyramid, polar Bond dipoles do not cancel.

IF_3 has 7 + 3(7) = 28 valence electrons.

:F—I—F:
:F:

T-shaped, polar Bond dipoles do not cancel.

Note: Each molecule contains the same number of atoms, but the structures are different because of differing numbers of lone pairs around each cental atom.

c. CF_4 has 4 + 4(7) = 32 valence electrons.

```
      :F:
   ..  |  ..
  :F — C — F:
   ..  |  ..
      :F:
       ..
```

tetrahedral, non-polar   Bond dipoles cancel.

SeF_4 has 6 + 4(7) = 34 valence electrons.

```
        ..
   ..  :F:
  :F    |
   ..\  |
      Se :
   ../  |
  :F    |
   ..  :F:
        ..
```

see-saw, polar   Bond dipoles do not cancel.

XeF_4 has 8 + 4(7) = 36 valence electrons.

```
   ..          ..
  :F           F:
   ..\  ..   / ..
       Xe
   ../  ..   \ ..
  :F           F:
   ..          ..
```

square planar, non-polar   Bond dipoles cancel.

Again, each molecule has the same number of atoms, but a different structure because of differing numbers of electrons around the central atom.

d. IF_5 has 7 + 5(7) = 42 valence electrons.

```
       ..   ..
      :F:   F:
       |  / ..
   ..  | /   ..
  :F — I — F:
   ..  /..   ..
   ../
  :F
   ..
```

square pyramid, polar   Bond dipoles do not cancel.

AsF_5 has 5 + 5(7) = 40 valence electrons.

```
        ..
   ..  :F:
  :F    |
   ..\  |     ..
      As — F:
   ../  |     ..
  :F    |
   ..  :F:
        ..
```

trigonal bipyramid, non-polar   Bond dipoles cancel.

Yet again, the molecules have the same number of atoms, but different structures because of the presence of differing numbers of lone pairs.

ADDITIONAL EXERCISES

65. a.

```
     H   H   H
     |   |   |
 H — C — C — C — H  + 5 O═O ──▶ 3 O═C═O + 4 H—O—H
     |   |   |
     H   H   H
```

Bonds broken:		Bonds made:	
2 C—C	(347 kJ/mol)	3 × 2 C═O	(799 kJ/mol)
8 C—H	(413 kJ/mol)	4 × 2 H—O	(467 kJ/mol)
5 O═O	(495 kJ/mol)		

ΔH = 2(347) + 8(413) + 5(495) - [6(799) + 8(467)] = -2057 kJ

b.

```
          H
          |
      H — C — OH
          |
  H       C ——— O                                 H   H
    \   /  \     \    H                           |   |
     \ /    H     \  /      ──▶ 2 O═C═O + 2 H — C — C — OH
      C  OH    H   C                              |   |
     / \ |     |  / \                             H   H
   HO    C ——— C     OH
          \     \
           H     OH
```

The molecules are complicated enough that it will be easier to break all bonds in glucose and make all the bonds in CO_2 and the alcohol.

Bonds broken:		Bonds made:	
5 C—C	(347 kJ/mol)	2 × 2 C═O	(799 kJ/mol)
7 C—O	(358 kJ/mol)	2 × 5 C—H	(413 kJ/mol)
5 O—H	(467 kJ/mol)	2 C—O	(358 kJ/mol)
7 C—H	(413 kJ/mol)	2 O—H	(467 kJ/mol)
		2 C—C	(347 kJ/mol)

ΔH = 5(347) + 7(358) + 5(467) + 7(413)

- [4(799) + 10(413) + 2(358) + 2(467) + 2(347)] = -203 kJ

67.

$$\mathrm{H_2N{-}NH_2} \longrightarrow 2\ \mathrm{N(g)} + 4\ \mathrm{H(g)}$$

$\Delta H° = D_{N-N} + 4\ D_{N-H} = D_{N-N} + 4(388.9)$

$\Delta H° = 2\Delta H_f°(N) + 4\ \Delta H_f°(H) - \Delta H_f°(N_2H_4) = 2(472.7) + 4(216.0) - 95.4$

$\Delta H° = 1714.0\ kJ = D_{N-N} + 4(388.9)$

$D_{N-N} = 158.4$ kJ/mol (160 kJ/mol in Table 8.4)

69. a.

$HF(g) \rightarrow H(g) + F(g)$	$\Delta H = 565$ kJ
$H(g) \rightarrow H^+(g) + e^-$	$\Delta H = 1312$ kJ
$F(g) + e^- \rightarrow F^-(g)$	$\Delta H = -328$ kJ
$HF(g) \rightarrow H^+(g) + F(g)^-$	$\Delta H = 1549$ kJ

b.

$HCl(g) \rightarrow H(g) + Cl(g)$	$\Delta H = 427$ kJ
$H(g) \rightarrow H^+(g) + e^-$	$\Delta H = 1312$ kJ
$Cl(g) + e^- \rightarrow Cl^-(g)$	$\Delta H = -349$ kJ
$HCl(g) \rightarrow H^+(g) + Cl^-(g)$	$\Delta H = 1390.$ kJ

c.

$HI(g) \rightarrow H(g) + I(g)$	$\Delta H = 295$ kJ
$H(g) \rightarrow H^+(g) + e^-$	$\Delta H = 1312$ kJ
$I(g) + e^- \rightarrow I^-(g)$	$\Delta H = -295$ kJ
$HI(g) \rightarrow H^+(g) + I^-(g)$	$\Delta H = 1312$ kJ

d.

$H_2O(g) \rightarrow OH(g) + H(g)$	$\Delta H = 467$ kJ
$H(g) \rightarrow H^+(g) + e^-$	$\Delta H = 1312$ kJ
$OH(g) + e^- \rightarrow OH^-(g)$	$\Delta H = -180$ kJ
$H_2O(g) \rightarrow H^+(g) + OH^-(g)$	$\Delta H = 1599$ kJ

71. Ionic solids can be characterized as being held together by strong-omnidirectional forces.

I. For electrical conductivity, charged species must be free to move. In ionic solids the charged ions are held rigidly in place. Once the forces are disrupted (melting or dissolution) the ions can move about (conduct).

II. Melting and boiling disrupts the attractions of the ions for each other. If the forces are strong it will take a lot of energy (high temp.) to accomplish this.

III. If we try to bend a piece of material, the atoms/ions must slide about each other. For an ionic solid the following might happen:

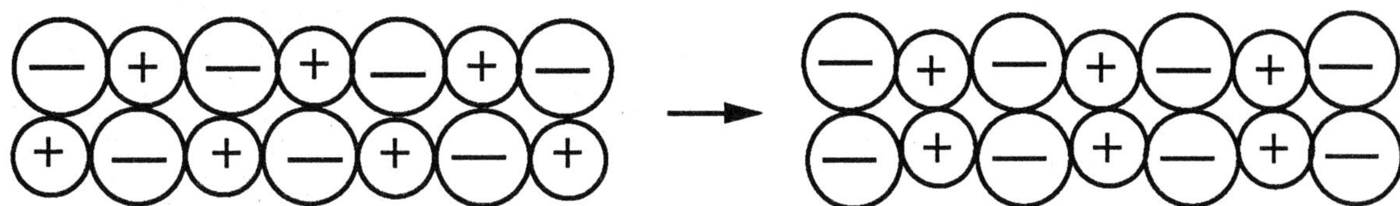

strong attraction strong repulsion

Just as the layers begin to slide, there will be very strong repulsions causing the solid to snap across a fairly clean plane.

IV. Polar molecules are attracted to ions and can break up the lattice.

These properties and their correlation to chemical forces will be discussed in detail in Chapter 10 and 11.

73. S_2Cl_2 has 2(6) + 2(7) = 26 valence electrons. SCl_2 has 6 + 2(7) = 20 valence electrons.

:Cl—S—S—Cl: :Cl—S—Cl:

75. a. Al_2Cl_6 has 2(3) + 6(7) = 48 valence electrons.

:Cl Cl Cl:
Al Al
:Cl Cl :Cl:

b. There are 4 pairs of electrons about each Al, we would predict the bond angles to be close to a tetrahedral angle of 109.5°.

c. non-polar

77. Resonance is possible in CO_3^{2-}. (See previous exercise) We would expect all C—O bonds to be equal and intermediate in strength and length to a single and a double bond. The bond length of 136 pm is consistent with this view of the structure.

79. a. BF_3 has 3 + 3(7) = 24 valence electrons.

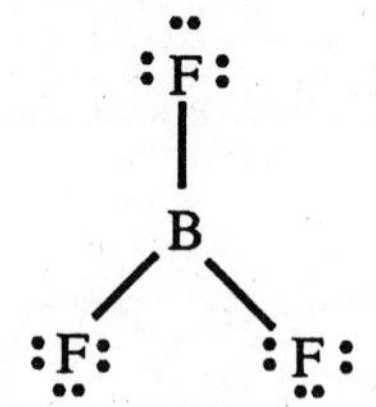

trigonal planar, non-polar

b. PF_3 has 5 + 3(7) = 26 valence electrons.

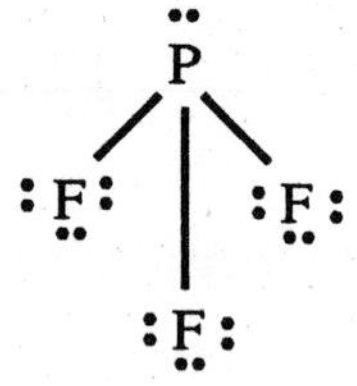

trigonal pyramid, polar

c. BrF_3 has 7 + 3(7) = 28 valence electrons.

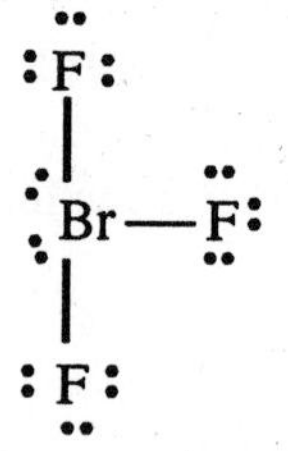

T-shaped, polar

Again we see a series of molecules with the same number of atoms, but different numbers of lone pairs: They have different structures.

81.

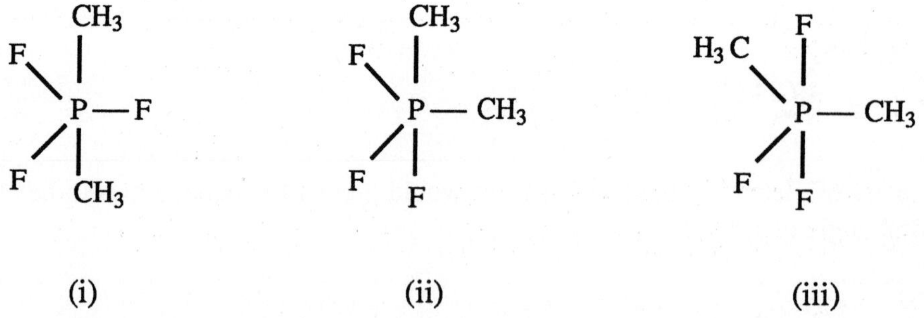

(i) (ii) (iii)

(i) is non-polar

(ii) and (iii) are polar with (ii) probably the more polar. Thus, dipole moments should help distinguish the three isomers; particularly in distinguishing (i) from the other two.

83. a.

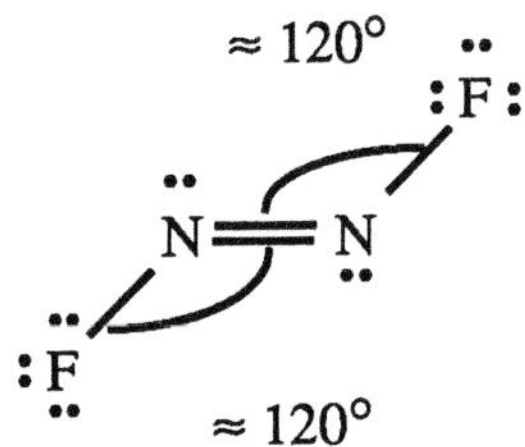

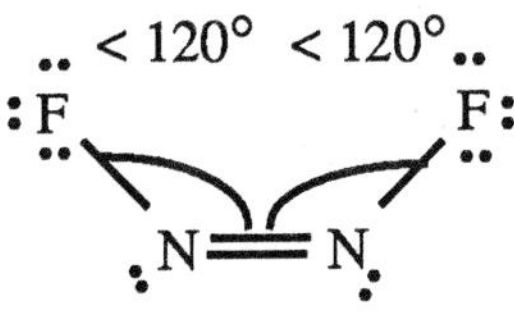

b. non-polar polar

85. This is a very weak bond. The structure

$$:\ddot{O}-\ddot{Xe}-\ddot{O}: \quad (\text{with } :\ddot{O}: \text{ bonded below Xe})$$

might be more accurate. It is difficult to think of a double bond as being so weak. This illustrates that no model is totally effective in describing the bonding in all molecules. We need to recognize that all models have limitations and that we can't apply all models to all molecules.

CHALLENGE PROBLEMS

87. Let us look at the complete cycle for Li_2S. (Assume all values are ± 1 kJ).

$2\,Li(s) \rightarrow 2\,Li(g)$	$2\,\Delta H_{sub}(Li) = 2(162)$ kJ
$2\,Li(g) \rightarrow 2\,Li^+(g) + 2\,e^-$	$2\,I = 2(520)$ kJ
$S(s) \rightarrow S(g)$	$\Delta H_{sub}(S) = 277$ kJ
$S(g) + e^- \rightarrow S^-(g)$	$EA_1 = -200$ kJ
$S^-(g) + e^- \rightarrow S^{2-}(g)$	$EA_2 = ?$
$2\,Li^+(g) + S^{2-}(g) \rightarrow Li_2S(s)$	$LE = -2472$ kJ
$2\,Li(s) + S(s) \rightarrow Li_2S(s)$	$\Delta H_f^\circ = -500$ kJ

$\Delta H_f^\circ = 2\,\Delta H_{sub}(Li) + 2I + \Delta H_{sub}(S) + EA_1 + EA_2 + LE$

$-500 = -1031 + EA_2$; $EA_2 = +531$ kJ

For each salt: $\Delta H_f^\circ = 2\,\Delta H_{sub}(M) + 2I + 277 - 200 + EA_2 + LE$

Na_2S: $-365 = 2(108) + 2(496) + 277 - 200 - 2203 + EA_2$; $EA_2 = +553$ kJ

K_2S: $-381 = 2(90) + 2(419) + 277 - 200 - 2052 + EA_2$; $EA_2 = +576$ kJ

Rb_2S: $-361 = 2(82) + 2(409) + 277 - 200 - 1949 + EA_2$; $EA_2 = +529$ kJ

Cs_2S: $-360 = 2(78) + 2(382) + 277 - 200 - 1850 + EA_2$; $EA_2 = +493$ kJ

We get values from 493 to 576 kJ

The mean value is $\frac{531 + 553 + 576 + 529 + 493}{5} = 536$ kJ

We can represent the results as $EA_2 = 540$ kJ $\pm$ 50 kJ

89. If no resonance:

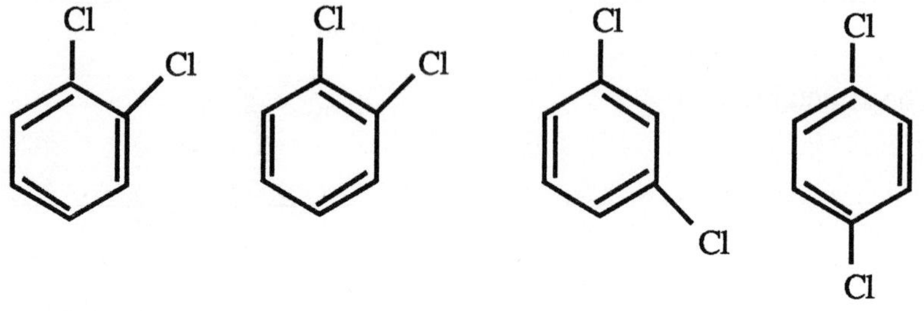

Cl Cl Cl Cl Cl Cl Cl Cl

4 different molecules

With resonance:

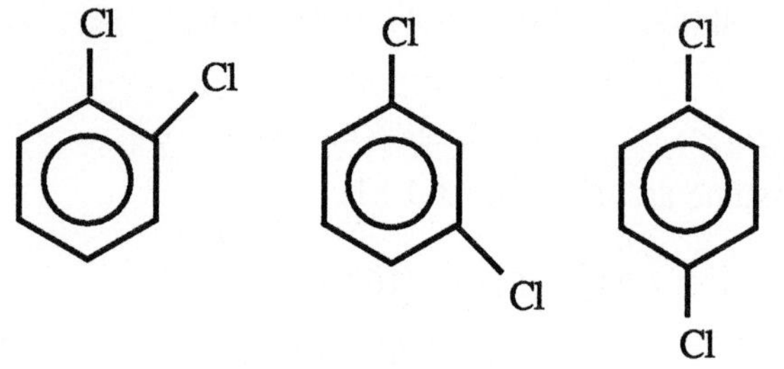

Cl Cl Cl Cl Cl Cl

3 different molecules

If resonance is present, we can't distinguish between a single and double bond between adjacent carbons that have a chlorine attached, since all carbon-carbon bonds are equivalent. That only 3 isomers are observed provides evidence for the existence of resonance.

91. SF_2 has 6 + 2(7) = 20 valence electrons. SF_4 has 6 + 4(7) = 34 valence electrons.

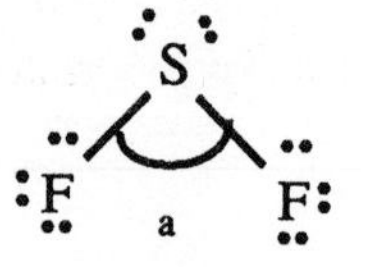

S F F a

bent, polar
a) < 109.5°

F F S F F a b

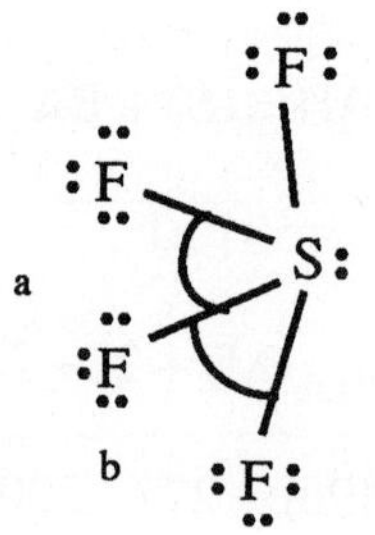

see-saw, polar
a) < 120°
b) < 90°

SF_6 has 6 + 6(7) = 48 valence electrons.

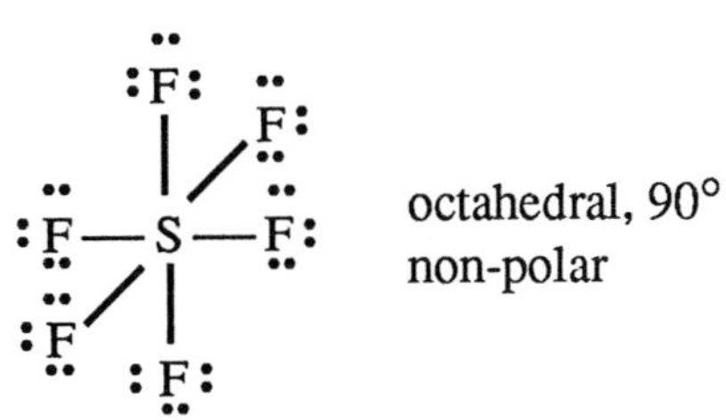

octahedral, 90°
non-polar

S_2F_4 has 2(6) + 4(7) = 40 valence electrons.

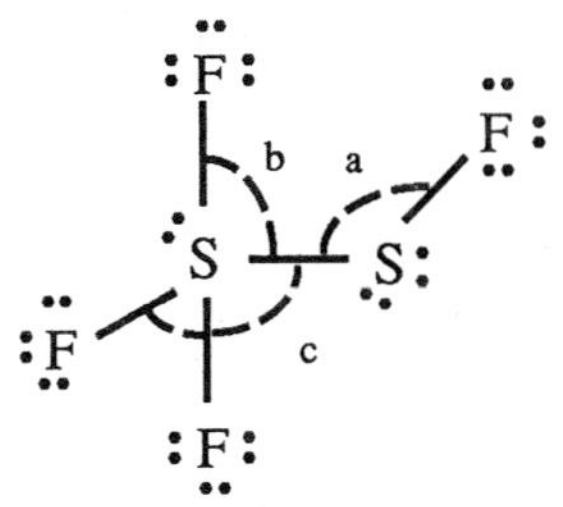

a) < 109.5°
b) < 90°
c) < 120°

There is no easy description for the S_2F_4 structure. There is a trigonal bipyramidal arrangement (3 F, 1 S, 1 lone pair) about one sulfur and a tetrahedral (1 F, 1 S, 2 lone pairs) arrangement of e^- pairs about the other sulfur. With this picture we could predict values for all bond angles. Polar.

93. a.

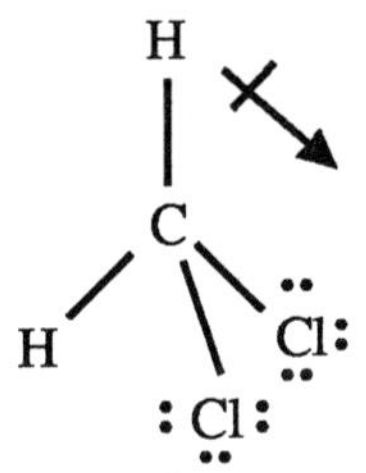

The C—H bond is not very polar since the electronegativities of C and H are about equal.

$\delta+$ $\delta-$
C—Cl is the charge distribution for each C—Cl bond. The two individual C—Cl bond dipoles add together to give an overall dipole moment for the molecule. The overall dipole will point from C (positive end) to the midpoint of the two Cl atoms (negative end).

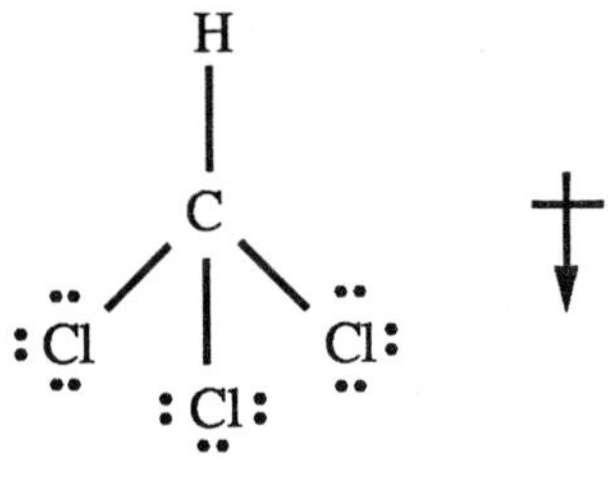

The C—H bond is essentially non-polar. The three C—Cl bond dipoles add together to give an overall dipole moment for the molecule. The overall dipole will have the negative end at the midpoint of the three chlorines and the positive end at the carbon.

CCl_4 is non-polar. CCl_4 is a tetrahedral molecule where all four C — Cl bond dipoles cancel when added together. Let's consider just the C and two of the Cl atoms. There will be a net dipole pointing in the direction of the middle of the two Cl atoms.

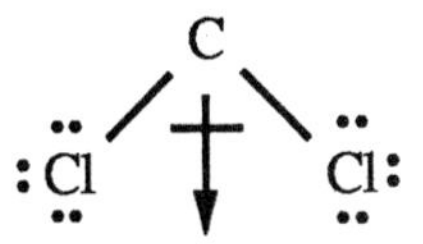

There will be an equal and opposite dipole arising from the other two Cl atoms. Combining:

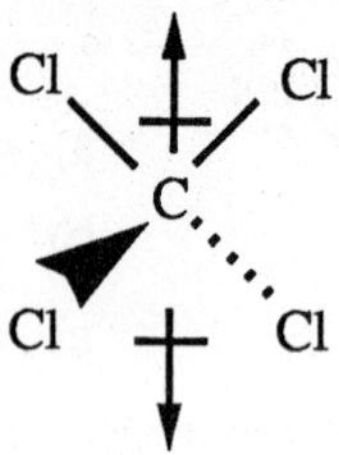

The two dipoles will cancel and CCl_4 is non-polar.

b. CO_2 is non-polar. CO_2 is a linear molecule with two equivalent bond dipoles that cancel. N_2O is polar since the bond dipoles do not cancel.

δ+ δ-

:N═N═O:

c. NH_3 is polar. The 3 N—H bond dipoles add together to give a net dipole in the direction of the lone pair. We would predict PH_3 to be non-polar on the basis of electronegativity, i.e., P—H bonds are non-polar. However, the presence of the lone pair makes the PH_3 molecule polar. The net dipole is in the direction of the lone pair and has a magntiude about one third that of the NH_3 dipole.

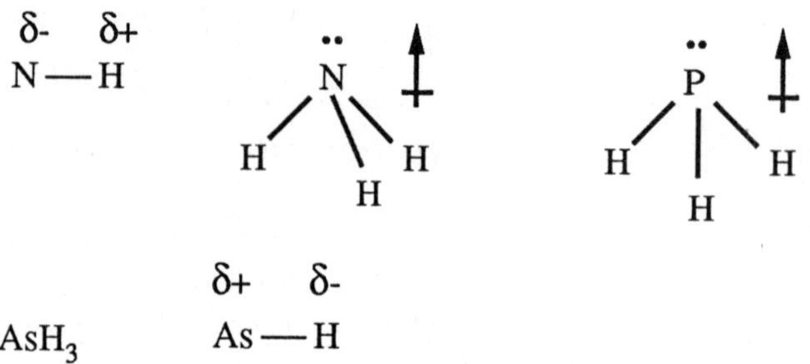

For AsH_3, the polarity arising from the lone pair is in the opposite direction as the bond dipoles. AsH_3 is slightly polar with a dipole moment one third that of PH_3 and in the direction of the lone pair.

As

H H H

95. a. EN is proportional to (IE - EA). The ionization energies of the noble gases are very high. This causes the EN to also be very high.

b. The electronegativity of F is greater than the electronegativity of Xe; it is also higher than the electronegativity of Kr. Thus, we might also expect F_2 to react with Kr and Rn.

CHAPTER NINE: COVALENT BONDING - ORBITALS

QUESTIONS

1. Bond energy is proportional to bond order. Bond length is inversely proportional to bond order. Bond energy and length can be measured.

3. Paramagnetic: Unpaired electrons are present.

   Measure the mass of a substance in the presence and absence of a magnetic field. A substance with unpaired electrons will be attracted by the magnetic field, giving an apparent increase in mass in the presence of the field. Greater number of unpaired electrons will give greater attraction and greater observed mass increase.

EXERCISES

Localized Electron Model: Hybrid Orbitals

5. a.

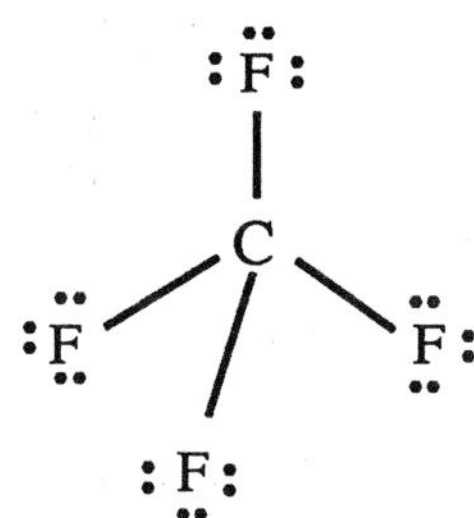

tetrahedral 109.5°
sp^3 non-polar

b.

trigonal pyramid < 109.5°
sp^3 polar

The angle should be slightly less than 109.5° because the lone pair requires more room than the bonding pairs.

c.

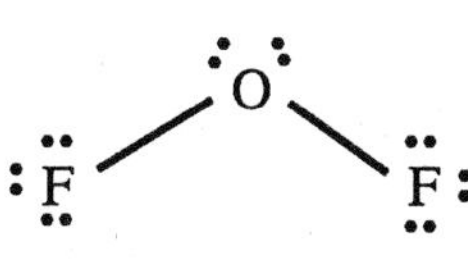

bent < 109.5°
sp^3 polar

d.

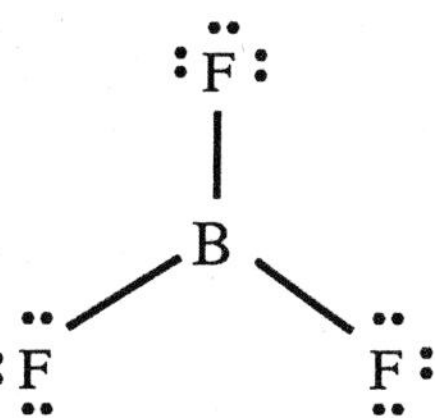

trigonal planar 120°
sp^2 non-polar

e.

H—Be—H

linear 180°
sp non-polar

f.

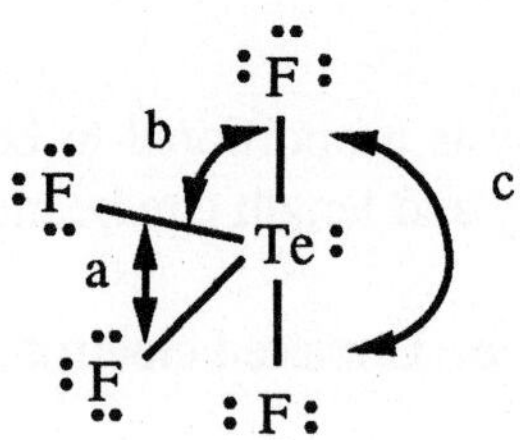

see-saw
a. < 120°, b. < 90°, c. ≈ 180°
dsp^3 polar

g.

F
a
F
As
F
b
F
F

trigonal bipyramid a. 90°, b. 120°
dsp^3 non-polar

h.

F—Kr—F

linear 180°
dsp^3 non-polar

i.

F F
Kr 90°
F F

square planar 90°
d^2sp^3 non-polar

j.

F
F
F—Se—F
F
F

octahedral 90°
d^2sp^3 non-polar

k.

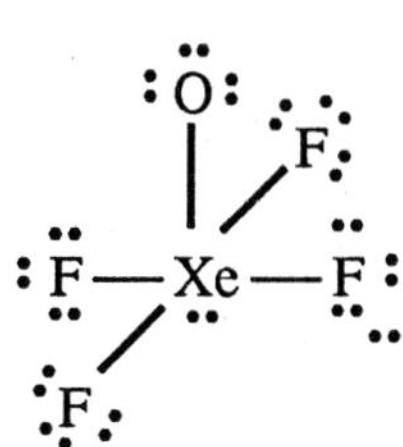

square pyramid $\approx 90°$

d^2sp^3 polar

l.

T-shaped

a. $< 90°$, b. $\approx 180°$

dsp^3 polar

m.

tetrahedral

109.5°

sp^3

non-polar

7.

For the p-orbitals to properly line up to form the π bond, all six atoms are forced into the same plane.

9. Biacetlyl ($C_4H_6O_2$) has 4(4) + 6(1) + 2(6) = 34 valence electrons.

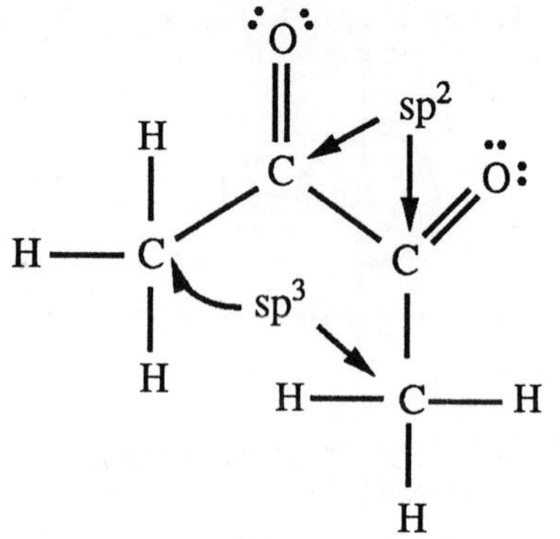

All CCO angles are 120°. The six atoms are not in the same plane because of free rotation about the carbon-carbon single bonds.

11 σ and 2 π bonds

Acetoin ($C_4H_8O_2$) has 4(4) + 8(1) + 2(6) = 36 valence electrons.

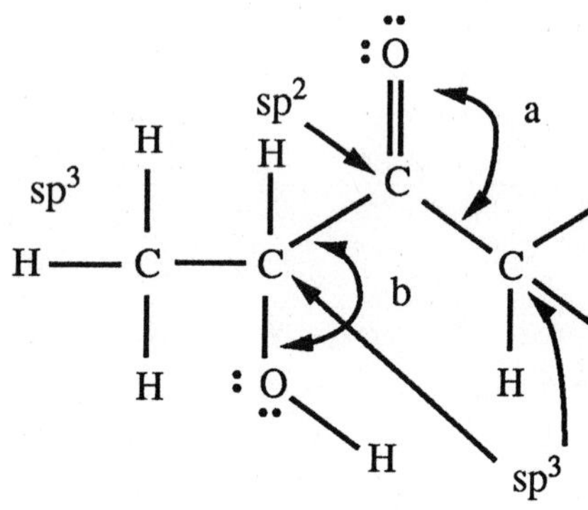

The carbon with doubly bonded O is sp^2 hybridized. The other 3 C-atoms are sp^3 hybridized.

Angle a) 120°

Angle b) 109.5°

13 σ and 1 π bonds

11.

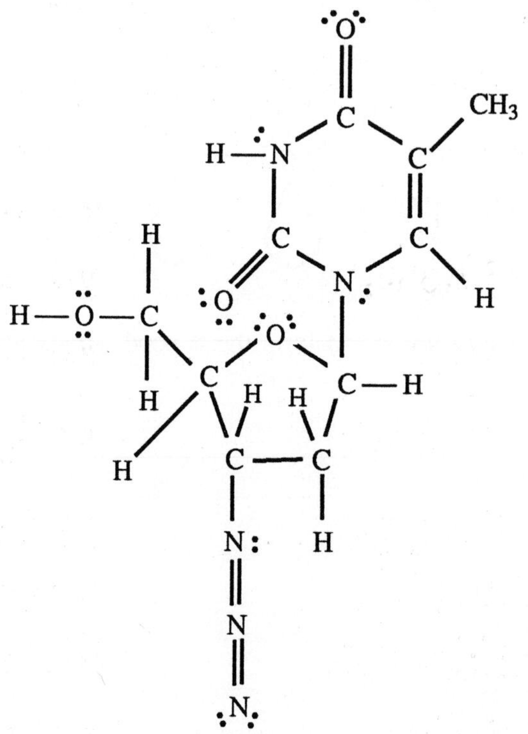

a. 6 b. 4 c. The center N in —N=N=N group

d. 33 σ e. 5 π bonds f. The six membered ring is planar.

g. 180° h. < 109.5° i. sp^3

13. a. 14 C are sp^3. 26 C are sp^2. None are sp.

b. 39 σ and 13 π bonds between carbon atoms.

The Molecular Orbital Model

15. If we calculate a non-zero bond order for a molecule, we predict that it can exist (is stable).

a. H_2^+ $(\sigma_{1s})^1$ B.O. = $\frac{1-0}{2}$ = 1/2, stable

H_2 $(\sigma_{1s})^2$ B.O. = $\frac{2-0}{2}$ = 1, stable

H_2^- $(\sigma_{1s})^2(\sigma_{1s}^*)^1$ B.O. = $\frac{2-1}{2}$ = 1/2, stable

H_2^{2-} $(\sigma_{1s})^2(\sigma_{1s}^*)^2$ B.O. = $\frac{2-2}{2}$ = 0, not stable

b. He_2^{2+} $(\sigma_{1s})^2$ B.O. = $\frac{2-0}{2}$ = 1, stable

He_2^+ $(\sigma_{1s})^2(\sigma_{1s}^*)^1$ B.O. = $\frac{2-1}{2}$ = 1/2, stable

He_2 $(\sigma_{1s})^2(\sigma_{1s}^*)^2$ B.O. = $\frac{2-2}{2}$ = 0, not stable

17. The electron configurations are:

a. H_2: $(\sigma_{1s})^2$, B.O. = 1, diamagnetic

b. B_2: $(\sigma_{2s})^2(\sigma_{2s}^*)^2(\pi_{2p})^2$, B.O. = 1, paramagnetic

c. F_2: $(\sigma_{2s})^2(\sigma_{2s}^*)^2(\pi_{2p})^4(\sigma_{2p})^2(\pi_{2p}^*)^4$, B.O. = 1, diamagnetic

d. CN^+: $(\sigma_{2s})^2(\sigma_{2s}^*)^2(\pi_{2p})^4$, B.O. = $\frac{6-2}{2}$ = 2 diamagnetic

e. CN: $(\sigma_{2s})^2(\sigma_{2s}^*)^2(\pi_{2p})^4(\sigma_{2p})^1$, B.O. $\frac{7-2}{2}$ = 2.5 paramagnetic

f. CN^-: $(\sigma_{2s})^2(\sigma_{2s}^*)^2(\pi_{2p})^4(\sigma_{2p})^2$, B.O. = 3, diamagnetic

g. N_2: $(\sigma_{2s})^2(\sigma_{2s}^*)^2(\pi_{2p})^4(\sigma_{2p})^2$, B.O. = 3, diamagnetic

h. N_2^+: $(\sigma_{2s})^2(\sigma_{2s}^*)^2(\pi_{2p})^4(\sigma_{2p})^1$, B.O. = 2.5, paramagnetic

i. N_2^-: $(\sigma_{2s})^2(\sigma_{2s}^*)^2(\pi_{2p})^4(\sigma_{2p})^2(\pi_{2p}^*)^1$, B.O. = 2.5, paramagnetic

19.

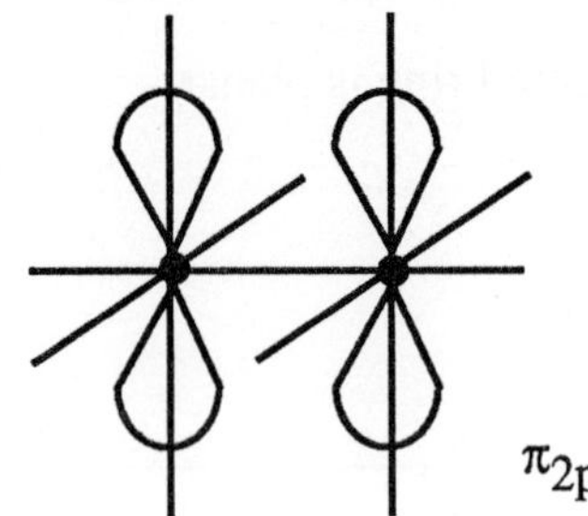

21. The bond orders are: CN^+, 2; CN, 2.5; CN^-, 3

Shortest → Longest bond length $CN^- < CN < CN^+$

Lowest → Highest bond energy $CN^+ < CN < CN^-$

23. C_2^{2-} has 10 valence electrons.

$[:C{\equiv}C:]^{2-}$ sp hybrid orbitals used.

$(\sigma_{2s})^2 (\sigma_{2s}{}^*)^2 (\pi_{2p})^4 (\sigma_{2p})^2$ B.O. = 3

Both give the same picture, a triple bond composed of a σ and two π-bonds. Both predict the ion to be diamagnetic. Lewis structures deal well with diamagnetic (all electrons paired) species. The Lewis model cannot really predict magnetic properties. C_2^{2-} is isoelectronic with the CO molecule.

ADDITIONAL EXERCISES

25. a. FClO, 7 + 7 + 6 = 20 e^-

:F̤̈—C̤̈l—Ö̤:

bent, sp^3 hybrid

b. $FClO_2$, 7 + 7 + 2(6) = 26 e^-

:F̤̈—C̤̈l—Ö̤:
|
:Ö̤:

trigonal pyramid, sp^3

c. $FClO_3$, 7 + 7 + 3(6) = 32 e^-

:F̈:
|
:Ö̤—Cl—Ö̤:
|
:Ö̤:

tetrahedral
sp^3

d. F_3ClO, 3(7) + 7 + 6 = 34 e^-

:F̤̈ :F̈:
\ |
Cl :
/ |
:Ö̤ :F̤:

see-saw
dsp^3

e. F_3ClO_2: 3(7) + 7 + 2(6) = 40 valence e^-

```
 ..    ..
:O    :F:
 ..\   |      ..
    Cl—F:
 ../   |      ..
:O    :F:
 ..    ..
```

trigonal bipyramid

dsp^3

27. SbF_5 monomer:

```
 ..    ..
:F    :F:
 ..\   |      ..
    Sb—F:
 ../   |      ..
:F    :F:
 ..    ..
```

trigonal bipyramid, dsp^3

SbF_5 polymer: Essentially there is an octahedral arrangement of fluorines about each antimony atom (d^2sp^3):

```
          b
Sb—F^a    F     F^c
      \   |   /
         Sb
      /   |   \
  c F     |     F b
          a
          F—Sb
```

The three types of fluorines are:

a. bridging, bonded to two Sb

b. 90° away from one type (a) and 180° away from the other type (a)

c. 90° away from both type (a)

29.

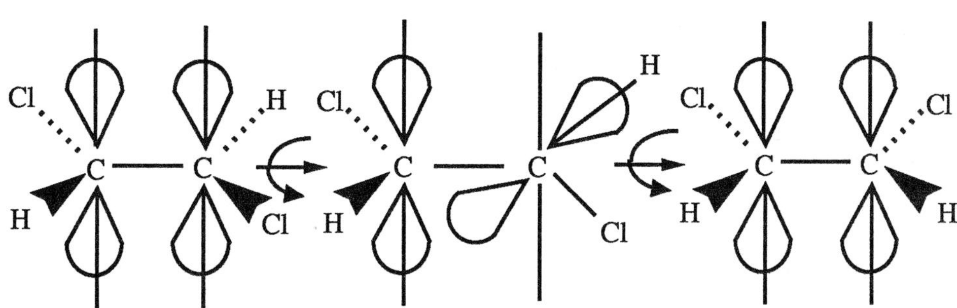

In order to rotate about the double bond, the molecule must go through the intermediate shown above. Hence, the π bonds must be broken, while the sigma bond remains intact. Bond energies are 347 kJ/mol for a C — C and 614 kJ/mol for a C $=$ C. If we take the single bond as the strength of the σ bond, then the strength of the π bond is (614 - 347) = 267 kJ/mol. Thus, 267 kJ/mol must be supplied to rotate about a carbon-carbon double bond.

31. a. P_4 4(5) = 20 valence electrons

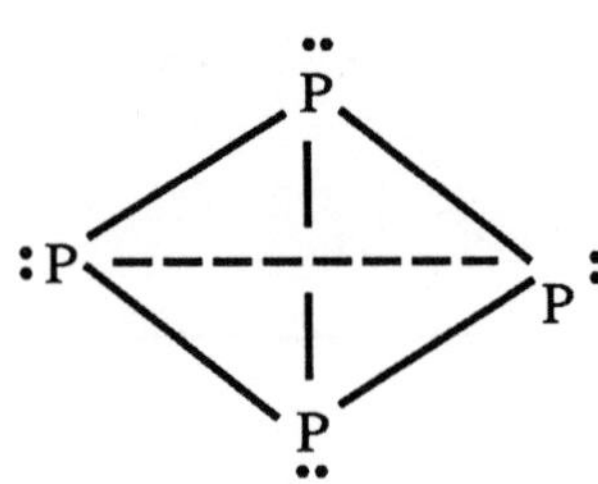

All P - sp^3

P—P—P bond angle 60° (Each P is at the corner of a tetrahedron. The faces of a tetrahedron are equilateral triangles.)

Note: We would normally predict 109.5° bond angles because there are 4 pair of e^- about each P. There is considerable strain in this molecule since each P is forced into 60° bond angles.

b. P_4O_6 4(5) + 6(6) = 56 valence electrons

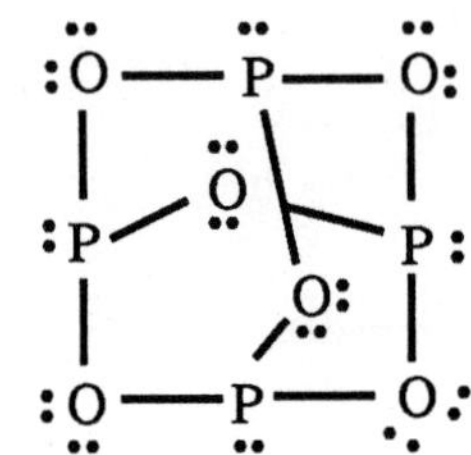

All P - sp^3

POP angle < 109.5°

OPO angle < 109.5°

c. P_4O_{10} 4(5) + 10(6) = 80 valence electrons

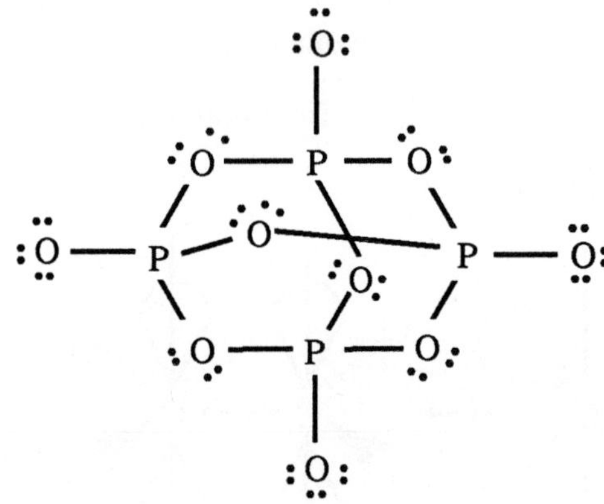

All P - sp^3

POP angle < 109.5°

OPO angle = 109.5°

33. a. $COCl_2$ has 4 + 6 +2(7) = 24 valence electrons.

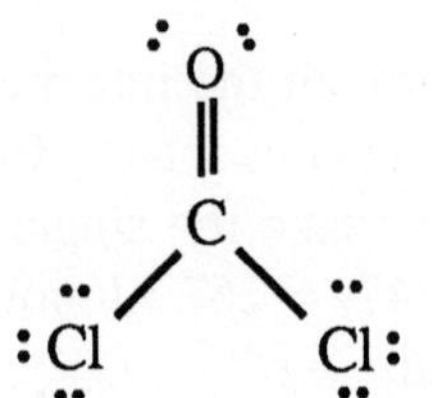

trigonal planar

polar

120°

sp^2

b. N_2F_2 has 2(5) + 2(7) = 24 valence electrons.

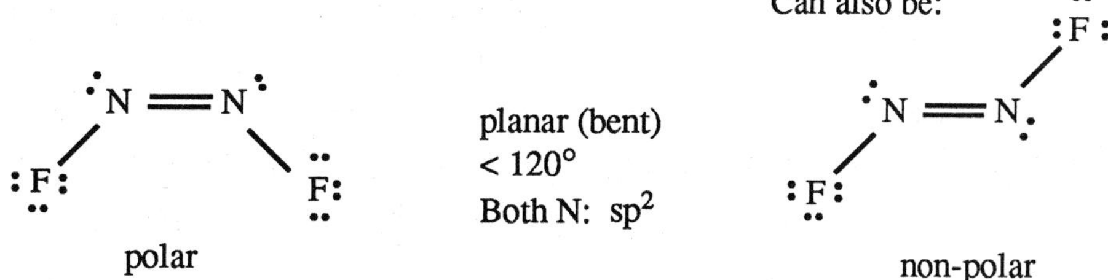

These are distinctly different molecules.

c. N_2O_3 has 2(5) + 3(6) = 28 valence electrons.

non-planar (bent and trigonal planar about nitrogens)
polar
$\approx 120°$
Both N: sp^2

d. COS has 4 + 6 + 6 = 16 valence electrons.

linear
polar
180°
sp

e. ICl_3 has 7 + 3(7) = 28 valence electrons.

T-shaped
polar
a. $< 90°$ b. $\approx 180°$
dsp^3

35. a. Yes, both have 4 effective pairs about the P.

b. sp^3 in each

c. P has to use one of its d orbitals to form the π bond.

d.

$$:\ddot{O}:^{-1}\ \text{(single bond, P +1)}\quad :\ddot{Cl}-P-\ddot{Cl}:\ \text{with}\ :\ddot{Cl}:\ \longleftrightarrow\ \ddot{O}^{\,0}=P^{0}\ \text{with}\ :\ddot{Cl}-P-\ddot{Cl}:,\ :\ddot{Cl}:$$

The structure with the P$=$O bond is favored on the basis of formal charge.

37. NO^+ has 5 + 6 - 1 = 10 valence electrons.

$$[:N\equiv O:]^+$$

NO has 5 + 6 = 11 valence electrons.

$$\dot{N}=\ddot{O} \longleftrightarrow \ddot{N}=\dot{O} \longleftrightarrow N\overset{\cdot}{=}O$$

NO^- has 12 valence electrons.

$$[\ddot{N}=\ddot{O}]^-$$

M.O. model:

NO^+ $(\sigma_{2s})^2 (\sigma_{2s}^*)^2 (\pi_{2p})^4 (\sigma_{2p})^2$, B.O. = 3, diamagnetic

NO $(\sigma_{2s})^2 (\sigma_{2s}^*)^2 (\pi_{2p})^4 (\sigma_{2p})^2 (\pi_{2p}^*)^1$, B.O. = 2.5, 1 unpaired e^-

NO^- $(\sigma_{2s})^2 (\sigma_{2s}^*)^2 (\pi_{2p})^2 (\sigma_{2p})^2 (\pi_{2p}^*)^2$, B.O. = 2, 2 unpaired e^-

Bond Energies: $NO^- < NO < NO^+$

Bond Lengths: $NO^+ < NO < NO^-$

The two models only give the same results for NO^+. The MO view is the correct one for NO and NO^-.

39. N_2: $(\sigma_{2s})^2 (\sigma_{2s}^*)^2 (\pi_{2p})^4 (\sigma_{2p})^2$ in ground state;

B.O. = 3; diamagnetic

1st excited state: $(\sigma_{2s})^2 (\sigma_{2s}^*)^2 (\pi_{2p})^4 (\sigma_{2p})^1 (\pi_{2p}^*)^1$

B.O. = $\frac{7 - 3}{2}$ = 2; paramagnetic (2 unpaired e^-)

41. Considering only the twelve valence electrons in O_2 we can put them in an energy level diagram.

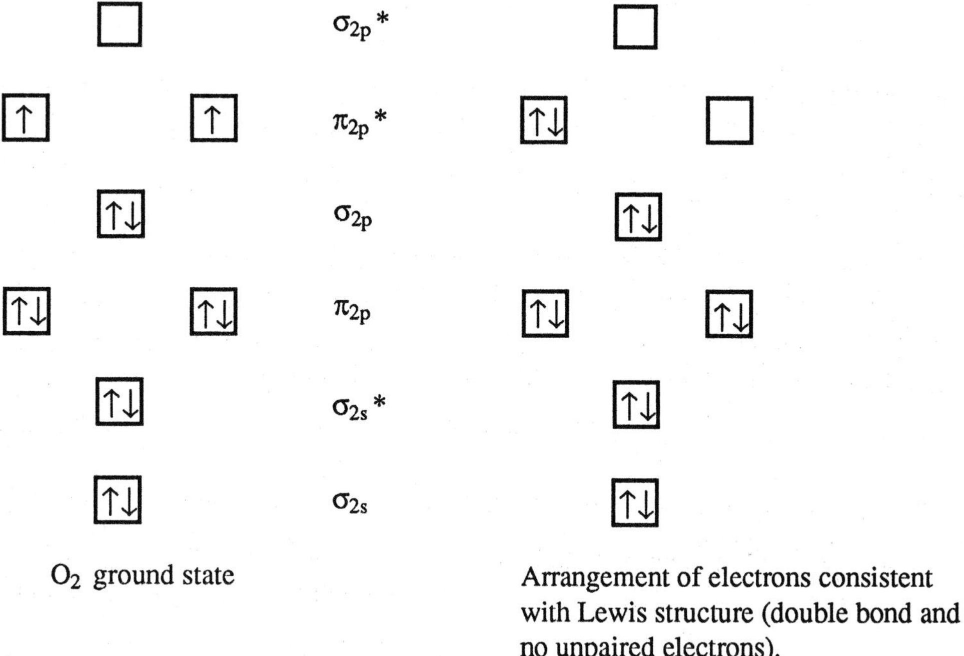

O_2 ground state

Arrangement of electrons consistent with Lewis structure (double bond and no unpaired electrons).

It takes energy to pair electrons in the same orbital. Thus, the structure with no unpaired electrons is at a higher energy; it is in an excited state.

CHALLENGE PROBLEMS

43.

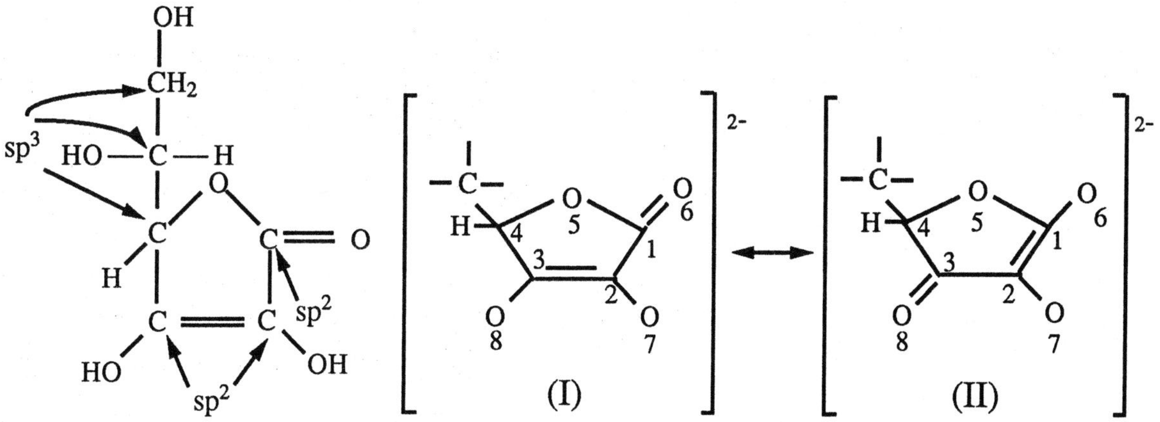

Two resonance forms can be drawn for the ring portion of the anion. In (I) the following groups of atoms must be coplanar because of the position of the double bonds:

$$C_1C_2O_5O_6 \text{ and } C_1C_2C_3C_4O_7O_8$$

In resonance form (II) the following two groups of atoms must be coplanar:

$$C_1C_2C_3O_5O_6O_7 \text{ and } C_2C_3C_4O_8$$

The only way to satisfy all of these requirements is for all eight atoms (the four carbons and one oxygen in the ring and the three oxygens bonded to the ring) to lie in the same plane.

45. The π bonds between S atoms and between C and S atoms are not as strong. The orbitals do not overlap with each other as well as the atomic orbitals of C and O.

47. a. The electron removed from N_2 is in the σ_{2p} molecular orbital which is lower in energy than the 2p atomic orbital from which the electron in atomic nitrogen is removed. Since the electron removed from N_2 is lower in energy than the electron in N, the ionization energy of N_2 should be greater than for N.

b. F_2 should have a lower first ionization energy than F. The electron removed from F_2 is in a $\pi_{2p}{}^*$ anti-bonding molecular orbital which is higher in energy than the 2p atomic orbitals. Thus, it is easier to remove an electron from F_2 than from F.

49. $$E = \frac{hc}{\lambda} = \frac{(6.626 \times 10^{-34}\ \text{J s})(2.998 \times 10^{8}\ \text{m/s})}{25 \times 10^{-9}\ \text{m}} = 7.9 \times 16^{-18}\ \text{J}$$

$$7.9 \times 10^{-18}\ \text{J} \times \frac{6.022 \times 10^{23}}{\text{mol}} \times \frac{1\ \text{kJ}}{1000\ \text{J}} = 4800\ \text{kJ/mol}$$

25 nm light has sufficient energy to ionize N_2 and N, and to break the triple bond. Thus, N_2, $N_2{}^+$, N, and N^+ will all be present.

To produce atomic nitrogen but no ions the range of energies of the light must be from 941 kJ/mol to 1402 kJ/mol.

$$941\ \text{kJ/mol} \times \frac{1\ \text{mol}}{6.022 \times 10^{23}} \times \frac{1000\ \text{J}}{\text{kJ}} = 1.56 \times 10^{-18}\ \text{J/photon}$$

$$\lambda = \frac{hc}{E} = \frac{(6.626 \times 10^{-34}\ \text{J s})(2.998 \times 10^{8}\ \text{m/s})}{1.56 \times 10^{-18}\ \text{J}} = 1.27 \times 10^{-7}\ \text{m} = 127\ \text{nm}$$

$$\frac{1402\ \text{kJ}}{\text{mol}} \times \frac{1\ \text{mol}}{6.022 \times 10^{23}} \times \frac{1000\ \text{J}}{\text{kJ}} = 2.328 \times 10^{-18}\ \text{J/photon}$$

$$\lambda = \frac{hc}{E} = \frac{(6.626 \times 10^{-34}\ \text{J s})(2.9979 \times 10^{8}\ \text{m/s})}{2.328 \times 10^{-18}\ \text{J}} = 8.533 \times 10^{-8}\ \text{m} = 85.33\ \text{nm}$$

Light with wavelengths in the range 85.33 nm $< \lambda <$ 127 nm will produce N but no ions.

CHAPTER TEN: LIQUIDS AND SOLIDS

QUESTIONS

1. There is an electrostatic attraction between the permanent dipoles of the polar molecules. The greater the polarity, the greater the attraction among molecules, the stronger the intermolecular forces.

3. As the size of the molecules increases, the strength of the London dispersion forces also increase. As the electron cloud gets larger it is easier for the electrons to be drawn away from the nucleus (more polarizable).

5. As the strengths of interparticle forces increase; surface tension, viscosity, melting point, and boiling point increase, while the vapor pressure decreases. See discussion in text.

7. Dipole forces are generally weaker than hydrogen bonding. They are similar in that they arise from an unequal sharing of electrons. We can look at hydrogen bonding as a particularly strong dipole force.

9. Liquids and solids both have characteristic volume and are not very compressible. Liquids and gases flow and assume the shape of their container.

11. As the interparticle forces increase the critical temperature increases.

13.

a.	crystalline solid:	regular, repeating structure.
	amorphous solid:	irregular arrangement of atoms or molecules.
b.	ionic solid:	made up of ions held together by ionic bonding.
	molecular solid:	made up of discrete covalently bonded molecules held together in the solid by weaker forces (LDF, dipole, or hydrogen bonds).
c.	molecular solid:	discrete molecules.
	covalent network solid:	no discrete molecules. A covalent network is like a large polymer. The interparticle forces are the covalent bonds between atoms.
d.	metallic solid:	completely delocalized electrons, conductor of electricity (ions in a sea of electrons).
	covalent network:	localized electrons, insulator or semiconductor.

15. No, an example is common glass which is primarily amorphous SiO_2 (a covalent network solid) as compared to ice (a crystalline solid held together by weaker H-bonds).

    The interparticle forces in the amorphous solid in this case are stronger than those in the crystalline solid. Whether or not a solid is amorphous or crystalline depends on the long range order in the solid and not on the strengths of the interparticle forces.

17. a. As the temperature is increased, more electrons in the valence band have sufficient kinetic energy, KE, to jump from the valence band to the conduction band.

b. A photon of light is absorbed by an electron which then goes from the valence band to the conduction band.

c. An impurity either adds electrons at an energy near that of the conduction band (n-type) or adds holes at an energy near that of the valence band (p-type).

19. a. condensation: vapor $\rightarrow$ liquid

b. evaporation: liquid $\rightarrow$ vapor

c. sublimation: solid $\rightarrow$ vapor

d. A supercooled liquid is a liquid which is at a temperature below its freezing point.

21. a. As intermolecular forces increase, the rate of evaporation decreases.

b. increase T: increase rate

c. increase surface area: increase rate

23. The phase change, $H_2O(g) \rightarrow H_2O(l)$ releases heat that can cause additional damage. Also steam can be at a temperature greater than 100°C.

EXERCISES

Interparticle Forces and Physical Properties

25. a. ionic b. LDF, dipole c. LDF only

d. LDF mostly (For all practical purposes, we consider a C — H bond to be non-polar even though there is a small difference in electronegativity.)

e. metallic f. metallic g. LDF only

h. H-bonding, LDF

27. a. OCS: OCS has dipole forces in addition to LDF. CO_2 is non-polar (only has LDF forces). The same reasoning is true for b-d answers.

b. PF_3 c. SF_2 d. SO_2

29. a. Neopentane is more compact than n-pentane. Thus, there is less area of contact between neopentane molecules: hence, weaker interparticle forces and a lower boiling point.

b. Ethanol is capable of H-bonding; dimethylether is not.

c. HF is capable of H-bonding.

d. LiCl, ionic: Ionic forces are much stronger than the intermolecular forces present in molecular solids.

$TiCl_4$: $4 + 4(7) = 32\ e^-$

$TiCl_4$ is a non-polar molecular substance with relatively weak London dispersion forces.

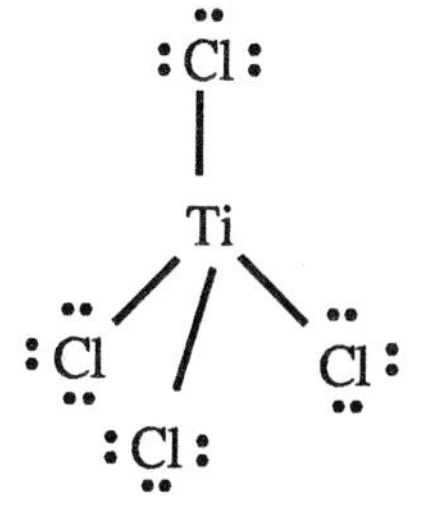

e. LiCl, ionic: HCl, molecular; Only dipole forces and LDF between HCl molecules. Ionic forces are much stronger than the molecular forces.

f. Both LiCl and CsCl are ionic. The lattice energy of LiCl is greater than that of CsCl because Li^+ ion is smaller than Cs^+ ion. Thus, stronger forces in LiCl.

31. As the electronegativity of the atoms bonded to H in a hydrogen bond increases, the strength of the hydrogen bond increases.

N····· H—N< < N····· H— O — < O····· H— O— etc.

weakest

33. a. NaCl — strong ionic bonding in lattice

b. CH_3CN — polar but no H-bonding

c. H_2 — non-polar like CH_4, smaller than CH_4, weaker LDF

d. SiO_2 — covalent network solid vs. a gas and a liquid

e. $HOCH_2CH_2OH$ — (ethylene glycol) greatest amount of H-bonding since two —OH groups are present.

f. NH_3 — Need N—H, O—H, or F—H for hydrogen bonding.

g. predict HF because of H-bonding - actually HI (see Exercise 10.77)

h. CO_2 — non-polar molecular substance, only weak LDF

Properties of Liquids

35. The attraction of H_2O for glass is stronger than H_2O — H_2O attraction. The miniscus is concave to increase the area of contact between glass and H_2O.

The Hg — Hg attraction is greater than the Hg — glass attraction. The miniscus is convex to minimize the Hg — glass contact. Polyethylene is a non-polar substance. The H_2O — H_2O attraction is stronger than the H_2O — polyethylene attraction; thus, the miniscus will have a convex shape.

37. The structure of H_2O_2 is H — O — O — H and produces greater hydrogen bonding than water. Long chains of hydrogen bonded H_2O_2 molecules get tangled together.

Structures and Properties of Solids

39.
a.	CO_2:	molecular	g.	KBr:	ionic
b.	SiO_2:	covalent network	h.	H_2O:	molecular
c.	Si:	atomic, covalent network	i.	NaOH:	ionic
d.	CH_4:	molecular	j.	U:	atomic, metallic
e.	Ru:	atomic, metallic	k.	$CaCO_3$:	ionic
f.	I_2:	molecular	l.	PH_3:	molecular

41. $n\lambda = 2d \sin\theta$

$$\lambda = \frac{2d \sin\theta}{n} = \frac{2 \times 201 \times 10^{-12} \text{ m} \sin 34.68°}{1}$$

$\lambda = 2.29 \times 10^{-10}$ m = 229 pm = 0.229 nm

43.

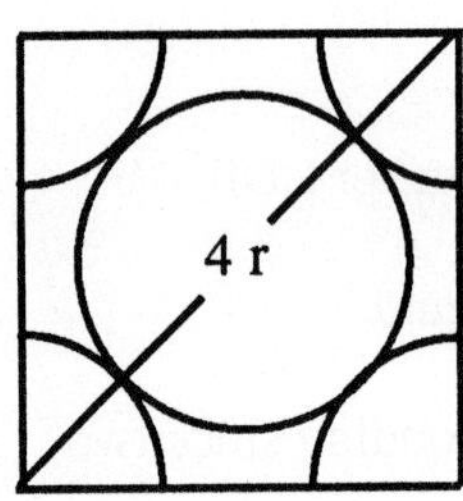

3.92×10^{-8} cm = length of cube edge

4r = length of diagonal

$$(4r)^2 = (3.92 \times 10^{-8})^2 + (3.92 \times 10^{-8})^2$$

$r = 1.39 \times 10^{-8}$ cm = 139 pm = 1.39 Å

In a face centered cubic unit cell:

$$8 \text{ corners} \times \frac{1/8 \text{ atom}}{\text{corner}} + 6 \text{ faces} \times \frac{1/2 \text{ atom}}{\text{face}} = 4 \text{ atoms}$$

$$\text{Density} = \frac{\text{mass}}{\text{volume}}$$

$$\text{Mass} = 4 \text{ atoms} \times \frac{\text{AW}}{\text{atom}} \times \frac{1 \text{ g}}{6.022 \times 10^{23} \text{ amu}}$$

Volume = $(\text{edge})^3 = (3.92 \times 10^{-8}\ \text{cm})^3$

$$21.45\ \text{g/cm}^3 = \frac{4 \times \text{AW} \times \dfrac{1\text{g}}{6.022 \times 10^{23}\ \text{amu}}}{(3.92 \times 10^{-8}\ \text{cm})^3}$$

AW = 194.5 amu ≈ 195 g/mol. The metal is platinum.

45. Body centered unit cell:

$$8 \text{ corners} \times \frac{1/8 \text{ Ti}}{\text{corner}} + 1 \text{ Ti at body center} = 2 \text{ Ti atoms}$$

$$4.50\ \text{g/cm}^3 = \frac{2(47.88\ \text{amu}) \times \dfrac{1\ \text{g}}{6.022 \times 10^{23}\ \text{amu}}}{l^3}$$

l = length of unit cell = 3.28×10^{-8} cm = 328 pm

Assume Ti atoms just touch along the body diagonal of the cube:

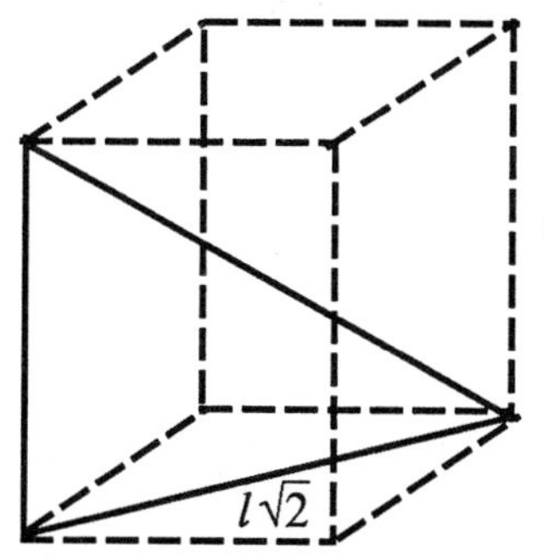

The triangle we need to solve is:

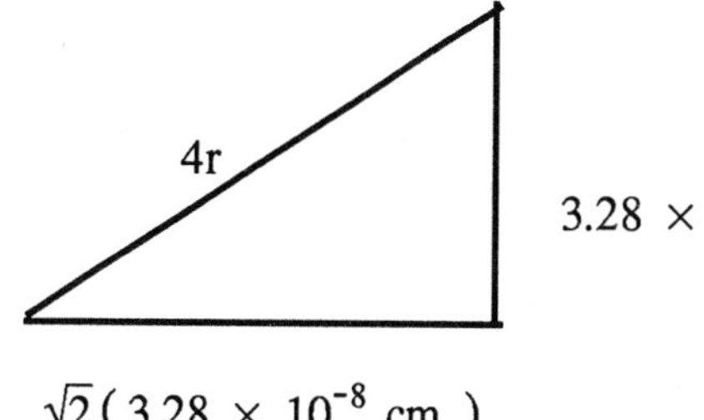

$$(4r)^2 = (3.28 \times 10^{-8})^2 + [\sqrt{2}(3.28 \times 10^{-8})]^2$$

$$r = 1.42 \times 10^{-8}\ \text{cm} = 142\ \text{pm} = 1.42\ \text{Å}$$

47. There are 4 Ni atoms in the unit cell.

$$\text{Density} = \frac{\text{Mass}}{\text{Volume}}$$

$$6.84\ \text{g/cm}^3 = \frac{4 \times 58.69\ \text{amu} \times \dfrac{1\ \text{g}}{6.022 \times 10^{23}\ \text{amu}}}{l^3}$$

$l = 3.85 \times 10^{-8}$ cm

For a face centered cube:

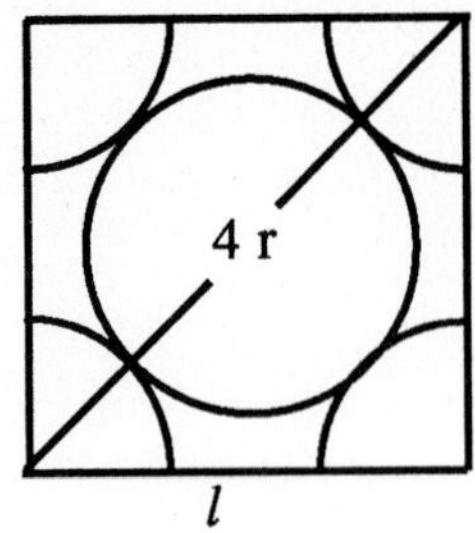

$(4r)^2 = l^2 + l^2 = 2\,l^2$

$16\,r^2 = 2(3.85 \times 10^{-8}\ \text{cm})^2$

$r = 1.36 \times 10^{-8}\ \text{cm} = 1.36\ \text{Å} = 136\ \text{pm}$

49. Al: $8\ \text{corners} \times \dfrac{1/8\ \text{Al}}{\text{corner}} = 1\ \text{Al}$

Ni: $6\ \text{face centers} \times \dfrac{1/2\ \text{Ni}}{\text{face center}} = 3\ \text{Ni}$

Composition: $AlNi_3$

51. $8\ \text{corners} \times \dfrac{1/8\ \text{Ca}}{\text{corner}} = 1\ \text{Ca}$

$6\ \text{faces} \times \dfrac{1/2\ \text{O}}{\text{face}} = 3$ oxygen atoms

1 Ti at body center. $CaTiO_3$

$8\ \text{corners} \times \dfrac{1/8\ \text{Ti}}{\text{corner}} = 1\ \text{Ti}$

$12\ \text{edges} \times \dfrac{1/4\ \text{O}}{\text{edge}} = 3\ \text{O}$

1 Ca at body center: $CaTiO_3$

Both pictures give the same formulas and six oxygen atoms coordinated to each titanium atom.

53. $E = 2.5\ \text{eV} \times 1.6 \times 10^{-19}\ \text{J/eV} = 4.0 \times 10^{-19}\ \text{J}$

$$E = \frac{hc}{\lambda}$$

$$\lambda = \frac{hc}{E} = \frac{(6.63 \times 10^{-34}\ \text{J s})(3.00 \times 10^{8}\ \text{m/s})}{4.0 \times 10^{-19}\ \text{J}} = 4.97 \times 10^{-7}\ \text{m} = 497\ \text{nm}$$

$\approx 5.0 \times 10^{2}\ \text{nm}$

55. a. $8\ \text{corners} \times \dfrac{1/8\ \text{Cl}}{\text{corner}} + 6\ \text{faces} \times \dfrac{1/2\ \text{Cl}}{\text{face}} = 4\ \text{Cl}$

$12\ \text{edges} \times \dfrac{1/4\ \text{Na}}{\text{edge}} + 1$ Na at body center = 4 Na

NaCl is the formula.

b. 1 Cs at body center

$8\ \text{corners} \times \dfrac{1/8\ \text{Cl}}{\text{corner}} = 1\ \text{Cl}$

CsCl is the formula.

c. There are 4 Zn inside the cube.

$$8 \text{ corners} \times \frac{1/8 \text{ S}}{\text{corner}} + 6 \text{ faces} \times \frac{1/2 \text{ S}}{\text{face}} = 4 \text{ S}$$

ZnS is the formula.

d. $$8 \text{ corners} \times \frac{1/8 \text{ Ti}}{\text{corner}} + 1 \text{ Ti at body center} = 2 \text{ Ti}$$

$$4 \text{ faces} \times \frac{1/2 \text{ O}}{\text{face}} + 2 \text{ O inside cube} = 4 \text{ O}$$

TiO_2 is the formula.

57. In has fewer valence electrons than Se, thus, Se doped with In would be a p-type semiconductor.

Phase Changes and Phase Diagrams

59. At 100°C (373 K) the vapor pressure of H_2O is 1.00 atm.

$$\ln\left(\frac{P_2}{P_1}\right) = \frac{\Delta H}{R}\left(\frac{1}{T_1} - \frac{1}{T_2}\right)$$

$$\ln\left(\frac{P_2}{1.00}\right) = \frac{40.7 \times 10^3 \text{ J/mol}}{8.3145 \text{ J/K•mol}}\left(\frac{1}{373 \text{ K}} - \frac{1}{388 \text{ K}}\right)$$

$\ln P_2 = 0.51$, $P_2 = e^{0.51} = 1.7$ atm (See Appendix One for a review of natural logarithms)

$$\ln\left(\frac{3.5}{1.0}\right) = \frac{40.7 \times 10^3 \text{ J/mol}}{8.3145 \text{ J/K•mol}}\left(\frac{1}{373} - \frac{1}{T_2}\right) = 1.253$$

$$2.6 \times 10^{-4} = \left(\frac{1}{373} - \frac{1}{T_2}\right)$$

$$2.6 \times 10^{-4} = 2.68 \times 10^{-3} - \frac{1}{T_2};\ \frac{1}{T_2} = 2.42 \times 10^{-3}$$

$$T_2 = \frac{1}{2.42 \times 10^{-3}} = 413 \text{ K or } 140.°\text{C}$$

61. $H_2O(s, -20°C) \longrightarrow H_2O(s, 0°C)$

$$q_1 = C_{ice} \times M \times \Delta T = 2.1 \frac{\text{J}}{\text{g °C}} \times 5.0 \times 10^2 \text{ g} \times 20.°\text{C} = 2.1 \times 10^4 \text{ J} = 21 \text{ kJ}$$

$H_2O(s, 0°C) \longrightarrow H_2O(l, 0°C)$

$$q_2 = 5.0 \times 10^2 \text{ g } H_2O \times \frac{1 \text{ mol}}{18.02 \text{ g}} \times 6.0 \frac{\text{kJ}}{\text{mol}} = 170 \text{ kJ}$$

$H_2O(l, 0°C) \longrightarrow H_2O(l, 100°C)$

$$q_3 = 4.2 \frac{\text{J}}{\text{g °C}} \times 5.0 \times 10^2 \text{ g} \times 100.°\text{C} = 2.1 \times 10^5 \text{ J} = 210 \text{ kJ}$$

$H_2O(l, 100°C) \longrightarrow H_2O(g, 100°C)$

$$q_4 = 5.0 \times 10^2 \text{ g} \times \frac{1 \text{ mol}}{18.02 \text{ g}} \times \frac{40.7 \text{ kJ}}{\text{mol}} = 1100 \text{ kJ}$$

$H_2O(g, 100°C) \longrightarrow H_2O(g, 250°C)$

$$q_5 = 1.8 \frac{\text{J}}{\text{g °C}} \times 5.0 \times 10^2 \text{ g} \times 150.°\text{C} = 1.4 \times 10^5 \text{ J} = 140 \text{ kJ}$$

$$q_{total} = q_1 + q_2 + q_3 + q_4 + q_5 = 21 + 170 + 210 + 1100 + 140 \approx 1600 \text{ kJ}$$

63. $\ln\left(\frac{P_1}{P_2}\right) = \frac{\Delta H_{vap}}{R}\left(\frac{1}{T_2} - \frac{1}{T_1}\right)$

At normal boiling point (1), $P_1 = 760.$ torr, $T_1 = 56.5\ °C = 329.7$ K

$T_2 = 25°C = 298.2$ K

$$\ln\left(\frac{760.}{P_2}\right) = \frac{32.0 \times 10^3 \text{ J/mol}}{8.3145 \text{ J/K•mol}}\left(\frac{1}{298.2} - \frac{1}{329.7}\right)$$

$6.633 - \ln P_2 = 1.23$

$\ln P_2 = 5.40$

$P_2 = e^{5.40} = 221$ torr

65. We want to graph ln P vs 1/T. The slope of the resulting straight line will be $-\Delta H/R$. Use Kelvin temperature.

P	ln P	T(Li)	1/T	T(Mg)	1/T
1	0	1023	9.775×10^{-4}	893	11.2×10^{-4}
10.	2.30	1163	8.598×10^{-4}	1013	9.872×10^{-4}
100.	4.61	1353	7.391×10^{-4}	1173	8.525×10^{-4}
400.	5.99	1513	6.609×10^{-4}	1313	7.616×10^{-4}
760.	6.63	1583	6.314×10^{-4}	1383	7.231×10^{-4}

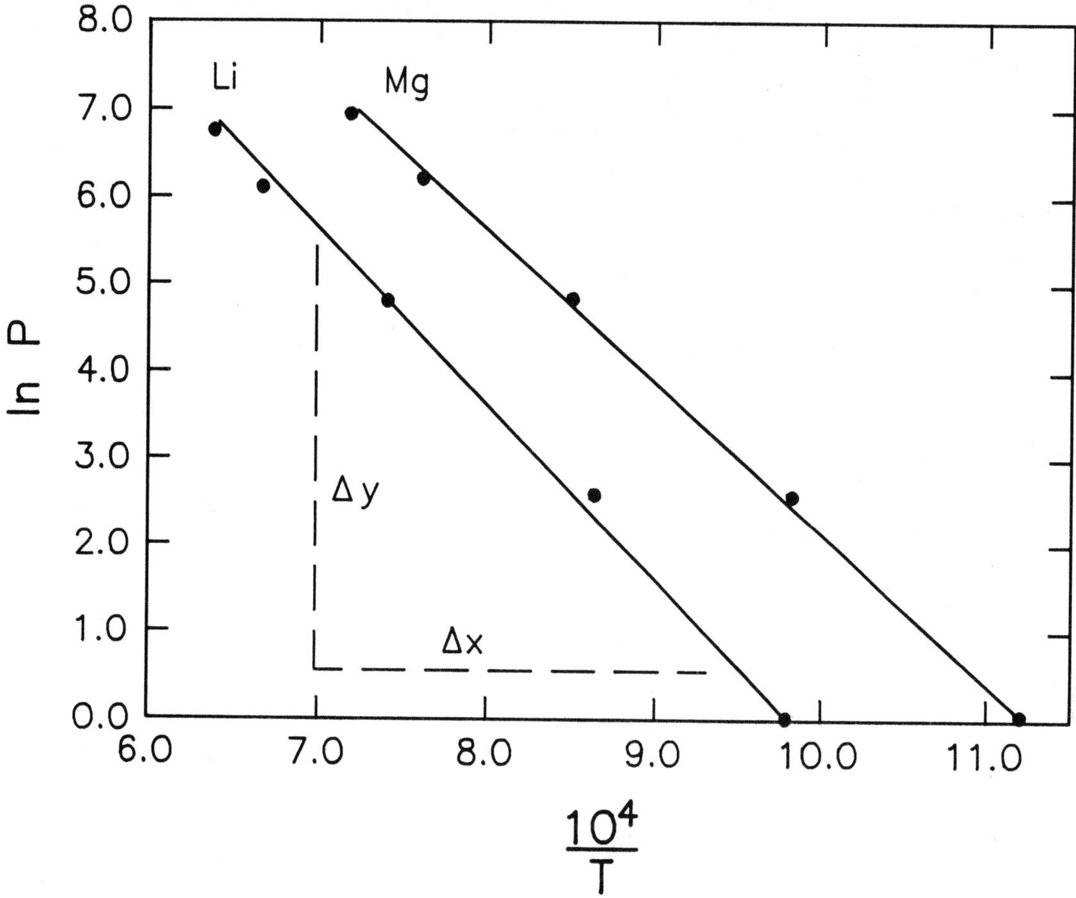

For Li, we get the slope by taking two points (x, y) that are on the line we draw, not data points. There may be experimental error and individual data points may not fall directly on the line. For a line:

$$\text{Slope} = \frac{\Delta y}{\Delta x} = \frac{y_2 - y_1}{x_2 - x_1}$$

or we can fit the straight line using a computer or calculator.

The equation of this line is:

$$\ln P = -1.90 \times 10^4 (1/T) + 18.6$$

$$\text{Slope} = -1.90 \times 10^4 \text{ K} = \frac{-\Delta H}{R}$$

$$\Delta H = -\text{Slope} \times R = 1.90 \times 10^4 \text{ K} \times 8.3145 \text{ J/mol•K}$$

$$\Delta H = 1.58 \times 10^5 \text{ J/mol} = 158 \text{ kJ/mol}$$

For Mg, the equation of the line is:

$$\ln P = -1.70 \times 10^4 (1/T) + 18.7$$

$$\text{Slope} = -1.70 \times 10^4 \text{ K} = \frac{-\Delta H}{R}$$

$\Delta H = -\text{Slope} \times R = 1.70 \times 10^4 \text{ K} \times 8.3145 \text{ J/mol} \cdot \text{K}$

$\Delta H = 1.41 \times 10^5 \text{ J/mol} = 141 \text{ kJ/mol}$

The bonding is stronger in Li.

67. A typical heating curve looks like

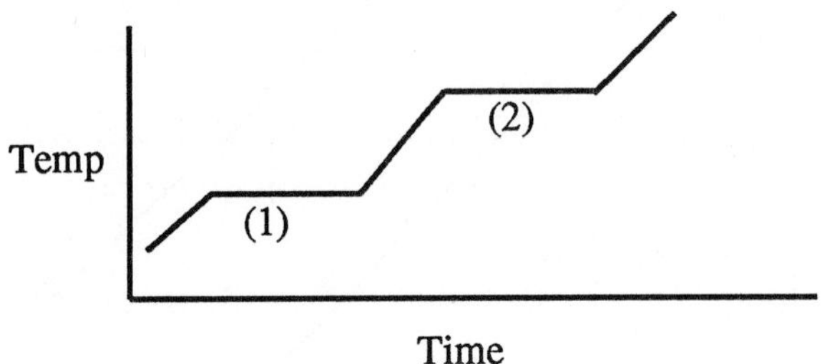

The plateau (1) gives a temperature at which the solid and liquid are in equilibrium. This temperature along with the pressure gives a point on the phase diagram. The plateau (2) will give a temperature on the liquid-vapor line.

69. A - solid; B - liquid; C - vapor; D - solid + vapor; E - solid + liquid + vapor; F - liquid + vapor; G - liquid + vapor; H - vapor

A - solid	E - solid + liquid + vapor
B - liquid	F - liquid + vapor
C - vapor	G - liquid + vapor
D - solid + vapor	H - vapor
triple point: E	critical point: G

normal freezing point: temperature at which solid - liquid line is at 1 atm.

normal boiling point: temperature at which liquid - vapor line is at 1 atm.

ADDITIONAL EXERCISES

71.

CH_3CO_2H	H-bonding
CH_2ClCO_2H	H-bonding + larger electronegative atom replacing H (greater dipole)
$CH_3CO_2CH_3$	polar

We predict $CH_3CO_2CH_3$ to have the weakest interparticle forces and CH_2ClCO_2H to have the strongest. The boiling points are consistent with this view.

73. $n\lambda = 2d \sin \theta$, $1.54 \text{ Å} = 1.54 \times 10^{-10} \text{ m} = 154 \text{ pm}$

$$d = \frac{n\lambda}{2 \sin \theta} = \frac{1 \times 154 \text{ pm}}{2 \times \sin 14.22°} = 313 \text{ pm} = 3.13 \times 10^{-10} \text{ m} = 3.13 \text{ Å}$$

75. If TiO_2 conducts electricity as a liquid it would be ionic.

$$\Delta EN = (3.5 - 1.5) = 2.0$$

From this difference in electronegativities we might predict TiO_2 to be ionic. It is actually a covalent network solid. The 1.5 is the value of the EN for Ti^{2+}. Ti^{4+} should have a higher electronegativity. See Question 8.3.

77. The boiling point of HF is higher than might be expected because of especially strong hydrogen bonding in HF. ΔH_{vap} should be high also, but is lower than expected because many of the H-bonds reform in the gas phase forming dimers, H—F···H—F.

79.

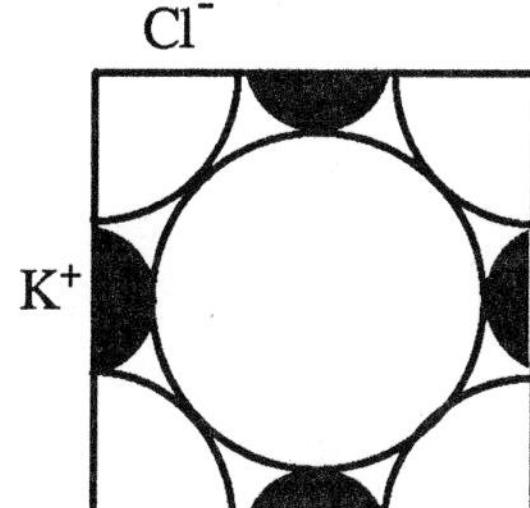

Assuming K^+ and Cl^- just touch along the edge.

$l = 2(314) = 628 \text{ pm} = 6.28 \times 10^{-8} \text{ cm}$

The unit cell contains 4 K^+ and 4 Cl^-

$$D = \frac{4(39.10)\text{ amu} + 4(35.45)\text{ amu}}{(6.28 \times 10^{-8}\text{ cm})^3} \times \frac{1\text{ g}}{6.022 \times 10^{23}\text{ amu}}$$

$$D = \frac{2.00\text{ g}}{\text{cm}^3}$$

81. $1.00\text{ lb} \times 454\ \frac{\text{g}}{\text{lb}} = 454\text{ g } H_2O$

A change of 1.00°F is equal to a change of 5/9°C.

The amount of heat in J in 1 Btu is:

$$4.18\ \frac{\text{J}}{\text{g °C}} \times 454\text{ g} \times \frac{5}{9}\text{ °C} = 1.05 \times 10^3\text{ J or } 1.05\text{ kJ}$$

It takes 40.7 kJ to vaporize 1 mol H_2O (ΔH_{vap}).

Combining these:

$$1.00 \times 10^4\ \frac{\text{Btu}}{\text{hr}} \times \frac{1.05\text{ kJ}}{\text{Btu}} \times \frac{1\text{ mol } H_2O}{40.7\text{ kJ}} = 258\text{ mol/hr}$$

$$\text{or: } \frac{258\text{ mol}}{\text{hr}} \times \frac{18.02\text{ g}}{\text{mol}} = 4650\text{ g/hr} = 4.65\text{ kg/hr}$$

83. a. two

b. higher pressure triple point: graphite, diamond, and liquid

lower pressure triple point: graphite, liquid, and vapor

c. It is converted to diamond.

d. diamond

CHALLENGE PROBLEMS

85. Both molecules are capable of H-bonding. However, in oil of wintergreen the hydrogen bonding is intramolecular.

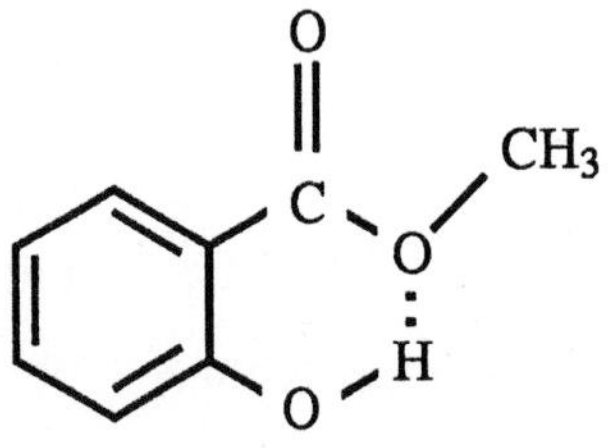

O
||
C
O
CH_3
H
O

In methyl-4-hydroxybenzoate the H-bonding is intermolecular, resulting in greater intermolecular forces and a higher melting point.

87. NaCl, $MgCl_2$, NaF, MgF_2, AlF_3 all have very high melting points indicative of strong interparticle forces. They are all ionic solids. $SiCl_4$, SiF_4, Cl_2, F_2, PF_5 and SF_6 are non-polar covalent molecules. Only LDF are present. PCl_3 and SCl_2 are polar molecules. LDF and dipole forces are present. In these 8 molecular substances the interparticle forces are weak and the melting points low. $AlCl_3$ doesn't seem to fit in as well. From the melting point, there are much stronger forces present than in the non-metal halides, but they aren't as strong as we would expect for an ionic solid. $AlCl_3$ illustrates a gradual transition from ionic to covalent bonding; from an ionic solid to discrete molecules.

89. $n\lambda = 2d \sin \theta$; $1 \text{ Å} = 10^{-8} \text{ cm} = 10^{-10} \text{ m} = 100 \text{ pm}$

$$d = \frac{n\lambda}{2\sin\theta} = \frac{1 \times 71.2 \text{ pm}}{2 \sin 5.564} = 367.2 \text{ pm} = 3.672 \times 10^{-8} \text{ cm}$$

$$13.28 \frac{\text{g}}{\text{cm}^3} = \frac{n \times 178.5 \text{ amu}}{(3.672 \times 10^{-8} \text{ cm})^3} \times \frac{1 \text{ g}}{6.022 \times 10^{23} \text{ amu}}$$

where n = the number of Hf atoms per unit cell

n = 2.2 This is most consistent with body centered cubes.

See Exercise 10.45 for geometry.

$$(4r)^2 = (367.2)^2 + (367.2 \sqrt{2})^2$$

$$r = 159 \text{ pm} = 1.59 \text{ Å}$$

91. There are atoms at each of the 8 corners (shared by eight unit cells) plus one atom inside the unit cell: 2/3 of one atom plus 1/6 from each of two others. Thus, there are two atoms in the unit cell.

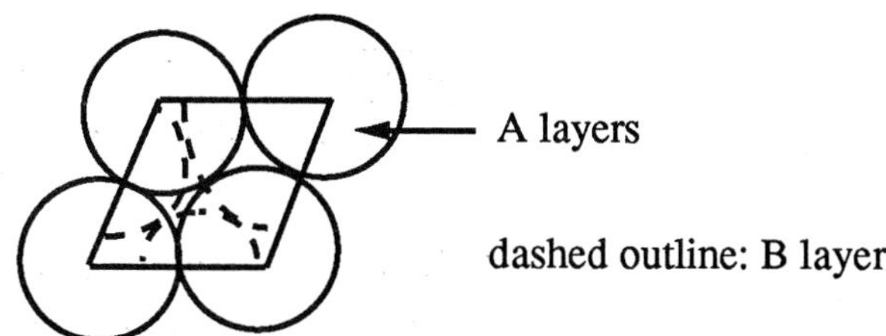

93. First, we need to get the empirical formula of spinel. Out of 100 g there are:

$$37.9 \text{ g Al} \times \frac{1 \text{ mol Al}}{26.98 \text{ g Al}} = 1.40 \ (2)$$

$$17.1 \text{ g Mg} \times \frac{1 \text{ mol Mg}}{24.31 \text{ g Mg}} = 0.703 \ (1)$$

Empirical formula: Al_2MgO_4

$$45.0 \text{ g O} \times \frac{1 \text{ mol O}}{16.00 \text{ g O}} = 2.81 \ (4)$$

Each unit cell will contain an integral value (n) of empirical formula units. Each unit has a mass of:

$$(24.31) + 2(26.98) + 4(16.00) = 142.27 \text{ g/mol}$$

$$3.57 \text{ g/cm}^3 = \frac{n(142.27 \text{ g/mol})}{(6.022 \times 10^{23} \text{/mol})(8.09 \times 10^{-8} \text{ cm})^3}$$

n = 8.00

Each unit cell has: 16 Al; 8 Mg; 32 O

95. a)

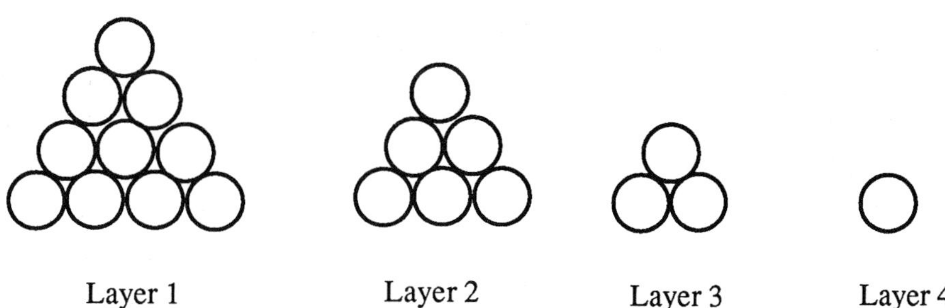

Layer 1 Layer 2 Layer 3 Layer 4

A total of 20 cannonballs will be needed.

b) Cubic closest packed, abc pattern of packing layers.

c) tetrahedron

97. Heat released:

$$0.250 \text{ g Na} \times \frac{1 \text{ mol}}{22.99 \text{ g}} \times \frac{368 \text{ kJ}}{2 \text{ mol}} = 2.00 \text{ kJ}$$

To melt 50.0 g of ice requires:

$$50.0 \text{ g ice} \times \frac{1 \text{ mol } H_2O}{18.02 \text{ g}} \times 6.02 \text{ kJ/mol} = 16.7 \text{ kJ}$$

The reaction doesn't release enough heat to melt all of the ice. The temperature will remain at 0°C.

CHAPTER ELEVEN: PROPERTIES OF SOLUTIONS

SOLUTION REVIEW

1. $125 \text{ g sucrose} \times \frac{1 \text{ mol}}{342.3 \text{ g}} = 0.365 \text{ mol}$

   $M = \frac{0.365 \text{ mol}}{1.00 \text{ L}} = 0.365 \text{ mol/L}$

2. $0.250 \text{ L} \times \frac{0.100 \text{ mol}}{\text{L}} \times \frac{134.0 \text{ g}}{\text{mol}} = 3.35 \text{ g } Na_2C_2O_4$

3. $25.00 \times 10^{-3} \text{ L} \times \frac{0.308 \text{ mol}}{\text{L}} = 7.70 \times 10^{-3} \text{ mol}$

   $\frac{7.70 \times 10^{-3} \text{ mol}}{0.500 \text{ L}} = 1.54 \times 10^{-2} \text{ mol/L}$

   $NiCl_2(s) \longrightarrow Ni^{2+}(aq) + 2\,Cl^-(aq)$

   $M_{Ni^{2+}} = 1.54 \times 10^{-2} \text{ mol/L}; \; M_{Cl^-} = 3.08 \times 10^{-2} \text{ mol/L}$

4. $158.5 \times 10^{-3} \text{ g Cu} \times \frac{1 \text{ mol}}{63.55 \text{ g}} = 2.494 \times 10^{-3} \text{ mol Cu}$

   $M_{Cu^{2+}} = \frac{2.494 \times 10^{-3} \text{ mol}}{1.00 \text{ L}} = 2.49 \times 10^{-3} \text{ mol/L}$

5. $\frac{2.8 \times 10^{-3} \text{ g}}{1.0 \times 10^{-3} \text{ L}} \times \frac{1 \text{ mol}}{112.4 \text{ g}} = 2.5 \times 10^{-2} \text{ mol/L}$

6. a. $Ca(NO_3)_2(s) \longrightarrow Ca^{2+}(aq) + 2\,NO_3^-(aq)$

      $M_{Ca^{2+}} = 1.06 \times 10^{-3} \text{ mol/L}; \; M_{NO_3^-} = 2.12 \times 10^{-3} \text{ mol/L}$

   b. $1 \times 10^{-3} \text{ L} \times \frac{1.06 \times 10^{-3} \text{ mol } Ca^{2+}}{\text{L}} \times \frac{40.08 \text{ g } Ca^{2+}}{\text{mol}} = 4 \times 10^{-5} \text{ g } Ca^{2+} \text{ ions}$

   c. $1.0 \times 10^{-6} \text{ L} \times \frac{2.12 \times 10^{-3} \text{ mol } NO_3^-}{\text{L}} \times \frac{6.02 \times 10^{23} \, NO_3^- \text{ ions}}{\text{mol } NO_3^-} = 1.3 \times 10^{15} \, NO_3^- \text{ ions}$

7. $1.00 \text{ L} \times \frac{0.040 \text{ mol HCl}}{\text{L}} = 0.040 \text{ mol HCl}$

   $0.040 \text{ mol HCl} \times \frac{1 \text{ L}}{0.25 \text{ mol HCl}} = 0.16 \text{ L} = 160 \text{ mL}$

8. a. $HNO_3(l) \rightarrow H^+(aq) + NO_3^-(aq)$

b. $Na_2SO_4(s) \rightarrow 2\,Na^+(aq) + SO_4^{2-}(aq)$

c. $AlCl_3(s) \rightarrow Al^{3+}(aq) + 3\,Cl^-(aq)$

d. $SrBr_2(s) \rightarrow Sr^{2+}(aq) + 2\,Br^-(aq)$

e. $KClO_4(s) \rightarrow K^+(aq) + ClO_4^-(aq)$

f. $NH_4Br(s) \rightarrow NH_4^+(aq) + Br^-(aq)$

g. $NH_4NO_3(s) \rightarrow NH_4^+(aq) + NO_3^-(aq)$

h. $CuSO_4(s) \rightarrow Cu^{2+}(aq) + SO_4^{2-}(aq)$

i. $NaOH(s) \rightarrow Na^+(aq) + OH^-(aq)$

QUESTIONS

9. $\text{Molarity} = \frac{\text{moles solute}}{\text{L solution}}$ $\text{Molality} = \frac{\text{moles solute}}{\text{kg solvent}}$

Since volume is temperature dependent and mass isn't, then molarity is temperature dependent and molality is temperature independent. In determining ΔT_f and ΔT_b, we are interested in how some temperature depends on composition. Thus, we don't want our expression of composition to also be dependent on temperature.

11. hydrophobic: water hating

hydrophilic: water loving

13. If solute-solvent attraction > solvent-solvent and solute-solute, there is a negative deviation from Raoult's Law. If solute-solvent < solvent-solvent and solute-solute attraction, then there is a positive deviation from Raoult's Law.

15. No, the solution is not ideal. For an ideal solution, this strength of interparticle forces in the solution are the same as in the pure solute and pure solvent. This results in $\Delta H_{soln} = 0$ for an ideal solution. ΔH_{soln} for methanol/water is not zero.

17. The osmotic pressure inside the fruit cells (and bacteria) is less than outside the cell. Water will leave the cells which will dehydrate bacteria that might be present causing them to die.

EXERCISES

Concentration Units:

19. $$\frac{3.0\text{ g }H_2O_2}{100\text{ g soln}} \times \frac{1.0\text{ g soln}}{\text{cm}^3\text{ soln}} \times \frac{1000\text{ cm}^3}{\text{L}} \times \frac{1\text{ mol }H_2O_2}{34.02\text{ g }H_2O_2} = 0.88\text{ mol/L}$$

$$\frac{3.0\text{ g }H_2O_2}{97.0\text{ g }H_2O} \times \frac{1000\text{ g}}{\text{kg}} \times \frac{1\text{ mol }H_2O_2}{34.02\text{ g }H_2O_2} = 0.91\text{ mol/kg}$$

$$3.0\text{ g }H_2O_2 \times \frac{1\text{ mol}}{34.02\text{ g}} = 8.8 \times 10^{-2}\text{ mol }H_2O_2$$

$$97.0\text{ g }H_2O \times \frac{1\text{ mol}}{18.02\text{ g}} = 5.38\text{ mol }H_2O, \qquad \chi = \frac{8.8 \times 10^{-2}}{5.38 + 0.088} = 1.6 \times 10^{-2}$$

21. Hydrochloric acid:

$$\frac{38 \text{ g HCl}}{100. \text{ g soln}} \times \frac{1.19 \text{ g soln}}{\text{cm}^3 \text{ soln}} \times \frac{1000 \text{ cm}^3}{\text{L}} \times \frac{1 \text{ mol HCl}}{36.46 \text{ g}} = 12 \text{ mol/L}$$

$$\frac{38 \text{ g HCl}}{62 \text{ g solvent}} \times \frac{1000 \text{ g}}{\text{kg}} \times \frac{1 \text{ mol HCl}}{36.46 \text{ g}} = 17 \text{ mol/kg}$$

$$38 \text{ g HCl} \times \frac{1 \text{ mol}}{36.46 \text{ g}} = 1.0 \text{ mol HCl}$$

$$62 \text{ g } H_2O \times \frac{1 \text{ mol}}{18.02 \text{ g}} = 3.4 \text{ mol } H_2O$$

$$\chi = \frac{1.0}{3.4 + 1.0} = 0.23$$

Nitric acid:

$$\frac{70. \text{ g } HNO_3}{100. \text{ g soln}} \times \frac{1.42 \text{ g soln}}{\text{cm}^3 \text{ soln}} \times \frac{1000 \text{ cm}^3}{\text{L}} \times \frac{1 \text{ mol } HNO_3}{63.02 \text{ g}} = 16 \text{ mol/L}$$

$$\frac{70. \text{ g } HNO_3}{30. \text{ g solvent}} \times \frac{1000 \text{ g}}{\text{kg}} \times \frac{1 \text{ mol } HNO_3}{63.02 \text{ g}} = 37 \text{ mol/kg}$$

$$70. \text{ g } HNO_3 \times \frac{1 \text{ mol}}{63.02 \text{ g}} = 1.1 \text{ mol } HNO_3$$

$$30. \text{ g } H_2O \times \frac{1 \text{ mol}}{18.02 \text{ g}} = 1.7 \text{ mol } H_2O$$

$$\chi = \frac{1.1}{1.7 + 1.1} = 0.39$$

Sulfuric acid:

$$\frac{95 \text{ g } H_2SO_4}{100. \text{ g soln}} \times \frac{1.84 \text{ g soln}}{\text{cm}^3 \text{ soln}} \times \frac{1000 \text{ cm}^3}{\text{L}} \times \frac{1 \text{ mol } H_2SO_4}{98.09 \text{ g } H_2SO_4} = 18 \text{ mol/L}$$

$$\frac{95 \text{ g } H_2SO_4}{5 \text{ g } H_2O} \times \frac{1000 \text{ g}}{\text{kg}} \times \frac{1 \text{ mol}}{98.09 \text{ g}} = 194 \text{ mol/kg} \approx 200 \text{ mol/kg}$$

$$95 \text{ g } H_2SO_4 \times \frac{1 \text{ mol}}{98.09 \text{ g}} = 0.97 \text{ mol } H_2SO_4$$

$$5 \text{ g } H_2O \times \frac{1 \text{ mol}}{18.02 \text{ g}} = 0.3 \text{ mol } H_2O$$

$$\chi = \frac{0.97}{0.97 + 0.3} = 0.76$$

Acetic acid:

$$\frac{99 \text{ g } CH_3CO_2H}{100. \text{ g soln}} \times \frac{1.05 \text{ g soln}}{\text{cm}^3 \text{ soln}} \times \frac{1000 \text{ cm}^3}{\text{L}} \times \frac{1 \text{ mol}}{60.05 \text{ g}} = 17 \text{ mol/L}$$

$$\frac{99 \text{ g } CH_3CO_2H}{1 \text{ g } H_2O} \times \frac{1000 \text{ g}}{\text{kg}} \times \frac{1 \text{ mol}}{60.05 \text{ g}} = 1600 \text{ mol/kg} \approx 2000 \text{ mol/kg}$$

$$99 \text{ g HOAc} \times \frac{1 \text{ mol}}{60.05 \text{ g}} = 1.6 \text{ mol HOAc}$$

$$1 \text{ g } H_2O \times \frac{1 \text{ mol}}{18.02 \text{ g}} = 0.06 \text{ mol } H_2O$$

$$\chi = \frac{1.6}{1.6 + 0.06} = 0.96$$

Ammonia:

$$\frac{28 \text{ g } NH_3}{100. \text{ g soln}} \times \frac{0.90 \text{ g}}{\text{cm}^3} \times \frac{1000 \text{ cm}^3}{\text{L}} \times \frac{1 \text{ mol}}{17.03 \text{ g}} = 15 \text{ mol/L}$$

$$\frac{28 \text{ g } NH_3}{72 \text{ g } H_2O} \times \frac{1000 \text{ g}}{\text{kg}} \times \frac{1 \text{ mol}}{17.03 \text{ g}} = 23 \text{ mol/kg}$$

$$28 \text{ g } NH_3 \times \frac{1 \text{ mol}}{17.03 \text{ g}} = 1.6 \text{ mol } NH_3$$

$$72 \text{ g } H_2O \times \frac{1 \text{ mol}}{18.02 \text{ g}} = 4.0 \text{ mol } H_2O$$

$$\chi = \frac{1.6}{4.0 + 1.6} = 0.29$$

23. $50.0 \text{ mL toluene} \times \frac{0.867 \text{ g}}{\text{mL}} = 43.4 \text{ g toluene}$

$$125 \text{ mL benzene} \times \frac{0.874 \text{ g}}{\text{mL}} = 109 \text{ g benzene}$$

$$\% \text{ toluene} = \frac{43.4}{43.4 + 109} \times 100 = 28.5\%$$

$$\frac{43.4 \text{ g toluene}}{175 \text{ mL soln}} \times \frac{1000 \text{ mL}}{\text{L}} \times \frac{1 \text{ mol toluene}}{92.13 \text{ g toluene}} = 2.69 \text{ mol/L}$$

$$\frac{43.4 \text{ g toluene}}{109 \text{ g benzene}} \times \frac{1000 \text{ g}}{\text{kg}} \times \frac{1 \text{ mol toluene}}{92.13 \text{ g toluene}} = 4.32 \text{ mol/kg}$$

$$43.4 \text{ g toluene} \times \frac{1 \text{ mol}}{92.13 \text{ g}} = 0.471 \text{ mol toluene}$$

$$109 \text{ g benzene} \times \frac{1 \text{ mol benzene}}{78.11 \text{ g benzene}} = 1.40 \text{ mol benzene}$$

$$\chi = \frac{0.471}{0.471 + 1.40} = 0.252$$

25. If we have 1.00 L of solution we have:

$$1.37 \text{ mol citric acid} \times \frac{192.1 \text{ g}}{\text{mol}} = 263 \text{ g citric acid}$$

$$1.00 \times 10^3 \text{ mL solution} \times \frac{1.10 \text{ g}}{\text{mL}} = 1.10 \times 10^3 \text{ g solution}$$

$$\% = \frac{263}{1.10 \times 10^3} \times 100 = 23.9\%$$

In 1.00 L of solution we have 263 g citric acid and (1.10×10^3 - 263) = 840 g of the solvent, H_2O.

$$\frac{1.37 \text{ mol citric acid}}{0.84 \text{ g}} = 1.6 \text{ mol/kg}$$

$$840 \text{ g } H_2O \times \frac{1 \text{ mol}}{18.02 \text{ g}} = 47 \text{ mol } H_2O$$

$$\chi = \frac{1.37}{47 + 1.37} = 0.028$$

27. If there are 100.0 mL of wine, there are

$$12.5 \text{ mL } C_2H_5OH \times \frac{0.79 \text{ g}}{\text{mL}} = 9.9 \text{ g } C_2H_5OH$$

$$87.5 \text{ mL } H_2O \times \frac{1.0 \text{ g}}{\text{mL}} = 87.5 \text{ g } H_2O$$

$$\% \text{ ethanol} = \frac{9.9}{87.5 + 9.9} \times 100 = 10.\% \text{ by mass}$$

$$m = \frac{9.9 \text{ g } C_2H_5OH}{0.0875 \text{ kg } H_2O} \times \frac{1 \text{ mol}}{46.07 \text{ g}} = 2.5 \text{ molal}$$

Energetics of Solutions and Solubility

29. Water is a polar molecule capable of hydrogen bonding. Polar molecules, molecules capable of hydrogen bonding, and ions can be hydrated.

a. CH_3CH_2OH b. $CHCl_3$: $CHCl_3$ is polar, CCl_4 is non-polar.

c. CH_3CO_2H d. Na_2S (Most sulfides are insoluble, Na_2S is an exception.)

e. $AlCl_3$: Al_2O_3 is an insoluble covalent network solid.

f. CO_2: CO_2 is non-polar and not very soluble, but SiO_2 is a covalent network solid and virtually insoluble in water.

31.

$$KCl(s) \rightarrow K^+(g) + Cl^-(g) \qquad \Delta H = -(-715 \text{ kJ/mol})$$

$$K^+(g) + Cl^-(g) \rightarrow K^+(aq) + Cl^-(aq) \qquad \Delta H = -684 \text{ kJ/mol}$$

$$KCl(s) \rightarrow K^+(aq) + Cl^-(aq) \qquad \Delta H_{soln} = 31 \text{ kJ/mol}$$

33. a. Mg^{2+}, smaller, higher charge b. Be^{2+}, smaller

c. Fe^{3+}, smaller size, higher charge

35. a. Water, $Cu(NO_3)_2$ is an ionic solid.

b. CCl_4, CS_2 is a non-polar molecule.

c. Water

d. CCl_4, the long non-polar chain favors a non-polar solvent

e. Water

f. CCl_4

37. As the length of the hydrocarbon chain increases, the solubility decreases. The —OH end of the alcohols can hydrogen bond with water. The hydrocarbon chain, however, cannot form H-bonds. As the chain gets longer, a greater portion of the molecule cannot interact with the water molecules; i.e. the effect of the —OH group decreases as the alcohol gets larger.

39. $CO_2 + OH^- \longrightarrow HCO_3^-$

No, the reaction of CO_2 with OH^- greatly increases the solubility of CO_2 by forming the soluble bicarbonate ion.

Vapor Pressure of Solutions

41. $P = \chi P^\circ$

In solution:

$$50.0\text{ g } C_6H_{12}O_6 \times \frac{1\text{ mol } C_6H_{12}O_6}{180.16\text{ g } C_6H_{12}O_6} = 0.278\text{ mol glucose}$$

$$600.0\text{ g } H_2O \times \frac{1\text{ mol}}{18.02\text{ g}} = 33.30\text{ mol } H_2O$$

$$\chi_{H_2O} = \frac{33.30}{33.58} = 0.9917$$

$$P = 0.9917 \times 23.8 = 23.6\text{ torr}$$

43. $25.8\text{ g } CH_4N_2O \times \frac{1\text{ mol}}{60.06\text{ g}} = 0.430\text{ mol}$

$$275\text{ g } H_2O \times \frac{1\text{ mol}}{18.02\text{ g}} = 15.3\text{ mol}$$

$$\chi_{H_2O} = \frac{15.3}{15.3 + 0.430} = 0.973$$

$$P_{H_2O} = \chi_{H_2O}P^\circ_{H_2O} = 0.973\ (23.8\text{ torr}) = 23.2\text{ torr at } 25°C$$

$$P_{H_2O} = 0.973\ (71.9\text{ torr}) = 70.0\text{ torr at } 45°C$$

45. a. $25\text{ mL } C_5H_{12} \times \frac{0.63\text{ g}}{\text{mL}} \times \frac{1\text{ mol}}{72.15\text{ g}} = 0.22\text{ mol } C_5H_{12}$

$$45\text{ mL } C_6H_{14} \times \frac{0.66\text{ g}}{\text{mL}} \times \frac{1\text{ mol}}{86.17\text{ g}} = 0.34\text{ mol } C_6H_{14}$$

In liquid: $\chi_{pentane} = \frac{0.22}{0.56} = 0.39$ and $\chi_{hexane} = 0.61$

$P_{pen} = 0.39(511 \text{ torr}) = 2.0 \times 10^2$ torr and $P_{hex} = 0.61(150 \text{ torr}) = 92$ torr

$P_{total} = 2.0 \times 10^2 + 92 = 292 \approx 290$ torr

b. In the vapor phase:

$$\chi_{pen} = \frac{P_{pen}}{P_{total}} = \frac{2.0 \times 10^2 \text{ torr}}{290 \text{ torr}} = 0.69$$

47. $0.15 = \dfrac{P_{pen}}{P_{pen} + P_{hex}}$ (in vapor) and $P_{pen} + P_{hex} = P_{total}$

Thus, $P_{pen} = 0.15\ P_{tot}$

In the liquid:

$P_{pen} + P_{hex} = P_{tot} = \chi_{pen}(511) + \chi_{hex}(150.)$

$P_{tot} = \chi_{pen}(511) + (1 - \chi_{pen})\ (150.)$

$P_{tot} = 150. + 361\ \chi_{pen}$

$P_{pen} = \chi_{pen}(511) = 0.15\ P_{tot}$

$0.15\ P_{tot} = 0.15(150.) + 0.15(361)\ \chi_{pen}$
$-0.15\ P_{tot} = -511\ \chi_{pen}$

$0 = 22.5 + 54.2\ \chi_{pen} - 511\ \chi_{pen}$ (carry extra significant figure)

$22.5 = 457\ \chi_{pen}$

$\chi_{pen} = 0.0492 \approx 0.049$ in the liquid

49. Compared to H_2O, d (methanol/water) will have a higher vapor pressure since methanol is more volatile than water. Both b (glucose/water) and c (NaCl/water) will have a lower vapor pressure than water. NaCl dissolves to give Na^+ ions and Cl^- ions; glucose is a nonelectrolyte. Since there are more particles, the vapor pressure of (c) will be the lowest.

Colligative Properties

51. $m = \dfrac{4.9 \text{ g sucrose}}{175 \text{ g solvent}} \times \dfrac{1000 \text{ g}}{\text{kg}} \times \dfrac{1 \text{ mol } C_{12}H_{22}O_{11}}{342.3 \text{ g } C_{12}H_{22}O_{11}} = 0.082$ molal

$\Delta T_b = K_b m = \dfrac{0.51°C}{\text{molal}} \times 0.082 \text{ molal} = 0.042°C$

$T_b = 100.042°C$ at 1.0 atm barometric pressure.

$\Delta T_f = K_f m = \frac{1.86°C}{molal} \times 0.082\ molal = 0.15°C$

$T_f = -0.15°C$

$\Pi = MRT$

In dilute aqueous solutions the density of the solution is about 1.0 g/mL, thus we can assume M = molality = 0.082.

$$\Pi = \frac{0.082\ mol}{L} \times \frac{0.08206\ L\ atm}{mol\ K} \times 298\ K = 2.0\ atm$$

53. a. Assume molarity and molality are equal.

$$M = m = \frac{1.0\ g}{L} \times \frac{1\ mol}{9.0 \times 10^4\ g} = 1.1 \times 10^{-5}\ M$$

$$\text{At 298 K, } \Pi = \frac{1.1 \times 10^{-5}\ mol}{L} \times \frac{0.08206\ L\ atm}{mol\ K} \times 298\ K \times \frac{760\ mm\ Hg}{atm}$$

$\Pi = 0.20$ mm Hg

$$\Delta T_f = K_f m = \frac{1.86°C}{molal} \times 1.1 \times 10^{-5}\ molal = 2.0 \times 10^{-5}\ °C$$

b. Osmotic pressure is better for determining the molecular weight of large molecules. A temperature change of 10^{-5} °C is very difficult to measure. A change in height of a column of mercury by 0.2 mm is not as hard to measure precisely.

55. $\Delta T_f = K_f m$; $2.63 = (40.)\ m$

$m = 6.6 \times 10^{-2}$ mol/kg and $m = \frac{moles\ reserpine}{kg\ solvent}$

$$6.6 \times 10^{-2} = \frac{1.00\ g\ reserpine}{25.0\ g\ solvent} \times \frac{1}{MW} \times \frac{1000\ g\ solvent}{kg\ solvent}$$

$$MW = \frac{1000}{25.0 \times 6.6 \times 10^{-2}} = 606\ g/mol \approx 610\ g/mol\ (2\ S.\ F.\ for\ K_f)$$

Properties of Electrolyte Solutions

57. $NaCl(s) \longrightarrow Na^+(aq) + Cl^-(aq)$

So the total concentration of particles (colligative molarity) is 2 (0.5) = 1.0.

$$\Pi = MRT = \frac{1.0\ mol}{L} \times \frac{0.08206\ L\ atm}{mol\ K} \times 298\ K = 24\ atm$$

A pressure greater than 24 atm must be applied.

59. $MgCl_2$ > NaCl > HOCl > glucose

61. $Ca(NO_3)_2(s) \rightarrow Ca^{2+}(aq) + 2\ NO_3^-(aq)$; i = 3

$$\Delta T_f = 3 \times \frac{1.86°C}{molal} \times 0.5\ molal = 2.8°C \approx 3°C,\ T_f = -3°C$$

The measured freezing point should be higher. Because of ion association the depression of the freezing point will be less than expected, resulting in a higher freezing point.

63. $\Delta T_f = iK_f m$

$$i = \frac{\Delta T_f}{K_f m} = \frac{0.440}{1.86 \times 0.091} = 2.6 \text{ for } 0.091 \text{ molal } CaCl_2$$

$$i = \frac{1.330}{1.86 \times 0.279} = 2.56 \text{ for } 0.279 \text{ molal } CaCl_2$$

$$i = \frac{2.345}{1.86 \times 0.475} = 2.65 \text{ for } 0.475 \text{ molal } CaCl_2$$

ADDITIONAL EXERCISES

65.

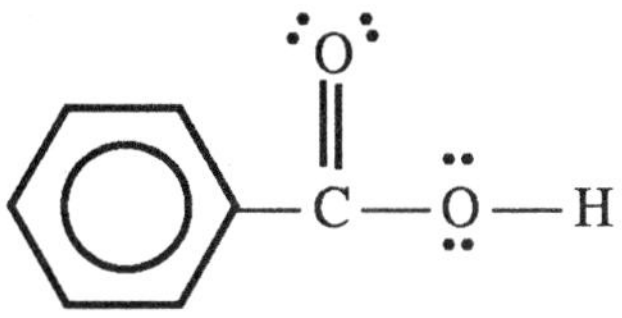

Benzoic acid is capable of hydrogen bonding. However, it is more soluble in nonpolar benzene than in water. In benzene, a non-polar hydrogen bonded dimer forms.

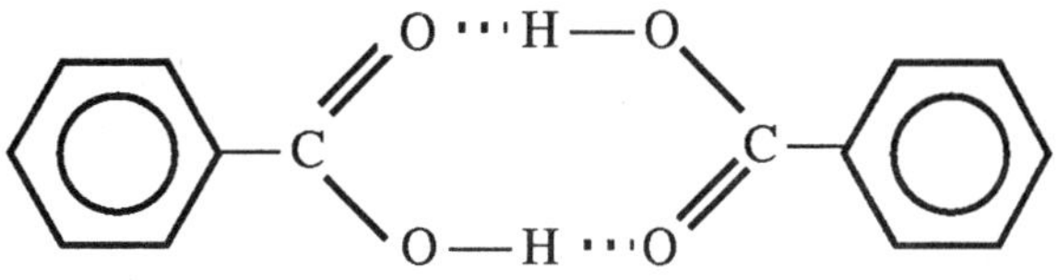

The dimer is nonpolar and thus more soluble in benzene than in water.

67. $$\frac{40.0\ g\ EG}{60.0\ g\ H_2O} \times \frac{1000\ g}{kg} \times \frac{1\ mol\ EG}{62.07\ g} = 10.7\ mol/kg$$

$$\frac{40.0\ g\ EG}{100.0\ g\ solution} \times \frac{1.05\ g}{cm^3} \times \frac{1000\ cm^3}{L} \times \frac{1\ mol}{62.07\ g} = 6.77\ mol/L$$

$$40.0\ g\ EG \times \frac{1\ mol}{62.07\ g} = 0.644\ mol\ EG$$

$$60.0\ g\ H_2O \times \frac{1\ mol}{18.02\ g} = 3.33\ mol\ H_2O$$

$$\chi = \frac{0.644}{3.33 + 0.644} = 0.162$$

69. Both $Al(OH)_3$ and NaOH are ionic. Since the lattice energy is proportiuonal to charge, the lattice energy of aluminum hydroxide is greater than that of sodium hydroxide. The attraction of water molecules for Al^{3+} and OH^- cannot overcome the larger lattice energy and $Al(OH)_3$ is insoluble.

71. a. CF_3CF_3 and $CF_3CF_2CF_3$. Solutions will be ideal when only dispersion forces are involved and both substances are nearly the same size. This is true for the two fluorocarbons. Hydrogen bonding will be important in water/acetone solutions.

b. C_7H_{16} and C_6H_{14}. Both non-polar, similar size. Dipole forces are present in both CHF_3 and CH_3OCH_3. It is unlikely that the forces in the solution will be the same as in the two pure substances.

c. CCl_4 and CF_4 since both are non-polar substances. There will be a hydrogen bonding in aqueous solutions of phosphoric acid.

73. $P_{total} = P_{methanol} + P_{propanol}$

$$= 0.50(303) + 0.50(44.6) = 150 + 22 = 172 \text{ mm Hg} \approx 170 \text{ mm Hg}$$

In the vapor: $\chi_{methanol} = \dfrac{P_{methanol}}{P_{total}} = \dfrac{150}{170} = 0.88$

$$\chi_{methanol} = 0.88 \text{ and } \chi_{propanol} = 0.12$$

75. $14.22 \text{ mg } CO_2 \times \dfrac{12.01 \text{ mg C}}{44.01 \text{ mg } CO_2} = 3.881 \text{ mg C}$

$$\% \text{ C} = \frac{3.881}{4.80} \times 100 = 80.9\% \text{ C}$$

$$1.66 \text{ mg } H_2O \times \frac{2.016 \text{ mg H}}{18.02 \text{ mg } H_2O} = 0.186 \text{ mg H}$$

$$\% \text{ H} = \frac{0.186}{4.80} \times 100 = 3.88\% \text{ H}$$

$$\% \text{ O} = 100 - (80.9 + 3.88) = 15.2\%$$

Out of 100 g:

$$80.9 \text{ g C} \times \frac{1 \text{ mol}}{12.01 \text{ g}} = 6.74 \qquad \frac{6.74}{0.950} = 7.09 \approx 7$$

$$3.88 \text{ g H} \times \frac{1 \text{ mol}}{1.008 \text{ g}} = 3.85 \qquad \frac{3.85}{0.950} = 4.05 \approx 4$$

$$15.2 \text{ g O} \times \frac{1 \text{ mol}}{16.00 \text{ g}} = 0.950 \qquad \frac{0.950}{0.950} = 1$$

Therefore, the empirical formula is C_7H_4O

$$\Delta T_f = K_f m$$

$m = \frac{22.3°C}{40.°C/molal} = 0.56$ molal

$0.56 \text{ molal} = \frac{1.32 \text{ g anthraquinone}}{0.0114 \text{ kg camphor}} \times \frac{1}{MW}$, $MW = \frac{1.32}{0.0114 \times 0.56} = 210$ g/mol

C_7H_4O: 7(12) + 4(1) + 16 ≈ 104 g/mol, so, molecular formula is $C_{14}H_8O_2$

77. $m = \frac{40.0 \text{ g } C_2H_6O_2}{60.0 \text{ g } H_2O} \times \frac{1000 \text{ g}}{\text{kg}} \times \frac{1 \text{ mol}}{62.07 \text{ g}} = 10.7$ mol/kg

$\Delta T_f = 1.86 \times 10.7 = 19.9°C$ and $T_f = -19.9°C$

$\Delta T_b = 0.51 \times 10.7 = 5.5°C$ and $T_b = 105.5°C$

79. $\Delta T_f = 3.00°C = \frac{1.86°C}{\text{molal}} \times m$; $m = 1.61$ mol/kg

$\frac{1.61 \text{ mol urea}}{\text{kg } H_2O} \times 0.150 \text{ kg } H_2O \times \frac{60.06 \text{ g urea}}{\text{mol urea}} = 14.5$ g

81. Out of 100 g, there are

$31.57 \text{ g C} \times \frac{1 \text{ mol}}{12.01 \text{ g}} = 2.629$ $\qquad \frac{2.629}{2.629} = 1.000$

$5.30 \text{ g H} \times \frac{1 \text{ mol}}{1.008 \text{ g}} = 5.26$ $\qquad \frac{5.26}{2.629} = 2.00$

$63.13 \text{ g O} \times \frac{1 \text{ mol}}{16.00 \text{ g}} = 3.946$ $\qquad \frac{3.946}{2.629} = 1.501$

Empirical formula: $C_2H_4O_3$

$\Delta T_f = K_f m$; $m = \frac{5.20°C}{1.86°C/molal} = 2.80$ molal

$\frac{2.80 \text{ mol}}{\text{kg}} = \frac{10.56 \text{ g}}{0.0250 \text{ kg } H_2O} \times \frac{1}{MW}$, MW = 151 g/mol (expt) and mass of $C_2H_4O_3$ = 76.05 g/mol

Molecular formula = $C_4H_8O_6$ and molar mass = 152.10 g/mol

Note: We can use the experimental molar mass to get the molecular formula. Knowing this, we can calculate the molar mass precisely from the formula.

CHALLENGE PROBLEMS

83. $\Delta T_f = 5.5 - 2.8 = 2.7°C$

$\Delta T_f = K_f m$

$m = \frac{\Delta T_f}{K_f} = \frac{2.7°C}{5.12°C/molal} = 0.527 \text{ molal} \approx 0.53$ molal

Let x = mass of naphthalene (molar mass: 128 g/mol)

Then 1.60 - x = mass of anthracene (molar mass: 178 g/mol)

$\frac{x}{128}$ = moles naphthalene and $\frac{1.60 - x}{178}$ = moles anthracene

$$\frac{\frac{x}{128} + \frac{1.60 - x}{178}}{0.0200 \text{ kg solvent}} = 0.527 \text{ (carry extra significant figure)}$$

$$\frac{178x + 1.60(128) - 128x}{(128)(178)} = 1.05 \times 10^{-2}$$

$50.x + 204.8 = 239.2$; $50.x = 239.2 - 204.8$; $50.x = 34.4$

x = 0.688 g naphthalene. So mixture is:

$$\frac{0.688 \text{ g}}{1.60 \text{ g}} \times 100 = 43\% \text{ naphthalene by mass and } 57\% \text{ anthracene by mass}$$

85. a. $\Delta T_f = K_f m$; $m = \frac{1.32}{5.12} = 0.258$ mol/kg

$$0.258 \text{ mol/kg} = \frac{1.22}{0.01560} \times \frac{1}{\text{MW}}$$

MW = 303 g/mol

Uncertainty in temp. = $\frac{0.04}{1.32} \times 100 = 3\%$

A 3% uncertainty of 303 g/mol = 9 g/mol

So, MW = 303 $\pm$ 9 g/mol

b. No, codeine could not be eliminated since it is in the possible range.

c. We would really like the uncertainty to be $\pm$ 1 g/mol. We need a T_f to be about 10 times what it was in this problem. Two possibilities:

1. make solution ten times as concentrated (possible solubility problem)

2. use camphor (K_f = 40) as the solvent.

Even then it may be difficult to distinguish between the two. Using a mass spectrometer (see Ch. 3 in textbook) would be a much better technique.

CHAPTER TWELVE: CHEMICAL KINETICS

QUESTIONS

1. The rate of a chemical reaction varies with time. Consider a reaction:

$$A \rightarrow \text{Products} \qquad \text{rate} = \frac{-\Delta[A]}{\Delta t}$$

If we graph [A] vs. t, it would roughly look like

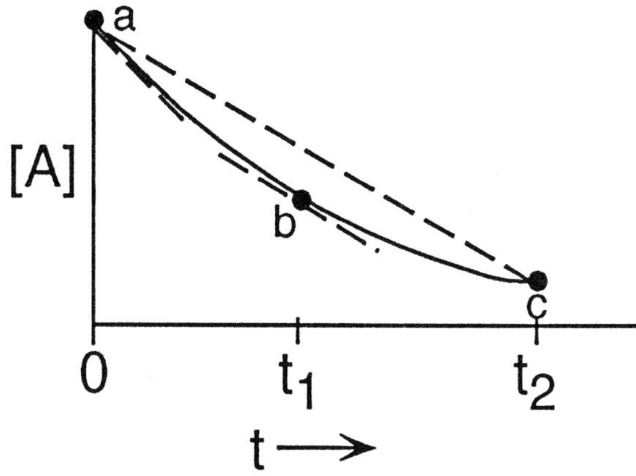

An instantaneous rate is the slope of the tangent line to the graph of [A] vs. t. We can determine the instantaneous rate at any point.

The initial rate is the instantaneous rate very near t = 0 (point a). The slope of the tangent at point b is the instantaneous rate at time, t_1.

The average rate is measured over a period of time. For example, the slope of the line connecting points a and c is the average rate of the reaction over the entire length of time, 0 to t_2.

3. a. The greater the frequency of collisions, the greater the opportunities for molecules to react, and, hence, the greater the rate.

   b. Chemical reactions involve the making and breaking of chemical bonds. The kinetic energy of the collision can be used to break bonds.

   c. For a reaction to occur, it is the reactive portion of each molecule that must be involved in a collision. Thus, only some collisions have the right orientation.

5. The probability of the simultaneous collision of three molecules with the correct energy and orientation is exceedingly small.

7. A catalyst increases the rate of a reaction by providing an alternate pathway (mechanism) with a lower activation energy.

9. No, the catalyzed reaction has a different mechanism and hence, a different rate law.

EXERCISES

Reaction Rates

11. $\text{Rate} = \frac{-\Delta[S_2O_3^{2-}]}{\Delta t} = \frac{0.0080 \text{ mol}}{\text{L s}}$

I_2 is consumed at half this rate.

$$\frac{-\Delta[I_2]}{\Delta t} = 0.0040 \text{ mol/L•s}$$

$$\frac{\Delta[S_4O_6^{2-}]}{\Delta t} = 0.0040 \text{ mol/L•s} \qquad \frac{\Delta[I^-]}{\Delta t} = 0.0080 \text{ mol/L•s}$$

13. a. mol/L•s

b. Rate = k; k has units of mol/L•s

c. Rate = k[A]

$$\frac{\text{mol}}{\text{L s}} = k\left(\frac{\text{mol}}{\text{L}}\right)$$

k must have units s^{-1}

d. Rate = $k[A]^2$

$$\frac{\text{mol}}{\text{L s}} = k\left(\frac{\text{mol}}{\text{L}}\right)^2$$

k must have units L/mol•s

e. L^2/mol^2•s

15. $$\frac{1.24 \times 10^{-12} \text{ cm}^3}{\text{molecules s}} \times \frac{1 \text{ L}}{1000 \text{ cm}^3} \times \frac{6.022 \times 10^{23} \text{ molecules}}{\text{mol}}$$

$k = 7.47 \times 10^8$ L/mol•s

Rate Laws from Experimental Data: Initial Rates Method

17. a. In the first two experiments, [NO] is held constant and $[Cl_2]$ is doubled. The rates also double. Thus, the reaction is first order with respect to Cl_2.

Mathematically: Rate = $k[NO]^x [Cl_2]^y$

$$\frac{0.35}{0.18} = \frac{k(0.10)^x (0.20)^y}{k(0.10)^x (0.10)^y} = \frac{(0.20)^y}{(0.10)^y}$$

$1.9 = 2.0^y \qquad y \approx 1$

We get the dependence on NO from the second and third experiments.

$$\frac{1.45}{0.35} = \frac{k(0.20)^x (0.20)}{k(0.10)^x (0.20)} = \frac{(0.20)^x}{(0.10)^x}$$

$4.1 = 2^x$; $\log 4.1 = x \log 2$; $0.62 = x(0.30)$

$x = 2.1 \approx 2$

So rate = $k[NO]^2 [Cl_2]$

b. $\frac{0.18 \text{ mol}}{\text{L min}} = k\left(\frac{0.10 \text{ mol}}{\text{L}}\right)^2\left(\frac{0.10 \text{ mol}}{\text{L}}\right)$

$k = 180 \text{ L}^2/\text{mol}^2\text{•min}$ (1st exp.)

$k = 175 \text{ L}^2/\text{mol}^2\text{•min}$ (2nd exp.)

$k = 181 \text{ L}^2/\text{mol}^2\text{•min}$ (3rd exp.)

To two significant figures, $k_{mean} = 1.8 \times 10^2 \text{ L}^2/\text{mol}^2\text{•min}$

19. a. Rate = $K[I^-]^x [OCl^-]^y$

$$\frac{7.91 \times 10^{-2}}{3.95 \times 10^{-2}} = \frac{k(0.12)^x (0.18)^y}{k(0.060)^x (0.18)^y} = 2.0^x$$

$2.00 = 2.0^x$, $x = 1$

$$\frac{3.95 \times 10^{-2}}{9.88 \times 10^{-3}} = \frac{k(0.060)(0.18)^y}{k(0.030)(0.090)^y}, \quad x = 1$$

$4.00 = 2.0 \times 2.0^y$; $2.0 = 2.0^y$; $y = 1$

Rate = $k[I^-][OCl^-]$

b. From the first experiment:

$$\frac{7.91 \times 10^{-2} \text{ mol}}{\text{L s}} = k\left(\frac{0.12 \text{ mol}}{\text{L}}\right)\left(\frac{0.18 \text{ mol}}{\text{L}}\right)$$

$k = 3.7 \text{ L/mol•s}$

All four experiments give the same value of k to two significant figures.

21. Rate = $k[ClO_2]^x[OH^-]^y$

From the first two experiments:

$2.30 \times 10^{-1} = k(0.100)^x(0.100)^y$ and $5.75 \times 10^{-2} = k(0.0500)^x(0.100)^y$

Dividing the two rate laws:

$$4.00 = \frac{(0.100)^x}{(0.0500)^x} = 2.00^x; \; x = 2$$

Comparing the second and third experiments:

$2.30 \times 10^{-1} = k(0.100)(0.100)^y$ and $1.15 \times 10^{-1} = k(0.100)(0.0500)^y$

Dividing: $2.00 = \frac{(0.100)^y}{(0.0500)^y} = 2.00^y$; $y = 1$

The rate law is:

Rate = $k[ClO_2]^2[OH^-]$

2.30×10^{-1} mol/L•s = k(0.100 mol/L)^2(0.100 mol/L)

k = 2.30×10^2 L^2/mol^2•s, k_{mean} = 2.30×10^2 L^2/mol^2•s

Integrated Rate Laws from Experimental Data

23. The first guess to make is that the reaction is first order. For a first order reaction a graph of ln[C_4H_6] vs. t should yield a straight line.

Time (s)	195	604	1246	2180	6210
[C_4H_6] (mol/L)	1.6×10^{-2}	1.5×10^{-2}	1.3×10^{-2}	1.1×10^{-2}	0.68×10^{-2}
ln[C_4H_6]	-4.14	-4.20	-4.34	-4.51	-4.99
1/[C_4H_6]	62.5	66.7	76.9	90.9	147

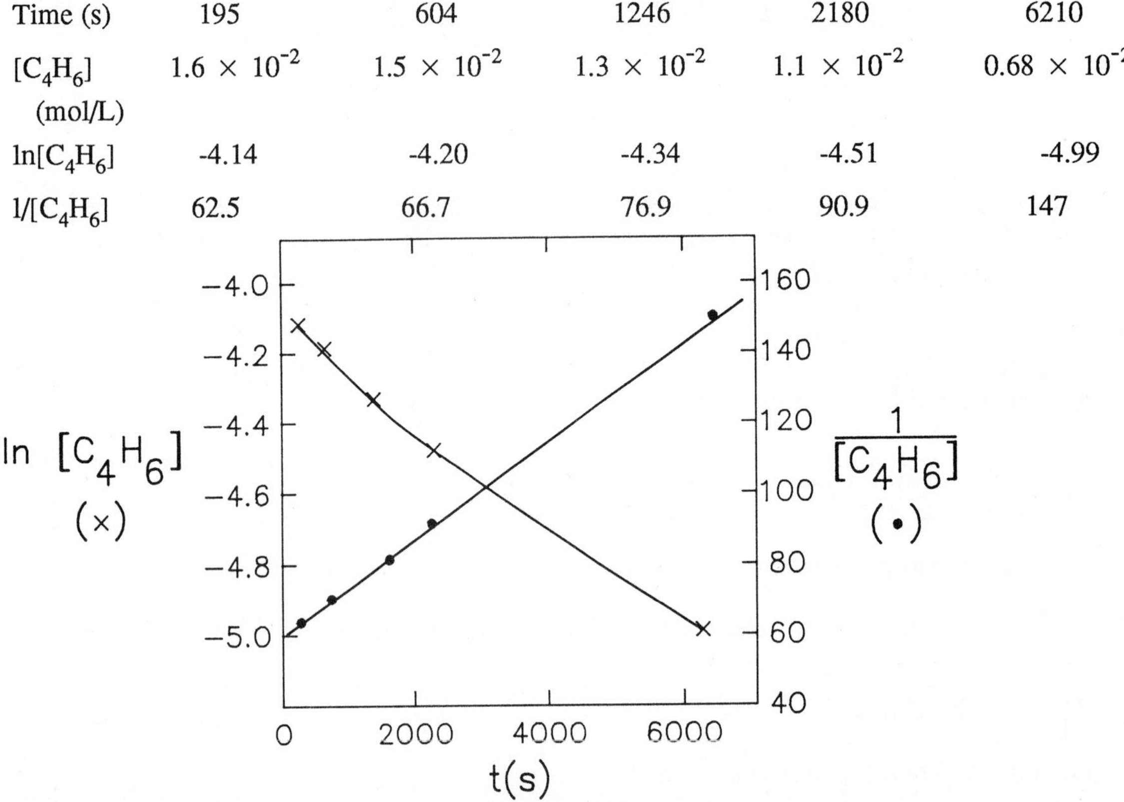

The natural log plot is not linear. The next guess is that the reaction is second order. For a second order reaction a plot of 1/[C_4H_6] vs. t should yield a straight line. Since we get a straight line for this graph, we conclude the reaction is second order in butadiene. Rate = $k[C_4H_6]^2$

For a second order reaction: $\frac{1}{[A]_t} = \frac{1}{[A]_o} + kt$

Thus the slope of the straight line equals the value of the rate constant. Using the points on the line at 1000 and 6000 s

$$k = \text{Slope} = \frac{144 \text{ L/mol} - 73 \text{ L/mol}}{6000 \text{ s} - 1000 \text{ s}} = 1.4 \times 10^{-2} \text{ L/mol}\bullet\text{s}$$

This value was obtained from reading the graph and "eyeballing" a straight line with a ruler.
Rate = $k[C_6H_5N_2Cl]$ with $k = 6.9 \times 10^{-2}\ s^{-1}$

25.

Time (s)	$[H_2O_2]$ (mol/L)	$\ln[H_2O_2]$
0	1.0	0
120	0.91	-0.094
300	0.78	-0.248
600	0.59	-0.528
1200	0.37	-0.994
1800	0.22	-1.51
2400	0.13	-2.04
3000	0.082	-2.50
3600	0.050	-3.00

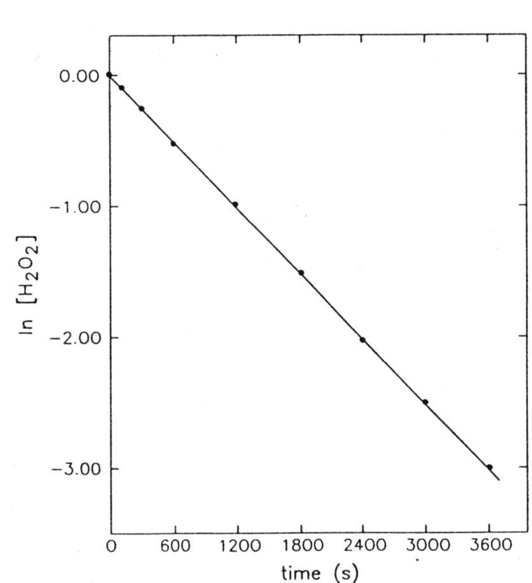

The plot $\ln[H_2O_2]$ vs t is linear.
Thus, the reaction is first order.

$\ln[H_2O_2] = -kt + \ln[H_2O_2]_o$.

Rate = $k[H_2O_2]$

$$\text{Slope} = -k = \frac{0 - (-3.00)}{0 - 3600} = -8.3 \times 10^{-4}\ s^{-1}$$

$k = 8.3 \times 10^{-4}\ s^{-1}$

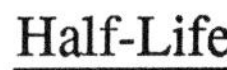

Half-Life

27. a. $k = \dfrac{\ln 2}{t_{1/2}} = \dfrac{0.69315}{t_{1/2}}$

For ^{239}Pu, $k = \dfrac{0.69315}{24{,}360\ \text{yr}} = 2.845 \times 10^{-5}\ \text{yr}^{-1}$

$$\frac{2.845 \times 10^{-5}}{\text{yr}} \times \frac{1\ \text{yr}}{365\ \text{d}} \times \frac{1\ \text{d}}{24\ \text{hr}} \times \frac{1\ \text{hr}}{60\ \text{min}} \times \frac{1\ \text{min}}{60\ \text{s}} = 9.021 \times 10^{-13}\ s^{-1}$$

For ^{241}Pu, $k = \dfrac{0.693}{13\ \text{yr}} = 0.0533\ \text{yr}^{-1} = 1.7 \times 10^{-9}\ s^{-1}$

b. The rate constant is larger for the decay of ^{241}Pu, hence, ^{241}Pu decays more rapidly.

c. For radioactive decay:

Rate = kN where N is the number of radioactive nuclei.

$$\text{Rate} = 1.7 \times 10^{-9}\ s^{-1} \times \left(5.0\ \text{g} \times \frac{1\ \text{mol}}{241\ \text{g}} \times \frac{6.02 \times 10^{23}\ \text{atoms}}{\text{mol}}\right)$$

Rate = 2.1 × 10^{13} disintegrations/s

$\ln \frac{[A]_t}{[A]_o} = -kt$

$\ln\left(\frac{N}{5.0}\right) = -0.0533\ yr^{-1}\ (1.0\ yr)$ (carry extra sig. fig.)

$\ln N = -0.0533 + \ln 5.0 = -0.0533 + 1.609 = 1.556$

$N = e^{1.556} = 4.7$ g left after 1 yr

$\ln N = -0.0533\ yr^{-1}\ (10.\ yr) + \ln 5.0 = -0.533 + 1.609 = 1.076$

$N = e^{1.076} = 2.9$ g ^{241}Pu left after 10. years

$\ln N = -0.0533\ yr^{-1}\ (100.\ yr) + \ln 5.0 = -5.33 + 1.609 = -3.72$

$N = e^{-3.72} = 0.024$ g left after 100. yr

29. $\ln \frac{[A]_t}{[A]_o} = -kt$

If $[A]_o = 100.$, then after 65 s, 45% of A has decayed or $[A]_{65} = 55$

$\ln\left(\frac{55}{100.}\right) = -k(65\ s)$

$k = 9.2 \times 10^{-3}\ s^{-1}$; $t_{1/2} = \frac{0.693}{k} = 75\ s$

31. $t_{1/2} = \ln \frac{1}{k[A]_o}$ or $k = \frac{1}{t_{1/2}[A]_o}$

$k = \frac{1}{143\ s\ (0.060\ mol/L)} = 0.12\ L/mol\bullet s$

33. a. $\ln\left(\frac{[A]_t}{[A]_o}\right) = -kt$

If reaction is 38.5% complete, then 38.5% of the original concentration is consumed, leaving 61.5%.

$[A] = 61.5\%$ of $[A]_o$ or $[A] = 0.615[A]_o$

$\ln(0.615) = -k(480\ s)$; $-0.486 = -k(480\ s)$

$k = 1.0 \times 10^{-3}\ s^{-1}$

b. $t_{1/2} = 0.693/k = 0.693/1.0 \times 10^{-3}\ s^{-1} = 693\ s = 690\ s$

c. 25% complete: $[A] = 0.75[A]_o$

$\ln(0.75) = -1.0 \times 10^{-3}(t)$

$t = 288 \text{ s} \approx 2.9 \times 10^2 \text{ s}$

75% complete: $[A] = 0.25[A]_o$

$\ln(0.25) = -1.0 \times 10^{-3}(t)$

$t = 1.39 \times 10^3 \text{ s} = 1.4 \times 10^3 \text{ s}$

or we know it takes 2 $t_{1/2}$ for reaction to be 75% complete:

$t = 2(693 \text{ s}) = 1386 \text{ s} \approx 1.4 \times 10^3 \text{ s}$

95% complete: $[A] = 0.05[A]_o$, $\ln(0.05) = -1.0 \times 10^{-3}(t)$ $\qquad t = 3 \times 10^3 \text{ s}$

Reaction Mechanisms

35. a. Rate = $k[CH_3NC]$ b. Rate = $k[O_3]\,[NO]$

37. We know from Exercise 12.25 that Rate = $k[H_2O_2]$.

If the first step, $H_2O_2 \longrightarrow 2\ OH$, is the slow step, then the mechanism gives the correct rate law.

Temperature Dependence of Rate Constants and Collision Theory

39. $k = A \exp(-E_a/RT)$

$\ln k = \frac{-E_a}{RT} + \ln A$; so $\ln k_1 = \frac{-E_a}{RT_1} + \ln A$ and $\ln k_2 = \frac{-E_a}{RT_2} + \ln A$

Subtracting the two equations, $\ln k_1 - \ln k_2 = \frac{-E_a}{R}\left(\frac{1}{T_1} - \frac{1}{T_2}\right) = \ln\left(\frac{k_1}{k_2}\right)$

$\ln 2.0 \times 10^3 - \ln k_2 = \frac{-15.0 \times 10^3 \text{ J/mol}}{8.3145 \text{ J/K•mol}}\left(\frac{1}{298 \text{ K}} - \frac{1}{348 \text{ K}}\right)$

$7.60 - \ln k_2 = -0.870$

$\ln k_2 = 8.47$

$k_2 = e^{8.47} = 4.8 \times 10^3 \text{ s}^{-1}$

41. a. $\ln\left(\frac{k_1}{k_2}\right) = \frac{E_a}{R}\left(\frac{1}{T_2} - \frac{1}{T_1}\right)$, $\quad \ln\left(\frac{2.45 \times 10^{-4}}{0.950}\right) = \frac{E_a}{8.3145}\left(\frac{1}{781} - \frac{1}{575}\right)$

$-8.263 = \frac{E_a}{8.3145}\left(-4.6 \times 10^{-4}\right)$

$E_a = 1.5 \times 10^5$ J/mol = 150 kJ/mol

$\ln k_1 = -E_a/RT + \ln A$

$$\ln(2.45 \times 10^{-4}) = \frac{-1.5 \times 10^5 \text{ J/mol}}{8.3145 \text{ J/mol•K}}\left(\frac{1}{575 \text{ K}}\right) + \ln A$$

$-8.314 = -31.38 + \ln A$

$\ln A = 23.07$

$A = e^{23.07} = 1.0 \times 10^{10}$ L/mol•s

b. $$k = A \exp(-E_a/RT) = 1.0 \times 10^{10} \exp\left(\frac{-1.5 \times 10^5}{8.3145 \times 648}\right) = 8.1 \times 10^{-3} \text{ L/mol•s}$$

43. From the Arrhenius equation, $k = A \exp(-E_a/RT)$

or in logarithmic form, $\ln k = \frac{-E_a}{RT} + \ln A$

Hence, a graph of ln k vs. 1/T should yield a straight line with a slope equal to $-E_a/R$.

T (K)	1/T	k (L/mol•s)	ln k
195	5.13×10^{-3}	1.08×10^9	20.80
230	4.35×10^{-3}	2.95×10^9	21.82
260	3.85×10^{-3}	5.42×10^9	22.41
298	3.36×10^{-3}	12.0×10^9	23.21
369	2.71×10^{-3}	35.5×10^9	24.29

From the "eyeball" line on the graph:

$$\text{Slope} = \frac{20.95 - 23.65}{5.00 \times 10^{-3} - 3.00 \times 10^{-3}}$$

$$= \frac{-2.70}{2.00 \times 10^{-3}} = -1.35 \times 10^3 \text{ K} = \frac{-E_a}{R}$$

$$E_a = 1.35 \times 10^3 \text{ K} \times \frac{8.3145 \text{ J}}{\text{K mol}}$$

$= 1.12 \times 10^4$ J/mol or 11.2 kJ/mol

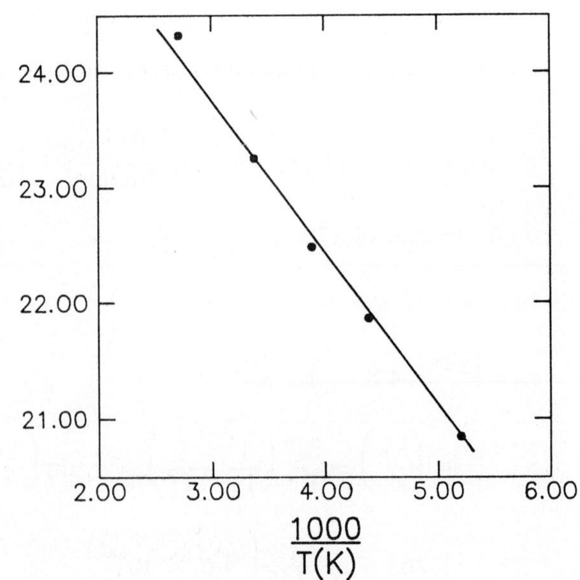

From the best straight line (by computer):

Slope = -1.43×10^3 K and E_a = 11.9 kJ/mol

45.

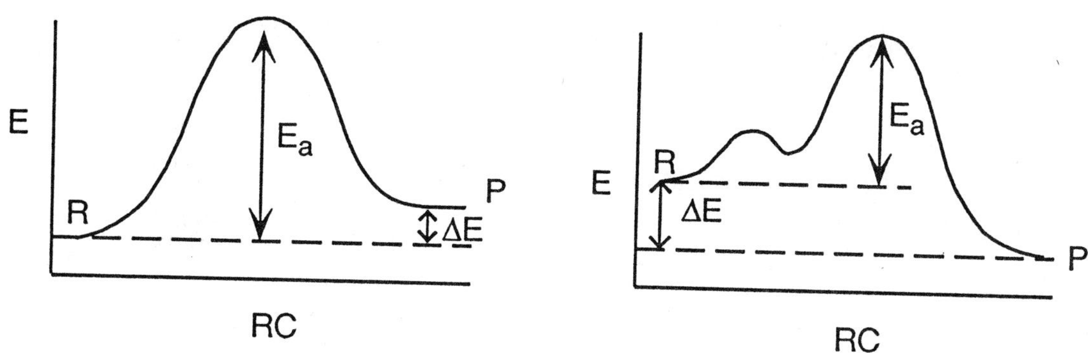

47.

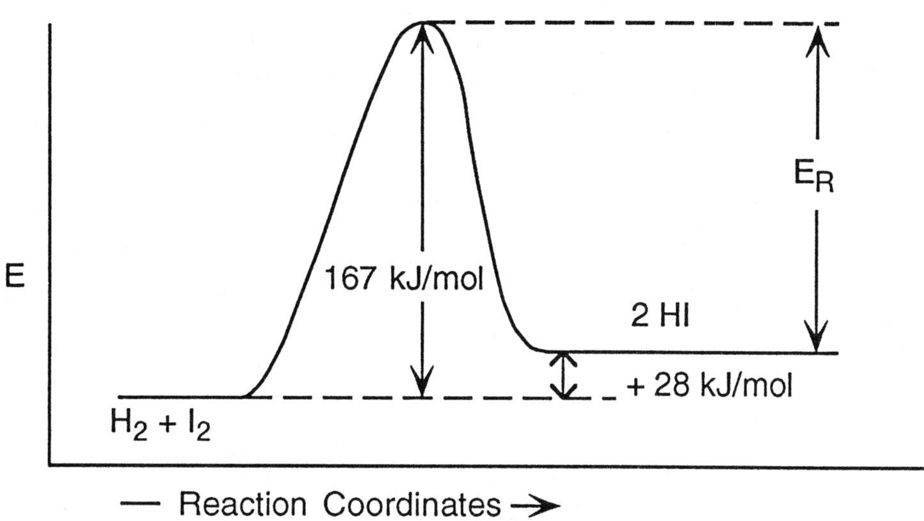

The activation energy for the reverse reaction is E_R in the diagram.

$$E_R = 167 - 28 = 139 \text{ kJ/mol}$$

Catalysis

49. a. NO is the catalyst.

b. NO_2 is an intermediate

c. $k = A \exp(-E_a/RT)$

$$\frac{k_{cat}}{k_{un}} = \frac{\exp[-E_a(\text{cat})/RT]}{\exp[-E_a(\text{un})/RT]}$$

$$\frac{k_{cat}}{k_{un}} = \exp\frac{E_a(\text{un}) - E_a(\text{cat})}{RT}$$

$$\frac{k_{cat}}{k_{un}} = \exp \frac{2100 \text{ J/mol}}{8.3145 \text{ J/mol•K} \times 298 \text{ K}} = e^{0.848} = 2.3$$

The catalyzed reaction is 2.3 times faster than the uncatalyzed reaction at 25°C.

51. The surface of the catalyst should be:

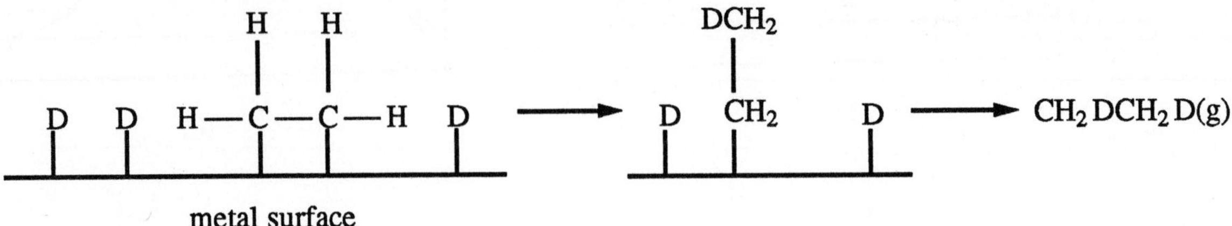

Thus, CH_2DCH_2D should be the product.

ADDITIONAL EXERCISES

53. Rate = $k[Cl]^{1/2}[CHCl_3]$

$$\frac{\text{mol}}{\text{L s}} = k\left(\frac{\text{mol}}{\text{L}}\right)^{1/2}\left(\frac{\text{mol}}{\text{L}}\right)$$

k must have units of $L^{1/2}/mol^{-1/2}$•s

55. Rate = $k[H_2SeO_3]^x[H^+]^y[I^-]^z$

Comparing the first and second experiments:

$$\frac{3.33 \times 10^{-7}}{1.66 \times 10^{-7}} = \frac{k(2.0 \times 10^{-4})^x\,(2.0 \times 10^{-2})^y\,(2.0 \times 10^{-2})^z}{k(1.0 \times 10^{-4})^x\,(2.0 \times 10^{-2})^y\,(2.0 \times 10^{-2})^z} = 2.0^x$$

$2.01 = 2.0^x,\ x = 1$

Comparing the first and fourth experiments:

$$\frac{6.66 \times 10^{-7}}{1.66 \times 10^{-7}} = \frac{k(1.0 \times 10^{-4})\,(4.0 \times 10^{-2})^y\,(2.0 \times 10^{-2})^z}{k(1.0 \times 10^{-4})\,(2.0 \times 10^{-2})^y\,(2.0 \times 10^{-2})^z}$$

$4.01 = 2.0^y,\ y = 2$

Comparing the first and sixth experiments:

$$\frac{13.2 \times 10^{-7}}{1.66 \times 10^{-7}} = \frac{k(1.0 \times 10^{-4})\,(2.0 \times 10^{-2})^2\,(4.0 \times 10^{-2})^z}{k(1.0 \times 10^{-4})\,(2.0 \times 10^{-2})^2\,(2.0 \times 10^{-2})^z}$$

$7.95 = 2.0^z$

$$z = \frac{\log 7.95}{\log 2} = 2.99 \approx 3$$

Rate: $k[H_2SeO_3][H^+]^2[I^-]^3$

Experiment #1:

$$\frac{1.66 \times 10^{-7}\ \text{mol}}{\text{L s}} = k\left(\frac{1.0 \times 10^{-4}\ \text{mol}}{\text{L}}\right)\left(\frac{2.0 \times 10^{-2}\ \text{mol}}{\text{L}}\right)^2 \left(\frac{2.0 \times 10^{-2}\ \text{mol}}{\text{L}}\right)^3$$

$k = 5.19 \times 10^5\ L^5/mol^5{\cdot}s \approx 5.2 \times 10^5\ L^5/mol^5{\cdot}s$

For all experiments:

Exp. #	$k(L^5/\ mol^5{\cdot}s)$	
1	5.19×10^5	
2	5.20×10^5	
3	5.20×10^5	
4	5.20×10^5	$k_{mean} = 5.2 \times 10^5\ L^5/mol^5{\cdot}s$
5	5.25×10^5	
6	5.16×10^5	
7	5.25×10^5	

57. $SO_2Cl_2(g) \longrightarrow SO_2(g) + Cl_2(g)$

Let P_O = initial partial pressure of SO_2Cl_2

If $x = P_{SO_2}$ at some time, then

$x = P_{SO_2} = P_{Cl_2}$

$P_{SO_2Cl_2} = P_O - x$

$P_{tot} = P_{SO_2Cl_2} + P_{SO_2} + P_{Cl_2} = P_O - x + x + x$

$P_{tot} = P_O + x$

So:

Time(hour)	0.00	1.00	2.00	4.00	8.00	16.00
P(atm)	4.93	5.60	6.34	7.33	8.56	9.52
$P_{SO_2Cl_2}$ (atm)	4.93	4.26	3.52	2.53	1.30	0.34
$\ln P_{SO_2Cl_2}$	1.60	1.45	1.26	0.928	0.262	-1.08

$P_{SO_2Cl_2} \propto [SO_2Cl_2]$ $\qquad P = \frac{n}{V}RT = [SO_2Cl_2]RT$

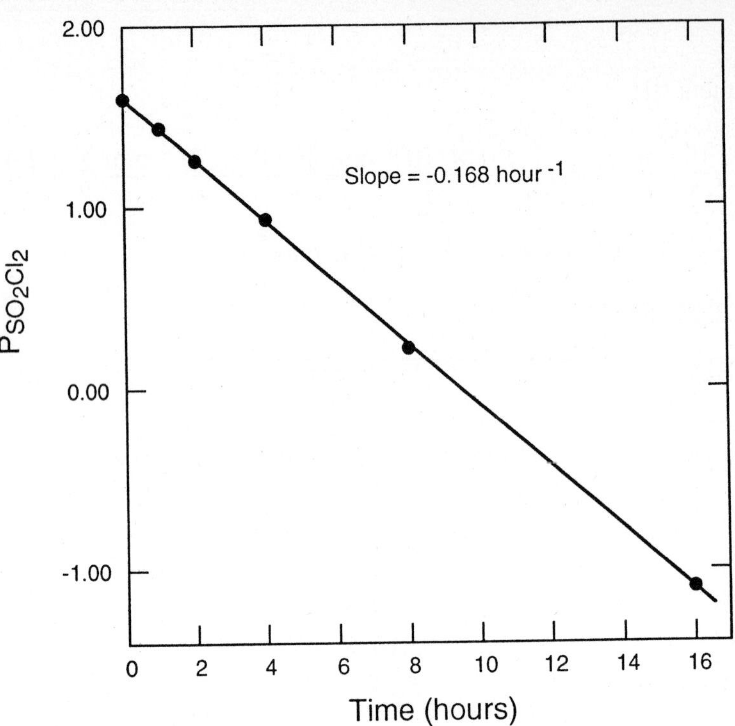

a. The reaction is first order in SO_2Cl_2

Slope of ln(P) vs. t is -0.168 $hour^{-1}$ = -k

$k = 0.168\ hour^{-1} = 4.67 \times 10^{-5}\ s^{-1}$

b. $t_{1/2} = \frac{0.69315}{k} = \frac{0.69315}{0.168} = 4.13$ hour

c. The total pressure increases at the same rate as the partial pressure of SO_2Cl_2 decreases.

$$\frac{\Delta P_{SO_2Cl_2}}{\Delta t} = \frac{-\Delta P_{tot}}{\Delta t} = -k\,P_{SO_2Cl_2}$$

$$\ln\left(\frac{P_{tot}}{P_O}\right) = kt;\ \ln\left(\frac{P_{tot}}{4.93}\right) = 0.168\ h^{-1}(0.500\ h)$$

$\ln P_{tot} = 0.0840 + 1.595 = 1.679;\ P_{tot} = e^{1.679} = 5.36$ atm

After 12.0 hours: From graph, $\ln P_{SO_2Cl_2} = -0.403$

$P_{SO_2Cl_2} = e^{-0.403} = 0.668$ atm

$P_{SO_2} = 4.93 - 0.668 = 4.26$; $P_{Cl_2} = 4.26$; $P_{tot} = 9.19$ atm

d. $$\ln\left(\frac{P_{SO_2Cl_2}}{P_O}\right) = -0.168(20.0\ hr) = -3.36$$

$\frac{P_{SO_2Cl_2}}{P_O} = e^{-3.36} = 3.47 \times 10^{-2}$, Fraction left = 0.0347 (3.47%)

59. a. W, lower activation energy than Os catalyst.

b. $k_w = A_w \exp(-E_a(W)/RT)$

$K_{uncat} = A_{uncat} \exp(-E_a(uncat)/RT)$. Assume $A_w = A_{uncat}$

$$\frac{k_w}{k_{uncat}} = \exp\left(\frac{-E_a(W)}{RT} + \frac{E_a(uncat)}{RT}\right)$$

$$\frac{k_w}{k_{uncat}} = \exp\left(\frac{-163{,}000 \text{ J/mol} + 335{,}000 \text{ J/mol}}{(8.3145 \text{ J/mol}\cdot\text{K})(298 \text{ K})}\right) = 1.41 \times 10^{30}$$

c. H_2 decreases the rate of the reaction. For the decomposition to occur, NH_3 molecules must be adsorbed on the surface of the catalyst. If H_2 that is produced from the reaction is also adsorbed on the catalyst, then there are fewer sites for NH_3 molecules to be adsorbed.

CHALLENGE PROBLEMS

61. $\text{Rate} = k[I^-]^x [OCl^-]^y [OH^-]^z$

Comparing the first and second experiments:

$$\frac{18.7 \times 10^{-3}}{9.4 \times 10^{-3}} = \frac{k(0.0026)^x (0.012)^y (0.10)^z}{k(0.0013)^x (0.012)^y (0.10)^z}$$

$2.0 = 2.0^x$, $x = 1$

Comparing the first and third experiments:

$$\frac{9.4 \times 10^{-3}}{4.7 \times 10^{-3}} = \frac{k(0.0013)(0.012)^y (0.10)^z}{k(0.0013)(0.006)^y (0.10)^z}$$

$2.0 = 2.0^y$, $y = 2$

Comparing the first and sixth experiments:

$$\frac{4.7 \times 10^{-3}}{9.4 \times 10^{-3}} = \frac{k(0.0013)(0.012)(0.20)^z}{k(0.0013)(0.012)(0.10)^z}$$

$1/2 = 2.0^z$, $z = -1$

$$\text{Rate} = \frac{k[I^-][OCl^-]}{[OH^-]}$$

For the first experiment:

$$\frac{9.4 \times 10^{-3} \text{ mol}}{\text{L s}} = k\,\frac{(0.0013 \text{ mol/L})(0.012 \text{ mol/L})}{(0.10 \text{ mol/L})}$$

$k = 60.3 \text{ s}^{-1} = 60. \text{ s}^{-1}$

For all experiments:

Exp. #	1	2	3	4	5	6	7
k (s^{-1})	60.3	59.9	60.3	59.8	59.9	60.3	59.8

$k_{mean} = 6.0 \times 10^1\ s^{-1}$

63. a. We check the first order dependence by graphing ln C vs. t for each set of data.

Dependence on NO

time (ms)	[NO] (molecules/cm^3)	ln [NO]
0	6.0×10^8	20.21
100.	5.0×10^8	20.03
500.	2.4×10^8	19.30
700.	1.7×10^8	18.95
1000.	9.9×10^7	18.41

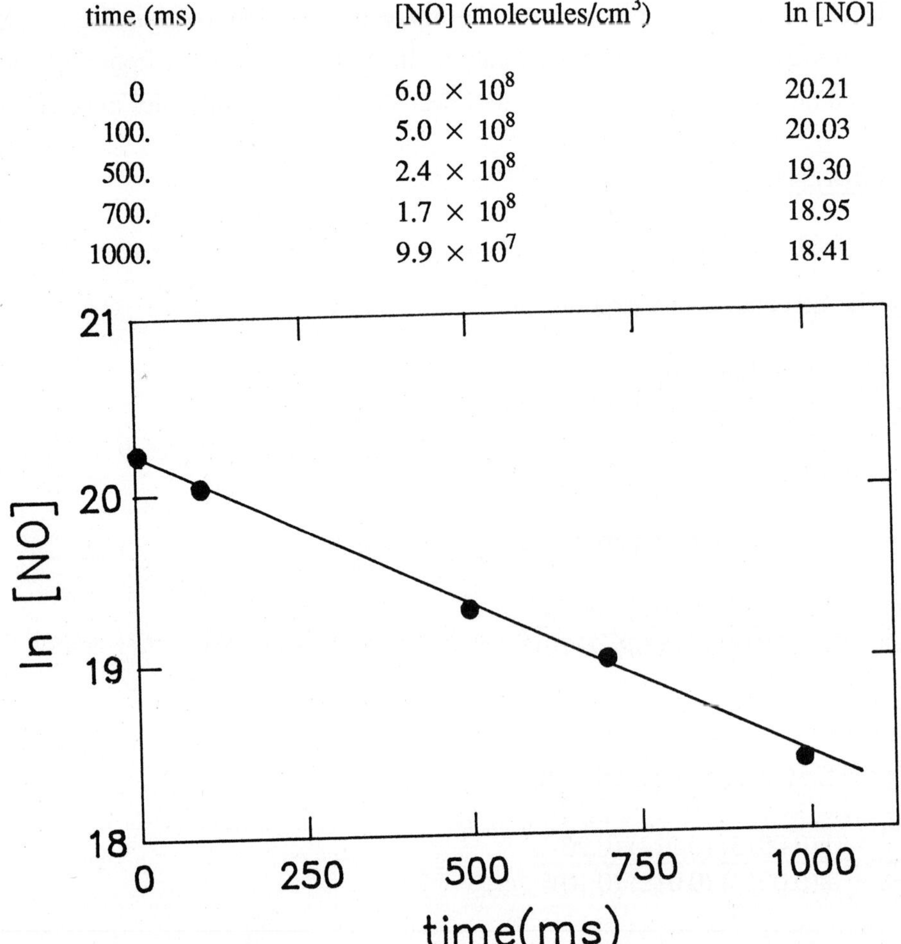

Since ln [NO] vs. t is linear, the reaction is first order with respect to nitric oxide.

We follow the same procedure for ozone. The data are:

time (ms)	$[O_3]$ (molecules/cm^3)	ln $[O_3]$
0	1.0×10^{10}	23.03
50.	8.4×10^{9}	22.85
100.	7.0×10^{9}	22.67
200.	4.9×10^{9}	22.31
300.	3.4×10^{9}	21.95

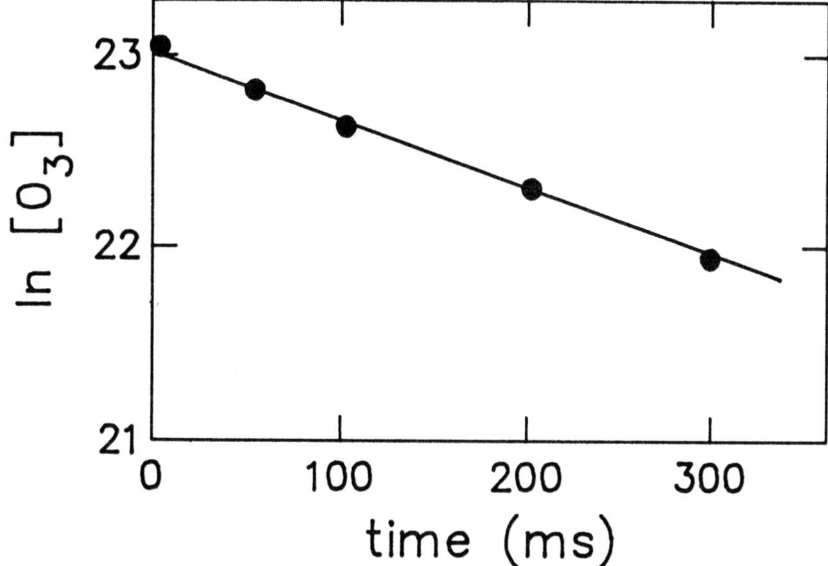

The plot of ln$[O_3]$ vs. t is linear. Hence, the reaction is first order with respect to ozone.

b. Rate = k[NO]$[O_3]$

c. Rate = k´[NO]

k´ = -(Slope from graph of ln [NO] vs. t)

$$k' = -\text{Slope} = -\frac{18.41 - 20.21}{(1000. - 0) \times 10^{-3}\ \text{s}} = 1.8\ \text{s}^{-1}$$

For ozone,

Rate = k´´$[O_3]$

and k´´ = -(Slope from ln$[O_3]$ vs. t)

$$k'' = -\text{Slope} = -\frac{(21.95 - 23.03)}{(300. - 0) \times 10^{-3}\ \text{s}}$$

$k'' = 3.6\ \text{s}^{-1}$

d. Rate = k[NO][O_3] = k′[NO], so k′ = k[O_3]

$k' = 1.8\ s^{-1} = k(1.0 \times 10^{14}$ molecules/cm^3)

$k = 1.8 \times 10^{-14}\ cm^3$/molecules•s

We can check this from the ozone data:

Rate = k″ [O_3] = k[NO][O_3], so k″ = k[NO]

$k'' = 3.6\ s^{-1} = k(2.0 \times 10^{14}$ molecules/cm^3)

$k = 1.8 \times 10^{-14}\ cm^3$/molecules• s

Both values agree.

CHAPTER THIRTEEN: CHEMICAL EQUILIBRIUM

QUESTIONS

1. a. The rates of the forward and reverse reactions are equal.

   b. There is no net change in the composition.

3. The equilibrium constant is a number that tells us the relative concentrations (pressures) of reactants and products at equilibrium.

   An equilibrium position is a set of concentrations that satisfy the equilibrium constant expression. More than one equilibrium position can satisfy the same equilibrium constant expression.

5. When we change the pressure by adding an unreactive gas, we do not change the partial pressure of any of the substances in equilibrium with each other. In this case the equilibrium will not shift. If we change the pressure by changing the volume, we will change the partial pressures of all the substances in equilibrium by the same factor. If there are unequal numbers of gaseous particles on the two sides of the equation, the equilibrium will shift.

EXERCISES

Characteristics of Chemical Equilibrium

7. No, equilibrium is a dynamic process. Both reactions:

$$H_2O + CO \longrightarrow H_2 + CO_2$$

and

$$H_2 + CO_2 \longrightarrow H_2O + CO$$

are occurring. Thus, ^{14}C atoms will be distributed between CO and CO_2.

The Equilibrium Constant

9. a. $K = \dfrac{[NO_2]\,[O_2]}{[NO]\,[O_3]}$ b. $K = \dfrac{[O_2]\,[O]}{[O_3]}$

   c. $K = \dfrac{[ClO]\,[O_2]}{[Cl]\,[O_3]}$ d. $K = \dfrac{[O_2]^3}{[O_3]^2}$

11. $K_p = K(RT)^{\Delta n}$, $\Delta n = 2$

$$K_p = \frac{2.6 \times 10^{-5}\ \text{mol}^2}{\text{L}^2}\left(\frac{0.08206\ \text{L atm}}{\text{mol K}} \times 400.\ \text{K}\right)^2$$

$$K_p = 2.8 \times 10^{-2}\ \text{atm}^2$$

13. a. $K = \dfrac{1}{[O_2]^5}$ b. $K = [N_2O]\,[H_2O]^2$

   c. $K = \dfrac{1}{[CO_2]}$ d. $K = \dfrac{[SO_2]^8}{[O_2]^8}$

15. $K = 278 = \frac{[SO_3]^2}{[SO_2]^2[O_2]}$ for $2\ SO_2 + O_2 \rightleftharpoons 2\ SO_3$

a. $SO_2(g) + 1/2\ O_2(g) \rightleftharpoons SO_3(g)$

$$K_{eq} = \frac{[SO_3]}{[SO_2][O_2]^{1/2}} = K^{1/2} = 16.7$$

b. $2\ SO_3(g) \rightleftharpoons 2\ SO_2(g) + O_2(g)$

$$K_{eq} = \frac{[SO_2]^2[O_2]}{[SO_3]^2} = \frac{1}{K} = 3.60 \times 10^{-3}$$

c. $SO_3(g) \rightleftharpoons SO_2(g) + 1/2\ O_2(g)$

$$K_{eq} = \frac{[SO_2][O_2]^{1/2}}{[SO_3]} = \left(\frac{1}{K}\right)^{1/2} = 6.00 \times 10^{-2}$$

d. $4\ SO_2(g) + 2\ O_2(g) \rightleftharpoons 4\ SO_3(g)$

$$K_{eq} = \frac{[SO_3]^4}{[SO_2]^4[O_2]^2} = K^2 = 7.73 \times 10^4$$

17. $CO_2(g) + H_2(g) \rightleftharpoons CO(g) + H_2O(g)$

$$K = \frac{[CO]\,[H_2O]}{[CO_2]\,[H_2]} = \frac{(5.9\text{ mol/L})(12.0\text{ mol/L})}{(18.0\text{ mol/L})(20.0\text{ mol/L})} = 0.20$$

19. $C(s) + CO_2(g) \rightleftharpoons 2\ CO(g)$

$$K_p = \frac{P^2_{CO}}{P_{CO_2}} = \frac{(2.6\text{ atm})^2}{2.9\text{ atm}} = 2.3\text{ atm}$$

21.

	$PCl_5(g)$	$\rightleftharpoons$	$PCl_3(g)$ +	$Cl_2(g)$
Initial	0.50 atm		0	0
	x atm PCl_5 reacts to reach equilibrium			
Change	$-x$	$\rightarrow$	$+x$	$+x$
Equil.	$0.50 - x$		x	x

$$P_{total} = P_{PCl_5} + P_{PCl_3} + P_{Cl_2} = 0.50 - x + x + x = 0.50 + x$$

$0.84\text{ atm} = 0.50 + x,\ x = 0.34\text{ atm}$

$P_{PCl_5} = 0.16\text{ atm},\quad P_{PCl_3} = P_{Cl_2} = 0.34\text{ atm}$

$$K_p = \frac{P_{PCl_3} \times P_{Cl_2}}{P_{PCl_5}} = \frac{(0.34)(0.34)}{(0.16)} = 0.72\text{ atm}$$

$$K = \frac{K_p}{(RT)^{\Delta n}} = \frac{0.72}{(0.08206)(523)} = 0.017\text{ mol/L}$$

Equilibrium Calculations

23. $H_2O + Cl_2O(g) \rightleftharpoons 2\ HOCl(g)$ $K_p = K = 0.0900 = \dfrac{[HOCl]^2}{[H_2O][Cl_2O]}$

a. $Q = \dfrac{(21.0)^2}{(200.)(49.8)} = 4.43 \times 10^{-2} < K$

Reaction will proceed to the right to reach equilibrium.

$H_2O(g) + Cl_2O(g) \rightarrow 2\ HOCl(g)$

b. $Q = \dfrac{(20.0)^2}{(296)(15.0)} = 0.0901 \approx K$ (at equilibrium)

c. $Q = \dfrac{\left(\dfrac{0.084\ mol}{2.0\ L}\right)^2}{\left(\dfrac{0.98\ mol}{2.0\ L}\right)\left(\dfrac{0.080\ mol}{2.0\ L}\right)} = \dfrac{(0.084)^2}{(0.98)(0.080)} = 0.090 = K$ (at equilibrium)

d. $Q = \dfrac{\left(\dfrac{0.25\ mol}{3.0\ L}\right)^2}{\left(\dfrac{0.56\ mol}{3.0\ L}\right)\left(\dfrac{0.0010\ mol}{3.0\ L}\right)} = \dfrac{(0.25)^2}{(0.56)(0.0010)} = 110 > K$

Reaction will proceed to the left to reach equilibrium.

$2\ HOCl \rightarrow H_2O + Cl_2O$ or $H_2O + Cl_2O \leftarrow 2\ HOCl$

25. $CaCO_3(s) \rightleftharpoons CaO(s) + CO_2(g)$

$K_p = P_{CO_2} = 1.04$

We only need to calculate the partial pressure of CO_2. At this temperature all CO_2 will be in the gas phase.

a. $PV = nRT$

$$P = \frac{nRT}{V} = \frac{\dfrac{58.4\ g}{44.01\ g/mol}\left(\dfrac{0.08206\ L\ atm}{mol\ K}\right)1173\ K}{(50.0\ L)} = 2.55 > K_p$$

Reaction will proceed to the left, the mass of CaO will decrease.

b. $P = \dfrac{(23.76)\ (0.08206)(1173)}{(44.01)(50.0)}\ atm = 1.04\ atm = K_p$

At equilibrium; mass of CaO will not change.

c. Mass of CO_2 same as in part b.

$P = 1.04\ atm = K_p$

At equilibrium; mass of CaO will not change.

d. $P = \frac{(4.82)\ (0.08206)(1173)}{(44.01)(50.0)}$ atm = 0.211 < K_p

Reaction will proceed to the right; the mass of CaO will increase.

27.

	$N_2(g)$ +	$O_2(g)$	$\rightleftharpoons$	2 NO(g)	$K = K_p = 0.050$
Initial	0.80 atm	0.20 atm		0	
	x atm N_2 reacts to reach equilibrium				
Change	$-x$	$-x$	$\longrightarrow$	$+2x$	
Equil.	$0.80 - x$	$0.20 - x$		$2x$	

$$K = 0.050 = \frac{P^2_{NO}}{P_{N_2}P_{O_2}} = \frac{(2x)^2}{(0.80 - x)\ (0.20 - x)}$$

$(0.050)\ (0.80 - x)\ (0.20 - x) = 4x^2$

$8.0 \times 10^{-3} - 0.050x + 0.050\,x^2 = 4x^2$

$3.95x^2 + 0.050x - 8.0 \times 10^{-3} = 0$

For a quadratic equation in the form: $a\,x^2 + b\,x + c = 0$

the solutions are $x = \frac{-b \pm \sqrt{b^2 - 4ac}}{2a}$

For this equation:

$$x = \frac{-0.050 \pm \sqrt{2.5 \times 10^{-3} + 0.1264}}{7.9}$$

$$x = \frac{-0.050 \pm \sqrt{0.1289}}{7.9}$$

Note: Rounding off intermediate answers in solving the quadratic formula and in the approximations method described below, leads to excessive round off error. In these problems we will discontinue the usual practice in this Solutions Guide and carry extra significant figures and round at the end.

$x = \frac{-0.050 \pm 0.359}{7.9}$, $x = 0.039$ or $x = -0.051$

A negative answer makes no physical sense, so $x = 0.039$

$P_{NO} = 2x = 0.078$ atm

There is an easier way to solve this problem in terms of algebra. There are a lot of manipulations necessary to get a solution using the quadratic formula. If we make an early mistake, it takes a lot of work to get the final answer and find out it is wrong. As a case in point it took me (KCB) about 20 minutes to find the mistake I made in initially solving this problem. In an early step I copied an exponent wrong. We can use our chemical sense to simplify the algebra. For this problem, we are trying to solve the equation:

$$\frac{4x^2}{(0.80 - x)\,(0.20 - x)} = 0.050$$

However, the value of the equilibrium constant, 0.050, is not very large; not much N_2 and O_2 will react to reach equilibrium. In this case, if we can say:

$$0.80 - x \approx 0.80$$

$$0.20 - x \approx 0.20$$

then the equation simplifies to:

$$\frac{4x^2}{(0.80)(0.20)} = 0.050 \text{ and } x = 0.045$$

When we check how good our assumptions are, we get:

$0.80 - x = 0.80 - 0.045 = 0.755$

$0.20 - x = 0.20 - 0.045 = 0.155$

These assumptions aren't very good (22.5% error in one of the assumptions). All is not lost! We can use this result as a further approximation. That is:

$$\frac{4x^2}{(0.80 - x)(0.20 - x)} \approx \frac{4x^2}{(0.80 - 0.045)(0.20 - 0.045)} = 0.050$$

Solving for x we get: $x = 0.038$. We repeat the process using 0.038 as a further guess for x.

$$\frac{4x^2}{(0.80 - 0.038)\,(0.20 - 0.038)} = 0.050,\ x = 0.039$$ Repeat using 0.039.

$$\frac{4x^2}{(0.80 - 0.039)\,(0.20 - 0.039)} = 0.050,\ x = 0.039$$

We just got the same answer in consecutive iterations; we have converged on the true answer. It is the same answer to two significant figures that we solved for using the quadratic formula. This method, called the method of successive approximations (See Appendix 1.4 in the text), at first appears to be more laborious. However, any one iteration is less time consuming than using the quadratic formula; it is easier to make an arithmetic mistake using the quadratic formula; and we can discover mistakes in a shorter period of time using successive approximations. Even more compelling, we're using our chemical sense to make approximations and, after all, this is a chemistry text. Try the method; once you get the hang of it you'll prefer successive approximation to the quadratic formula.

29.

	$2\,SO_2(g)$	+ $O_2(g)$	$\rightleftharpoons$	$2\,SO_3(g)$	$K_p = 0.25$
Initial	0.50 atm	0.50 atm		0	
	$2x$ atm SO_2 reacts to reach equilibrium				
Change	$-2x$	$-x$	$\rightarrow$	$+2x$	
Equil.	$0.50 - 2x$	$0.50 - x$		$2x$	

$$\frac{P^2_{SO_3}}{P^2_{SO_2}P_{O_2}} = \frac{(2x)^2}{(0.50 - 2x)^2(0.50 - x)} \approx \frac{4x^2}{(0.50)^3} = 0.25$$

$x = 0.088$ Assumption is not good (x is 20% of 0.5).

Using method of successive approximations (Appendix 1.4):

$$\frac{4x^2}{[0.50 - 2(0.088)]^2[0.50 - (0.088)]} = \frac{4x^2}{(0.324)^2(0.412)} = 0.25; \quad x = 0.052$$

$$\frac{4x^2}{[0.50 - 2(0.052)]^2[0.50 - (0.052)]} = \frac{4x^2}{(0.396)^2(0.448)} = 0.25; \quad x = 0.066$$

$$\frac{4x^2}{(0.368)^2(0.434)} = 0.25; \quad x = 0.061$$

$$\frac{4x^2}{(0.378)^2(0.439)} = 0.25; \quad x = 0.063$$

The next trials converge at 0.062.

$P_{SO_2} = 0.50 - 2x = 0.376 = 0.38$ atm

$P_{SO_3} = 2x = 0.124 = 0.12$ atm, $P_{O_2} = 0.50 - x = 0.438 = 0.44$ atm

31. a.

	$N_2O_4(g)$	$\rightleftharpoons$	$2\ NO_2(g)$	$K_p = 0.25$
Initial	0		0.050 atm	
	$2x$ atm NO_2 reacts to reach equilibrium			
Change	$+x$	$\leftarrow$	$-2x$	
Equil.	x		$0.050 - 2x$	

$$K_p = 0.25 = \frac{(0.050 - 2x)^2}{x}, \quad 0.25x = 2.5 \times 10^{-3} - 0.20x + 4x^2$$

$$4x^2 - 0.45x + 2.5 \times 10^{-3} = 0$$

$$x = \frac{+0.45 \pm [(-0.45)^2 - 4(4)(2.5 \times 10^{-3})]^{1/2}}{2(4)}$$

$$x = \frac{+0.45 - 0.403}{8} = 5.9 \times 10^{-3} \text{ (other root, 0.11 makes no sense)}$$

$P_{N_2O_4} = 5.9 \times 10^{-3}$ atm

$P_{NO_2} = 0.050 - 2(0.0059) = 3.8 \times 10^{-2}$ atm

b.

	$N_2O_4(g)$	$\rightleftharpoons$	$2\ NO_2(g)$	$K_p = 0.25$
Initial	0.040 atm		0	
	x atm N_2O_4 reacts to reach equilibrium			
Change	$-x$	$\rightarrow$	$+2x$	
Equil.	$0.040 - x$		$2x$	

$$\frac{(2x)^2}{0.040 - x} = 0.25$$

$$4x^2 = 0.010 - 0.25x$$

$$4x^2 + 0.25x - 0.010 = 0$$

$$x = \frac{-0.25 \pm [(-0.25)^2 - 4(4)(-0.010)]^{1/2}}{8}$$

Only way to get positive answer is:

$$x = \frac{-0.25 + 0.47}{8} = 0.028$$

$P_{NO_2} = 2x = 0.056$ atm

$P_{N_2O_4} = 0.040 - x = 0.012$ atm

c. $N_2O_4 \rightleftharpoons 2\ NO_2$ $K_p = 0.25$

$Q = 1.0$, so some NO_2 must go to N_2O_4 ($Q > K_p$).

	N_2O_4	$\rightleftharpoons$	$2\ NO_2$
Initial	1.0 atm		1.0 atm
	$2x$ atm NO_2 reacts to reach equilibrium		
Change	$+x$	$\leftarrow$	$-2x$
Equil.	$1.0 + x$		$1.0 - 2x$

$$\frac{(1.0 - 2x)^2}{1.0 + x} = 0.25$$

$$1.0 - 4.0x + 4x^2 = 0.25 + 0.25x$$

$$4x^2 - 4.25x + 0.75 = 0$$

$$x = \frac{4.25 \pm [(-4.25)^2 - 4(4)(0.75)]^{1/2}}{8}$$

$x = \dfrac{4.25 \pm 2.46}{8}$ $x = 0.22$ or 0.84

$x = 0.84$, $1 - 2x < 0$ Not possible

$x = 0.22$ is the chemically correct solution.

$P_{N_2O_4} = 1.0 + x = 1.2$ atm

$P_{NO_2} = 1.0 - 2x = 0.56$ atm ≈ 0.6 atm

LeChatelier's Principle

33. a. right b. right c. no effect

d. Reaction is exothermic: $\Delta H° = (-393.5) - (-110.5 - 242) = -41$ kJ

$CO(g) + H_2O(g) \longrightarrow H_2(g) + CO_2(g) + \text{Heat}$

Increase T; add heat; the equilibrium shifts to the left.

35. a. right b. left c. right d. no effect e. no effect

f. An increase in temperature will shift the equilibrium to the right, since the reaction is endothermic, $\Delta H° = 2(25.9) = 51.8$ kJ.

37. We can do two things in choosing a solvent. First we must avoid water. Any extra water we add from the solvent tends to push the equilibrium to the left. This eliminates water and 95% ethanol as solvent choices. Of the remaining two solvents, acetonitrile will not take part in the reaction, whereas ethanol is a reactant. If we use ethanol as the solvent it will drive the equilibrium to the right, thereby reducing the concentrations of the objectionable butyric acid to a minimum. Thus, the best solvent is 100% ethanol.

Kinetics and Equilibrium

39. The catalyst increased the rates of both reactions. Equilibrium was reached more rapidly with the catalyst. The uncatalyzed reaction is too slow to be practical.

A catalyst increases the rate of a reaction by providing an alternate pathway with a lower activation energy. Thus, in the new pathway, the activation energy will be lowered for both the forward and reverse reactions.

41. $$K = \frac{k_f}{k_r} = \frac{[Br^-][SO_4^{2-}]^3}{[BrO_3^-][SO_3^{2-}]^3}$$

Rate forward reaction = Rate reverse reaction

$$k_f [BrO_3^-] [SO_3^{2-}] [H^+] = k_r [Br^-]^a [SO_4^{2-}]^b [H^+]^c [SO_3^{2-}]^d [BrO_3^-]^e$$

$$\frac{k_f}{k_r} = \frac{[Br^-]^a [SO_4^{2-}]^b [H^+]^c [SO_3^{2-}]^d [BrO_3^-]^e}{[H^+] [BrO_3^-] [SO_3^{2-}]} = K = \frac{[Br^-] [SO_4^{2-}]^3}{[BrO_3^-] [SO_3^{2-}]^3}$$

For the two concentration expressions to be the same: a = 1, b = 3, c = 1, d = -2, e = 0

$$\text{Rate}_r = k_r \frac{[Br^-][SO_4^{2-}]^3 [H^+]}{[SO_3^{2-}]^2}$$

ADDITIONAL EXERCISES

43. $2\ O_3(g) \rightleftharpoons 3\ O_2(g) \quad K_p = \dfrac{P_{O_2}^3}{P_{O_3}^2}$

a. $O_3(g) \rightleftharpoons 3/2\ O_2(g)$

$$K_p' = \frac{P_{O_2}^{3/2}}{P_{O_3}} = K_p^{1/2}$$

b. $3\ O_2(g) \rightleftharpoons 2\ O_3(g)$

$$K_p'' = \frac{P_{O_3}^2}{P_{O_2}^3} = \frac{1}{K_p}$$

45. a.

$Na_2O(s) \rightleftharpoons 2\ Na(l) + 1/2\ O_2(g)$	K_1
$2\ Na(l) + O_2(g) \rightleftharpoons Na_2O_2(s)$	$1/K_3$
$Na_2O(s) + 1/2\ O_2(g) \rightleftharpoons Na_2O_2(s)$	$K_{eq} = (K_1)\ (1/K_3)$

$$K_{eq} = \frac{2 \times 10^{-25}}{5 \times 10^{-29}} = 4 \times 10^3$$

b.

$NaO(g) \rightleftharpoons Na(l) + 1/2\ O_2(g)$	K_2
$Na_2O(s) \rightleftharpoons 2\ Na(l) + 1/2\ O_2(g)$	K_1
$2\ Na(l) + O_2(g) \rightleftharpoons Na_2O_2(s)$	$1/K_3$
$NaO(g) + Na_2O(s) \rightleftharpoons Na_2O_2(s) + Na(l)$	

$$K_{eq} = \frac{K_1 K_2}{K_3} = 8 \times 10^{-2}$$

c.

$2\ NaO(g) \rightleftharpoons 2\ Na(l) + O_2(g)$	K_2^2
$2\ Na(l) + O_2(g) \rightleftharpoons Na_2O_2(s)$	$1/K_3$
$2\ NaO(g) \rightleftharpoons Na_2O_2(s)$	

$$K_{eq} = \frac{K_2^2}{K_3} = 8 \times 10^{18}$$

47.

	$Fe^{3+}(aq)$ +	$SCN^-(aq)$	$\rightleftharpoons$	$FeSCN^{2+}(aq)$	$K = 1.1 \times 10^3$
Before	0.10	2.0		0	
	0.10 mol/L Fe^{3+} reacts completely (large K, products dominate)				
Change	-0.10	-0.10	$\rightarrow$	+0.10	Reaction goes to completion.
After	0	1.9		0.10	New Initial
	x mol/L $FeSCN^{2+}$ reacts				
Change	$+x$	$+x$	$\leftarrow$	$-x$	Go back to equilibrium
Equil.	x	$1.9 + x$		$0.10 - x$	

$$K = 1.1 \times 10^3 = \frac{[FeSCN^{2+}]}{[Fe^{3+}][SCN^-]} = \frac{0.10 - x}{(x)(1.9 + x)} \approx \frac{0.10}{1.9x}$$

$x = 4.8 \times 10^{-5}$; $0.10 - x = 0.10 - 0.000048 = 0.099952 = 0.10$

Good Assumption!

$x = [Fe^{3+}] = 4.8 \times 10^{-5}\ M$ $\quad\quad [FeSCN^{2+}] = 0.10\ M$

$[SCN^-] = 1.9\ M$

49. When SO_2 is added, the equilibrium will shift to the left, producing more SO_2Cl_2 and some energy. The energy increases the temperature of the reaction mixture.

51. $H_2O(g) + Cl_2O(g) \rightleftharpoons 2\ HOCl(g)$ $\quad\quad K = 0.090$

a. The initial concentrations of H_2O and Cl_2O are:

$$\frac{1.0\ g\ H_2O}{1.0\ L} \times \frac{1\ mol}{18.02\ g} = 5.5 \times 10^{-2}\ mol/L$$

$$\frac{2.0\ g\ Cl_2O}{1.0\ L} \times \frac{1\ mol}{86.90\ g} = 2.3 \times 10^{-2}\ mol/L$$

	H_2O (g)	+	$Cl_2O(g)$	$\rightleftharpoons$	2 HOCl(g)
Initial	5.5×10^{-2}		2.3×10^{-2}		0
	x mol/L H_2O reacts				
Change	$-x$		$-x$	$\rightarrow$	$+2x$
Equil.	$5.5 \times 10^{-2} - x$		$2.3 \times 10^{-2} - x$		$2x$

$$K = 0.090 = \frac{(2x)^2}{(5.5 \times 10^{-2} - x)(2.3 \times 10^{-2} - x)}$$

Assume x is small, so:

$$\frac{4x^2}{(0.055)(0.023)} = 0.090,\ x = 0.0053$$

Assumption is not good. Refining our guess:

$$\frac{4x^2}{(0.055 - 0.0053))(0.023 - 0.0053)} = 0.090,\ x = 0.0041$$

$$\frac{4x^2}{(0.055 - 0.0041))(0.023 - 0.0041)} = 0.090,\ x = 0.0047$$

$$\frac{4x^2}{(0.055 - 0.0047))(0.023 - 0.0047)} = 0.090,\ x = 0.0046$$

$$\frac{4x^2}{(0.055 - 0.0046))(0.023 - 0.0046)} = 0.090,\ x = 0.0046$$

We have two identical successive answers, so we have converged on the answer.

$x = 4.6 \times 10^{-3}$

$[HOCl] = 2x = 9.2 \times 10^{-3}$ mol/L

$[H_2O] = 0.055 - x = 0.050$ mol/L

$[Cl_2O] = 0.023 - x = 0.018$ mol/L

b.

	$H_2O(g)$ +	$Cl_2O(g)$	$\rightleftharpoons$	$2\ HOCl(g)$
Initial	0	0		$\frac{1.0 \text{ mol}}{2.0 \text{ L}} = 0.50\ M$
	$2x$ mol/L HOCl reacts			
Change	$+x$	$+x$	$\leftarrow$	$-2x$
Equil.	x	x		$0.50 - 2x$

$$K = 0.090 = \frac{[HOCl]^2}{[H_2O]\,[Cl_2O]} = \frac{(0.50 - 2x)^2}{x^2}$$

We can solve this exactly very quickly. The expression is a perfect square, so we can take the square root of each side:

$$0.30 = \frac{0.50 - 2x}{x}$$

$0.30x = 0.50 - 2x$

$2.30x = 0.50$

$x = 0.217$ (carry extra significant figure)

$x = [H_2O] = [Cl_2O] = 0.217 = 0.22\ M$

$[HOCl] = 0.50 - 2x = 0.50 - 0.434 = 0.07\ M$

CHALLENGE PROBLEMS

53. a.

	$2\ NO(g)$	+	$Br_2(g)$	$\rightleftharpoons$	$2\ NOBr(g)$
Initial	98.4 torr		41.3 torr		0
Change	$-2x$		$-x$	$\rightarrow$	$2x$
Equilibrium	$98.4 - 2x$		$41.3 - x$		$2x$

$P_{total} = (98.4 - 2x) + (41.3 - x) + 2x = 139.7 - x$

$P_{total} = 110.5 = 139.7 - x$

$x = 29.2$ torr

$P_{NO} = 98.4 - 2(29.2) = 40.0$ torr $= 0.0526$ atm

$P_{Br_2} = 41.3 - 29.2 = 12.1$ torr $= 0.0159$ atm

$P_{NOBr} = 2(29.2) = 58.4 \text{ torr} = 0.0768 \text{ atm}$

$$K_p = \frac{P^2_{NOBr}}{P^2_{NO} P_{Br_2}} = \frac{(0.0768 \text{ atm})^2}{(0.0526 \text{ atm})^2 \ (0.0159 \text{ atm})} = 134 \text{ atm}^{-1}$$

b.

	$2\,NO(g)$	+	$Br_2(g)$	$\rightleftharpoons$	$2\,NOBr(g)$
Initial	0.30		0.30		0
Change	$-2x$		$-x$	$\rightarrow$	$+2x$
Equilibrium	$0.30 - 2x$		$0.30 - x$		$2x$

This would yield a cubic equation. There is no way we can simplify the task. K_p is pretty large, so we will solve this problem in two steps; assume reaction goes to completion then solve the back equilibrium problem.

	$2\,NO$	+	Br_2	$\rightleftharpoons$	$2\,NOBr$
Initial	0.30		0.30		0
Change (React Completely)	-0.30		-0.15	$\rightarrow$	+0.30
New Initial	0		0.15		0.30
Change	$+2y$		$+y$	$\leftarrow$	$-2y$
Equilibrium	$2y$		$0.15 + y$		$0.30 - 2y$

$$\frac{(0.30 - 2y)^2}{(2y)^2 \ (0.15 + y)} = 134$$

$$\frac{(0.30 - 2y)^2}{(0.15 + y)} = 134 \times 4y^2 = 536y^2$$

If y is small: $\frac{(0.30)^2}{0.15} = 536y^2$ and $y = 0.034$

But $0.15 + y = 0.184$; assumption is not good (23% error).

Use 0.034 as approximation for y.

$$\frac{(0.30 - 0.068)^2}{0.15 + 0.034} = 536y^2 ; \ y = 0.023$$

$$\frac{(0.30 - 0.046)^2}{0.173} = 536y^2; \ y = 0.026$$

$$\frac{(0.30 - 0.052)^2}{0.176} = 536y^2; \ y = 0.026$$

So: $P_{NO} = 2y = 0.052$ atm

$P_{Br_2} = 0.15 + y = 0.176 \text{ atm} \approx 0.18 \text{ atm}$

$P_{NOBr} = 0.30 - 2y = 0.25$ atm

55. a. $N_2(g) + O_2(g) \rightleftharpoons 2\ NO(g)$

$$K_p = 1 \times 10^{-31} = \frac{P_{NO}^2}{P_{N_2}P_{O_2}} = \frac{P_{NO}^2}{(0.8)(0.2)}$$

$P_{NO} = 1 \times 10^{-16}$ atm

PV = nRT

$$n = \frac{PV}{RT} = \frac{(1 \times 10^{-16}\ \text{atm})\ (1.0 \times 10^{-3}\ \text{L})}{\left(\frac{0.08206\ \text{L atm}}{\text{mol K}}\right)(298\ \text{K})} = 4 \times 10^{-21}\ \text{mol}$$

$$\frac{4 \times 10^{-21}\ \text{mol}}{\text{cm}^3} \times \frac{6.02 \times 10^{23}\ \text{molecules}}{\text{mol}} = \frac{2 \times 10^3\ \text{molecules}}{\text{cm}^3}$$

b. There is more NO in the atmosphere than we would expect from the value of K. The answer must lie in the rates of the reaction. At 25°C the rates of both reactions:

$$N_2 + O_2 \rightarrow 2\ NO \text{ and } 2\ NO \rightarrow N_2 + O_2$$

are essentially zero. Very strong bonds must be broken; the activation energy is very high. Nitric oxide is produced in high energy or high temperature environments. In nature some NO is produced by lightning. The primary manmade source is from automobiles. The production of NO is endothermic (ΔH_f° = + 90 kJ/mol). At high temperature, K will increase and the rates of the reaction increase; so NO is produced. Once the NO gets into a more normal environment in the atmosphere, it doesn't go back to N_2 and O_2 because of the slow rate.

57.

	$CCl_4(g) \rightleftharpoons$	$C(s)$ +	$2\ Cl_2(g)$	$K_p = 0.76$ atm
Initial	P_O		0	
Change	-p	$\rightarrow$	2p	
Equil.	P_O -p		2p	

$P_{total} = P_O - p + 2p = P_O + p = 1.2$ atm

$$\frac{(2p)^2}{P_O - p} = 0.76, \quad 4p^2 = 0.76\ P_O - 0.76\ p, \quad P_O = \frac{4p^2 + 0.76\ p}{0.76}$$

$$\frac{4p^2}{0.76} + p + p = 1.2\ \text{atm}, \quad 5.3p^2 + 2p - 1.2 = 0$$

$$p = \frac{-2 \pm (4 + 25)^{1/2}}{2(5.3)} = 0.32\ \text{atm and } P_O + 0.32 = 1.2. \text{ Thus, } P_O = 0.9\ \text{atm}$$

CHAPTER FOURTEEN: ACIDS AND BASES

QUESTIONS

1. a. A strong acid is 100% dissociated in water.
   b. A strong base is 100% dissociated in water.
   c. A weak acid is much less than 100% dissociated in water.
   d. A weak base is one that only a small percentage of the molecules react with water to form OH^-.

3. $H_2O \rightleftharpoons H^+ + OH^-$ $\qquad K_w = [H^+][OH^-] = 1.0 \times 10^{-14}$

   Neutral solution $[H^+] = [OH^-]$; $[H^+] = 1.0 \times 10^{-7}$ *M*; pH = 7.00

5. a. Arrhenius acid: generate H^+ in water
   b. Brönsted-Lowry acid: proton donor
   c. Lewis acid: electron pair acceptor

7. Nitrogen

9. a. weaker bond = weaker base (stronger bond = stronger base)
   b. greater electronegativity = weaker base
   c. more oxygen atoms = weaker base

EXERCISES

Nature of Acids and Bases

11. In deciding whether a substance is an acid, base, strong, or weak, we should keep in mind three things:

    1. There are only a few important strong acids and bases.

    2. All other acids and bases are weak.

    3. Structural features to look for to identify a substance as an acid or a base: $—CO_2H$ groups (acid) or the presence of N (base) in organic compounds.

    a. weak acid b. weak acid c. weak base

    d. strong base e. weak base f. weak acid

13. HNO_3: strong acid $\qquad$ HOCl: $K_a = 3.5 \times 10^{-8}$ $\qquad$ NH_4^+: $K_a = 5.6 \times 10^{-10}$

    H_3O^+ is the strongest acid that can exist in water.

$$H_3O^+ > HNO_3 > HOCl > NH_4^+$$

All are acidic in water, so all of these acids are stronger acids than H_2O.

$$H_3O^+ > HNO_3 > HOCl > NH_4^+ > H_2O$$

15. a. HI is a strong acid and $K_a = K_w = 10^{-14}$ for H_2O.

HI is a stronger acid than H_2O.

b. H_2O, $K_a = 10^{-14}$; $HClO_2$, $K_a = 1.2 \times 10^{-2}$

$HClO_2$ is a stronger acid than H_2O.

c. HF, $K_a = 7.2 \times 10^{-4}$; HCN, $K_a = 6.2 \times 10^{-10}$

HF is a stronger acid than HCN.

17.

	Acid	Base	Conjugate Base of Acid	Conjugate Acid of Base
a.	H_2O	H_2O	OH^-	H_3O^+
b.	CH_3COCH_3	CH_3O^-	$CH_3COCH_2^-$	CH_3OH
c.	H_2S	NH_3	HS^-	NH_4^+
d.	H_2SO_4	H_2O	HSO_4^-	H_3O^+
e.	H_3O^+	OH^-	H_2O	H_2O
f.	H_2O	$H_2PO_4^-$	OH^-	H_3PO_4
g.	HCN	H_2O	CN^-	H_3O^+

19. a. $H_3PO_4 + H_2O \rightleftharpoons H_3O^+ + H_2PO_4^-$ $\quad K_{a_1} = \frac{[H_3O^+][H_2PO_4^-]}{[H_3PO_4]}$

or $H_3PO_4 \rightleftharpoons H^+ + H_2PO_4^-$ $\quad K_{a_1} = \frac{[H^+][H_2PO_4^-]}{[H_3PO_4]}$

Only the first reaction will be written for the rest, but both are equivalent.

b. $H_2PO_4^- + H_2O \rightleftharpoons H_3O^+ + HPO_4^{2-}$ $\quad K_{a_2} = \frac{[H_3O^+][HPO_4^{2-}]}{[H_2PO_4^-]}$

c. $HPO_4^{2-} + H_2O \rightleftharpoons H_3O^+ + PO_4^{3-}$ $\quad K_{a_3} = \frac{[H_3O^+][PO_4^{3-}]}{[HPO_4^{2-}]}$

d. $HNO_2 + H_2O \rightleftharpoons H_3O^+ + NO_2^-$ $\quad K_a = \frac{[H_3O^+][NO_2^-]}{[HNO_2]}$

e. $Ti(H_2O)_6^{4+} + H_2O \rightleftharpoons Ti(H_2O)_5(OH)^{3+} + H_3O^+$ $\quad K_a = \frac{[H_3O^+][Ti(H_2O)_5(OH)^{3+}]}{[Ti(H_2O)_6^{4+}]}$

21. a. $PO_4^{3-} + H_2O \rightleftharpoons HPO_4^{2-} + OH^-$ $K_b = \dfrac{[OH^-][HPO_4^{2-}]}{[PO_4^{3-}]}$

b. $HPO_4^{2-} + H_2O \rightleftharpoons H_2PO_4^- + OH^-$ $K_b = \dfrac{[OH^-][H_2PO_4^-]}{[HPO_4^{2-}]}$

c. $H_2PO_4^- + H_2O \rightleftharpoons H_3PO_4 + OH^-$ $K_b = \dfrac{[OH^-][H_3PO_4]}{[H_2PO_4^-]}$

d. $NH_3 + H_2O \rightleftharpoons NH_4^+ + OH^-$ $K_b = \dfrac{[NH_4^+][OH^-]}{[NH_3]}$

e. $CN^- + H_2O \rightleftharpoons OH^- + HCN$ $K_b = \dfrac{[OH^-][HCN]}{[CN^-]}$

23. $H^- + H_2O \rightarrow H_2 + OH^-$; $OCH_3^- + H_2O \rightarrow CH_3OH + OH^-$

Autoionization of Water and the pH Scale

25. a. $pH = -\log(1.4 \times 10^{-3}) = 2.85$ b. $pH = -\log(2.5 \times 10^{-10}) = 9.60$

c. $pH = -\log(6.1) = -0.79$ Note: If $[H^+] > 1\ M$, pH is negative.

d. $[OH^-] = 3.5 \times 10^{-2}\ M$; $pOH = 1.46$

$pH + pOH = 14$; $pH = 12.54$

$$\text{or } [H^+] = \frac{1.0 \times 10^{-14}}{3.5 \times 10^{-2}} = 2.9 \times 10^{-13}\ M$$

$pH = -\log(2.9 \times 10^{-13})$; $= 12.54$

e. $[OH^-] = 8 \times 10^{-11}\ M$; $pOH = 10.1$; $pH = 3.9$

f. $[OH^-] = 5.0\ M$; $pOH = -0.70$; $pH = 14.70$

Note: if $[OH^-] > 1.0\ M$, then pOH is negative and $pH > 14$.

g. $pOH = 10.5$, $pH = 14.0 - 10.5 = 3.5$

h. $pOH = 2.3$, $pH = 14.0 - 2.3 = 11.7$

27. a. $pH = 7.41$; $[H^+] = 10^{-7.41} = 3.9 \times 10^{-8}\ M$

$pOH = 14 - 7.41 = 6.59$; $[OH^-] = 10^{-6.59} = 2.6 \times 10^{-7}\ M$

$$\text{or } [OH^-] = \frac{K_w}{[H^+]} = \frac{1.0 \times 10^{-14}}{3.9 \times 10^{-8}} = 2.6 \times 10^{-7}\ M$$

b. pH = 15.3; $[H^+] = 10^{-15.3} = 5 \times 10^{-16}$ *M*

pOH = 14 - pH = -1.3; $[OH^-] = 10^{1.3} = 20$ *M*

c. H = -1.0

$[H^+] = 10$ *M*

pOH = 14.0 - pH = 15.0

$[OH^-] = 1 \times 10^{-15}$ *M*

d. pH = 3.2

$[H^+] = 10^{-3.2} = 6 \times 10^{-4}$ *M*

pOH = 14.0 - pH = 10.8

$[OH^-] = 10^{-10.8} = 2 \times 10^{-11}$ *M*

e. pOH = 5.0; $[OH^-] = 1 \times 10^{-5}$ *M*

pH = 9.0; $[H^+] = 1 \times 10^{-9}$ *M*

f. pOH = 9.6

$[OH^-] = 10^{-9.6} = 2.5 \times 10^{-10}$ *M* $= 3 \times 10^{-10}$ *M* (1 significant figure)

pH = 4.4; $[H^+] = 10^{-4.4} = 4 \times 10^{-5}$ *M*

29. a. Since the value of the equilibrium constant increases as the temperature increases, the reaction is endothermic.

b. $H_2O \rightleftharpoons H^+ + OH^-$ $K_w = 5.47 \times 10^{-14} = [H^+][OH^-]$

In pure water $[H^+] = [OH^-]$:

$5.47 \times 10^{-14} = [H^+]^2$; $[H^+] = 2.34 \times 10^{-7}$ *M*

$pH = -\log [H^+] = -\log (2.34 \times 10^{-7}) = 6.631$

c. A neutral solution of water at 50°C has:

$[H^+] = [OH^-]$; $[H^+] = 2.34 \times 10^{-7}$ *M*; pH = 6.631

Obviously, the condition that $[H^+] = [OH^-]$ is the most general definition of a neutral solution.

d.

Temp (°C)	Temp(K)	1/T	K_w	ln K_w
0	273	3.66×10^{-3}	1.14×10^{-15}	-34.408
25	298	3.36×10^{-3}	1.00×10^{-14}	-32.236
35	308	3.25×10^{-3}	2.09×10^{-14}	-31.499
40	313	3.19×10^{-3}	2.92×10^{-14}	-31.165
50	323	3.10×10^{-3}	5.47×10^{-14}	-30.537

From the graph: $37°C = 310.\ K$; $1/T = 3.23 \times 10^{-3}$

$\ln K_w = -31.38$; $K_w = 2.35 \times 10^{-14}$

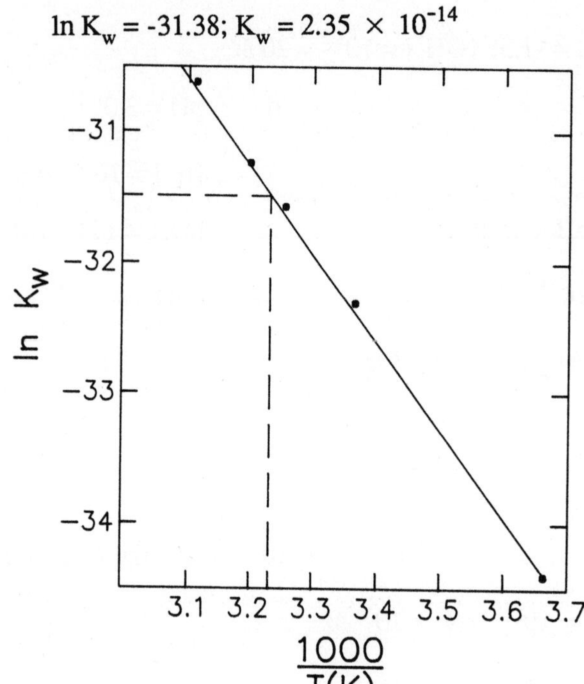

e. At 37°C, $2.35 \times 10^{-14} = [H^+][OH^-] = [H^+]^2$

$[H^+] = 1.53 \times 10^{-7}$; pH = 6.815

Solutions of Acids

31. a. $H^+(aq)$, $Cl^-(aq)$, and H_2O (HCl is a strong acid.) $[H^+] = 0.250$, pH = 0.602

b. $H^+(aq)$, $Br^-(aq)$, and H_2O (HBr is a strong acid.) pH = 0.602

c. H_2O, $H^+(aq)$, $ClO_4^-(aq)$ ($HClO_4$ is a strong acid.) pH = 0.602

d. H_2O, $H^+(aq)$, $NO_3^-(aq)$ (HNO_3 is a strong acid.) pH = 0.602

33. a. $HCl \rightarrow H^+ + Cl^-$ Strong acids dissociate completely.

$$\frac{0.1 \text{ mol HCl}}{\text{L}} \times \frac{1 \text{ mol H}^+}{\text{mol HCl}} = 0.1\ M = [H^+]$$

$$pH = -\log(0.1) = 1.0$$

b. $[H^+] = 0.1\ M$ pH = 1.0

c. $[H^+] = 0.1\ M$ pH = 1.0

35. $50.0 \text{ mL con. HCl soln} \times \frac{1.19 \text{ g}}{\text{mL}} \times \frac{38 \text{ g HCl}}{100. \text{ g con. HCl soln}} \times \frac{1 \text{ mol HCl}}{36.46 \text{ g}} = 0.62 \text{ mol HCl}$

$20.0 \text{ mL con. } HNO_3 \text{ soln} \times \frac{1.42 \text{ g}}{\text{mL}} \times \frac{70. \text{ g } HNO_3}{100. \text{ g soln}} \times \frac{1 \text{ mol } HNO_3}{63.02 \text{ g } HNO_3} = 0.32 \text{ mol } HNO_3$

$HCl + H_2O \longrightarrow H_3O^+ + Cl^-$ and $HNO_3 + H_2O \longrightarrow H_3O^+ + NO_3^-$

So we will have 0.62 + 0.32 = 0.94 mol of H_3O^+ in the final solution.

$[H^+] = \frac{0.94 \text{ mol}}{0.500 \text{ L}} = 1.9\ M$

$[OH^-] = \frac{1.0 \times 10^{-14}}{1.9} = 5.3 \times 10^{-15}\ M$; pH = -0.28

37. 0.050 mol/L CH_3CO_2H

Major species: CH_3CO_2H, H_2O

We will consider the acetic acid, $K_a = 1.8 \times 10^{-5}$, as a more important source of H^+ than water, i.e., we will ignore the H^+ contribution from water. For our purposes, this assumption will almost always hold true.

We will use the abbreviations HOAC for CH_3CO_2H and OAc^- for $CH_3CO_2^-$.

	HOAc	$\rightleftharpoons$	H^+	+	OAc^-
Initial	0.050 mol/L		$0(10^{-7})$		0
	x mol/L HOAc dissociates to reach equilibrium				
Change	$-x$	$\longrightarrow$	$+x$		$+x$
Equil	$0.050 - x$		x		x

$K_a = 1.8 \times 10^{-5} = \frac{[H^+][OAc^-]}{[HOAc]} = \frac{x^2}{0.050 - x} \approx \frac{x^2}{0.050}$ (assuming $x << 0.050$)

$x = [H^+] = 9.5 \times 10^{-4}\ M$ Assumptions good (~ 2% error).

pH = 3.02

Note: Assumption is good if % error in assumption is < 5%. Both assumptions we have made are good by the 5% rule, i.e., we can ignore H^+ contribution from water and x << 0.050. Always check your assumptions. If the assumptions fail, we must solve exactly using either the quadratic equation or the method of successive approximations (see Appendix 1.4 of text).

For 0.10 M HOAc, we follow the same procedure. The final equation is:

$1.8 \times 10^{-5} = \frac{x^2}{0.10 - x} \approx \frac{x^2}{0.10}$, $x = [H^+] = 1.34 \times 10^{-3}\ M \approx 1.3 \times 10^{-3}\ M$

Assumptions good. pH = 2.89

For 0.40 M HOAc:

$$1.8 \times 10^{-5} = \frac{x^2}{0.40 - x} \approx \frac{x^2}{0.40},\ x = [H^+] = 2.7 \times 10^{-3}\ M$$

Assumptions good. pH = 2.57

39. Major species: $HC_6H_2Cl_3O$ and H_2O. We will consider trichlorophenol as the major source of H^+, i.e., ignore H^+ contribution from H_2O.

	$HC_6H_2Cl_3O$	$\rightleftharpoons$	H^+	+	$C_6H_2Cl_3O^-$
Initial	0.05 M		$0(10^{-7})$		0
	x mol/L trichlorophenol dissociates to reach equilibrium				
Change	$-x$	$\rightarrow$	$+x$		$+x$
Equil	$0.05 - x$		x		x

$$K_a = 1 \times 10^{-6} = \frac{[H^+][C_6H_2Cl_3O^-]}{[HC_6H_2Cl_3O]} = \frac{x^2}{0.05 - x} \approx \frac{x^2}{0.05} \quad \text{(assuming } x << 0.05\text{)}$$

$x = [H^+] = 2 \times 10^{-4}\ M$ Assumptions good (0.4% error).

pH = 3.7

$[C_6H_2Cl_3O^-] = [H^+] = 2 \times 10^{-4}\ M,\ [OH^-] = 5 \times 10^{-11}\ M$

$[HC_6H_2Cl_3O] = 0.05 - x = 0.05 - 0.0002 = 0.04998 = 0.05\ M$

40. In all four parts of this problem the major species are the weak acid and water. In all we will consider the weak acid to be the major source of H^+, i.e. ignore H^+ contribution from H_2O.

a.

	$HC_2H_3O_2$ + H_2O	$\rightleftharpoons$	H_3O^+	+	$C_2H_3O_2^-$
Initial	0.20 M		$0(10^{-7})$		0
	x mol/L $HC_2H_3O_2$ dissociates to reach equilibrium				
Change	$-x$	$\rightarrow$	$+x$		$+x$
Equil	$0.20 - x$		x		x

$$K_a = 1.8 \times 10^{-5} = \frac{[H^+][C_2H_3O_2^-]}{[HC_2H_3O_2]} = \frac{x^2}{0.20 - x} \approx \frac{x^2}{0.20}$$

$x = [H^+] = 1.9 \times 10^{-3}\ M$

We have made two assumptions which we must check.

1. $0.20 - x \approx 0.20$

   $0.20 - x = 0.20 - 0.002 = 0.198 = 0.20$ Good assumption (1% error).
   If the percent error in assumption is < 5%, assumption is valid.

2. Acetic acid is the major source of H^+, i.e. we can ignore 10^{-7} *M* H^+ already present in neutral H_2O.

$[H^+]$ from $HC_2H_3O_2 = 2 \times 10^{-3} >> 10^{-7}$ This assumption is valid.

In future problems we will always begin the problem solving process by making these assumptions and we will always check them. However, we may not explicitly state that the assumptions are valid. We will always state when the assumptions are not valid and we have to use other techniques to solve the problem.

Remember, anytime we make an assumption, we must check its validity before the solution to the problem is complete.

$[H^+] = [C_2H_3O_2^-] = 1.9 \times 10^{-3}\ M,\ [OH^-] = 5.3 \times 10^{-12}\ M$

$[HC_2H_3O_2] = 0.20 - x = 0.198 \approx 0.20\ M$

$pH = -\log(1.9 \times 10^{-3}) = 2.72$

b.

	HNO_2	$\rightleftharpoons$	H^+	+	NO_2^-	$K_a = 4.0 \times 10^{-4}$
Initial	1.5 *M*		$0(10^{-7})$		0	
	x mol/L HNO_2 dissociates to reach equilibrium					
Change	$-x$	$\rightarrow$	$+x$		$+x$	
Equil	$1.5 - x$		x		x	

$$K_a = 4.0 \times 10^{-4} = \frac{[H^+][NO_2^-]}{[HNO_2]} = \frac{x^2}{1.5 - x} \approx \frac{x^2}{1.5}$$

$x = [H^+] = 2.4 \times 10^{-2}\ M$

Assumptions good: $10^{-7} << 2.4 \times 10^{-2} << 1.5$

$[H^+] = [NO_2^-] = 2.4 \times 10^{-2}\ M,\ [OH^-] = 4.2 \times 10^{-13}$

$[HNO_2] = 1.5 - x = 1.48 \approx 1.5\ M$; pH = 1.62

c.

	HF	$\rightleftharpoons$	H^+	+	F^-	$K_a = 7.2 \times 10^{-4}$
Initial	0.020 *M*		$0(10^{-7})$		0	
	x mol/L HF dissociates to reach equilibrium					
Change	$-x$	$\rightarrow$	$+x$		$+x$	
Equil	$0.020 - x$		x		x	

$$K_a = 7.2 \times 10^{-4} = \frac{[H^+][F^-]}{[HF]} = \frac{x^2}{0.020 - x} \approx \frac{x^2}{0.020}$$ (assuming $x << 0.020$)

$x = [H^+] = 3.8 \times 10^{-3} >> 10^{-7}$

But $0.020 - x = 0.020 - 0.004 = 0.016$ (x is 20% of 0.020)

The assumption $x << 0.020$ is not good (x is more than 5% of 0.020). We must continue. We can solve the equation by using the quadratic formula or by the method of successive approximations (see Appendix 1.4 of text). We let 0.016 *M* be a new approximation for [HF]. That is, try $x = 0.004$; so $0.020 - 0.004 = 0.016$, then solve for a new value of x.

$$\frac{x^2}{0.020 - x} \approx \frac{x^2}{0.016} = 7.2 \times 10^{-4},\ x = 3.4 \times 10^{-3}$$

We use this new value of x to further refine our estimate of [HF], i.e. $0.020 - x = 0.020 - 0.0034 = 0.0166$ (carry extra significant figure).

$$\frac{x^2}{0.020 - x} \approx \frac{x^2}{0.0166} = 7.2 \times 10^{-4},\ x = 3.5 \times 10^{-3}$$

We repeat, until we get a self-consistent answer. In this case it will be:

$$x = 3.5 \times 10^{-3}$$

So: $[H^+] = [F^-] = x = 3.5 \times 10^{-3}\ M$; $[OH^-] = 2.9 \times 10^{-12}\ M$

$[HF] = 0.020 - x = 0.020 - 0.0035 = 0.017\ M$; $pH = 2.46$

d.

$$CH_3CH(OH)CO_2H \rightleftharpoons CH_3CH(OH)CO_2^- + H^+$$

HLac Lac$^-$

	HLac	$\rightleftharpoons$	Lac$^-$	+	H^+	$K_a = 1.4 \times 10^{-4}$
Initial	0.83 *M*		0		$0(10^{-7})$	
	x mol/L HLac dissociates to reach equilibrium					
Change	$-x$	$\rightarrow$	$+x$		$+x$	
Equil	$0.83 - x$		x		x	

$$K_a = 1.4 \times 10^{-4} = \frac{[H^+][Lac^-]}{[HLac]} = \frac{x^2}{0.83 - x} \approx \frac{x^2}{0.83}$$

$x = [H^+] = 1.1 \times 10^{-2} >> 10^{-7}$

$0.83 - 0.01 = 0.82$ Assumptions good (1.3% error).

So: $[H^+] = [Lac^-] = 1.1 \times 10^{-2}$, $[OH^-] = 9.1 \times 10^{-13}$

$[HLac] = 0.83 - 0.01 = 0.82\ M$; $pH = 1.96$

41.

	$B(OH)_3$ + H_2O	$\rightleftharpoons$	$B(OH)_4^-$	+	H^+
Initial	0.50 *M*		0		$0(10^{-7})$
	x mol/L $B(OH)_3$ reacts to reach equilibrium				
Change	$-x$	$\rightarrow$	$+x$		$+x$
Equil	$0.50 - x$		x		x

$$K_a = 5.8 \times 10^{-10} = \frac{[H^+][B(OH)_4^-]}{[B(OH)_3]} = \frac{x^2}{0.50 - x}$$

If 0.50 - x ≈ 0.50, then $5.8 \times 10^{-10} \approx \frac{x^2}{0.50}$

$x = [H^+] = 1.7 \times 10^{-5}$ *M*; pH = 4.77 Assumptions good.

43. $0.56 \text{ g } C_6H_5CO_2H(HBz) \times \frac{1 \text{ mol HBz}}{122.1 \text{ g}} = 4.6 \times 10^{-3}$ mol

Initial concentration of benzoic acid is $\frac{4.6 \times 10^{-3} \text{ mol}}{\text{L}}$

	HBz	⇌	H^+	+	Bz^-
Initial	4.6×10^{-3} *M*		$0(10^{-7})$		0
	x mol/L HBz dissociates to reach equilibrium				
Change	$-x$	→	$+x$		$+x$
Equil	$4.6 \times 10^{-3} - x$		x		x

$$K_a = 6.4 \times 10^{-5} = \frac{[H^+][Bz^-]}{[HBz]} = \frac{x^2}{4.6 \times 10^{-3} - x} \approx \frac{x^2}{4.6 \times 10^{-3}}$$

$x = [H^+] = 5.4 \times 10^{-4} >> 10^{-7}$

$0.0046 - x = 0.0046 - 0.00054 = 0.0041$

Assumption is not good (0.00054 is greater than 5% of 0.0046).

When the assumption(s) fail, we must solve exactly using the quadratic formula or the method of successive approximations (see Appendix 1.4 of text).

Using successive approximations:

$$\frac{x^2}{0.0041} = 6.4 \times 10^{-5},\ x = 5.1 \times 10^{-4}$$

$$\frac{x^2}{(0.0046 - 0.00051)} = \frac{x^2}{0.00041} = 6.4 \times 10^{-5},\ x = 5.1 \times 10^{-4}$$

So $x = [H^+] = [Bz^-] = 5.1 \times 10^{-4}$ *M*

$[HBz] = 4.6 \times 10^{-3} - x = 4.1 \times 10^{-3}$ *M*

pH = 3.29; pOH = 10.71; $[OH^-] = 10^{-10.71} = 1.9 \times 10^{-11}$ *M*

45. Major species: HIO_3, H_2O

Major source of H^+: HIO_3

	HIO_3	$\rightleftharpoons$	H^+ +	IO_3^-
Initial	0.50 *M*		$0(10^{-7})$	0
	x mol/L HIO_3 dissociates to reach equilibrium			
Change	$-x$	$\rightarrow$	$+x$	$+x$
Equil	$0.50 - x$		x	x

$$K_a = 0.17 = \frac{[H^+][IO_3^-]}{[HIO_3]} = \frac{x^2}{0.50 - x}$$

If $0.50 - x \approx 0.50$, then:

$x = 0.29$ Assumption is bad (60% error). Use 0.29 as an estimate of x and use the method of successive approximations to solve (Appendix 1.4 of text).

$$0.17 = \frac{x^2}{0.50 - x} \approx \frac{x^2}{0.50 - 0.29},\ x = 0.20$$

$$0.17 = \frac{x^2}{0.50 - 0.20},\ x = 0.23;\ 0.17 = \frac{x^2}{0.50 - 0.23},\ x = 0.21$$

$$0.17 = \frac{x^2}{0.50 - 0.21},\ x = 0.22;\ 0.17 = \frac{x^2}{0.50 - 0.22},\ x = 0.22$$

$x = [H^+] = 0.22$ mol/L; pH = -log(0.22) = 0.66

47. HCl is a strong acid. It will give 0.10 *M* H^+, 0.10 M Cl^-. Major species are: H^+, Cl^-, HOCl and H_2O.

	HOCl	$\rightleftharpoons$	H^+	+	OCl^-
Initial	0.10 *M*		0.10 *M*		0
	x mol/L HOCl dissociates to reach equilibrium				
Change	$-x$	$\rightarrow$	$+x$		$+x$
Equil	$0.10 - x$		$0.10 + x$		x

$$K_a = 3.5 \times 10^{-8} = \frac{[H^+][OCl^-]}{[HOCl]} = \frac{(0.10 + x)(x)}{0.10 - x} \approx x$$

Assumption is good. We are really assuming that HCl is the only important source of H^+. The contribution to $[H^+]$ from the HOCl is negligible. Therefore,

$[H^+] = 0.10$ *M*; pH = 1.00

49.

	$HClO_2$	$\rightleftharpoons$	H^+	+	ClO_2^-	$K_a = 1.2 \times 10^{-2}$
Initial	0.22 *M*		$0(10^{-7})$		0	
	x mol/L $HClO_2$ dissociates to reach equilibrium					
Change	$-x$	$\rightarrow$	$+x$		$+x$	
Equil	$0.22 - x$		x		x	

$$K_a = 1.2 \times 10^{-2} = \frac{[H^+][ClO^-]}{[HClO_2]} = \frac{x^2}{(0.22 - x)} \approx \frac{x^2}{0.22}$$

$x = 5.1 \times 10^{-2}$

The assumption that x is small is not good (23% error). Using the method of successive approximations (see Appendix 1.4 of text):

$\frac{x^2}{0.169} = 1.2 \times 10^{-2}$, $x = 4.5 \times 10^{-2}$

$\frac{x^2}{0.175} = 1.2 \times 10^{-2}$, $x = 4.6 \times 10^{-2}$ We can stop here (4.6×10^{-2} will repeat).

$[H^+] = [ClO_2^-] = x = 4.6 \times 10^{-2}\ M$; % ionized = $\frac{4.6 \times 10^{-2}}{0.22} \times 100 = 21\%$

51. a.

	HOAc	$\rightleftharpoons$	H^+	+	OAc^-
Initial	0.50 M		$0(10^{-7})$		0
	x mol/L HOAc ($HC_2H_3O_2$) dissociates to reach equilibrium				
Change	$-x$	$\rightarrow$	$+x$		$+x$
Equil	$0.50 - x$		x		x

$$K_a = 1.8 \times 10^{-5} = \frac{[H^+][OAc^-]}{[HOAc]} = \frac{x^2}{(0.50 - x)} \approx \frac{x^2}{0.50}$$

$x = [H^+] = [OAc^-] = 3.0 \times 10^{-3}\ M$ Assumptions good.

% ionized = $\frac{[OAc^-]}{0.50\ M} \times 100 = 0.60\%$

b. The set-up for (b) and (c) is similar to (a) except the final equation is slightly different.

$$K_a = 1.8 \times 10^{-5} = \frac{x^2}{(0.050 - x)} \approx \frac{x^2}{0.050}$$

$x = [OAc^-] = 9.5 \times 10^{-4}$ Assumptions good.

% ionized = $\frac{9.5 \times 10^{-4}}{0.050} \times 100 = 1.9\%$

c. $K_a = 1.8 \times 10^{-5} = \frac{x^2}{(0.0050 - x)} \approx \frac{x^2}{0.0050}$

$x = [OAc^-] = 3.0 \times 10^{-4}$

$0.0050 - x = 0.0047$ Assumption that x is negligible is borderline (6% error).

Using successive approximations:

$1.8 \times 10^{-5} = \frac{x^2}{0.0047}$, $x = 2.9 \times 10^{-4}$, next trial gives 2.9×10^{-4}

$$\% \text{ ionized} = \frac{2.9 \times 10^{-4}}{5.0 \times 10^{-3}} \times 100 = 5.8\%$$

Note: As the solution is more dilute, the percent ionization increases. This is what we would predict using Le Chatelier's Principle.

53. $HOBr \rightleftharpoons H^+ + OBr^-$

From normal setup: $[H^+] = [OBr^-] = x$

$[HOBr] = 0.063 - x = 0.063 - [H^+]$

$pH = 4.95$, $[H^+] = 10^{-4.95} = 1.1 \times 10^{-5}$

$$K_a = \frac{[H^+][OBr^-]}{[HOBr]} = \frac{(1.1 \times 10^{-5})^2}{(0.063 - 1.1 \times 10^{-5})} = 1.9 \times 10^{-9}$$

55. a.

	H_3AsO_4	$\rightleftharpoons$	H^+	+	$H_2AsO_4^-$	$K_{a_1} = 5 \times 10^{-3}$
Initial	$0.10\,M$		$0(10^{-7})$		0	
	x mol/L H_3AsO_4 dissociates to reach equilibrium					
Change	$-x$	$\rightarrow$	$+x$		$+x$	
Equil	$0.10 - x$		x		x	

$$K_{a_1} = 5 \times 10^{-3} = \frac{[H^+][H_2AsO_4^-]}{[H_3AsO_4]} = \frac{x^2}{(0.10 - x)} \approx \frac{x^2}{0.10}$$

$x = 2 \times 10^{-2}$, Assumption bad (x is 20% of 0.10). Using successive approximations:

$\frac{x^2}{0.08} = 5 \times 10^{-3}$, $x = 2 \times 10^{-2}$

$x = [H^+] = 2 \times 10^{-2}\,M$; $pH = 1.7$

b.

	H_2CO_3	$\rightleftharpoons$	H^+	+	HCO_3^-	$K_{a_1} = 4.3 \times 10^{-7}$
Initial	$0.10\,M$		$0(10^{-7})$		0	
	x mol/L H_2CO_3 dissociates					
Change	$-x$	$\rightarrow$	$+x$		$+x$	
Equil	$0.10 - x$		x		x	

$$K_a = 4.3 \times 10^{-7} = \frac{[H^+][HCO_3^-]}{[H_2CO_3]} = \frac{x^2}{(0.10 - x)} \approx \frac{x^2}{0.10}$$

$x = [H^+] = 2.1 \times 10^{-4}$ Assumptions good. $pH = 3.68$

Solutions of Bases

57. NO_3^-: $K_b \approx 0$ since HNO_3 is a strong acid. H_2O: $K_b = 10^{-14}$

NH_3: $K_b = 1.8 \times 10^{-5}$; CH_3NH_2: $K_b = 4.38 \times 10^{-4}$

$CH_3NH_2 > NH_3 > H_2O >> NO_3^-$

59. a. NH_3 b. NH_3 c. OH^- d. CH_3NH_2

61. $2.48 \text{ g TlOH} \times \dfrac{1 \text{ mol TlOH}}{221.4 \text{ g}} = 1.12 \times 10^{-2} \text{ mol}$

TlOH is a strong base, so $[OH^-] = \dfrac{1.12 \times 10^{-2} \text{ mol}}{L}$

pOH = 1.951; pH = 12.049

63. a. Major species: K^+, OH^-, H_2O

$[OH^-] = 0.150$, pOH = -log(0.150) = 0.824

pH = 14.000 - pOH = 13.176

b. Major species: Cs^+, OH^-, H_2O

$[OH^-] = 0.150$, pOH = 0.824, pH = 13.176

c. Major species: NH_3, H_2O

We will consider NH_3 as the major source of OH^-, i.e. ignore OH^- contribution from H_2O.

	NH_3 + H_2O	$\rightleftharpoons$	NH_4^+	+	OH^-
Initial	0.150 M		0		$0(10^{-7})$
	x mol/L NH_3 reacts with H_2O to reach equilibrium				
Change	$-x$	$\rightarrow$	$+x$		$+x$
Equil	$0.150 - x$		x		x

$$K_b = \frac{[NH_4^+][OH^-]}{[NH_3]} = 1.8 \times 10^{-5} = \frac{x^2}{0.150 - x} \approx \frac{x^2}{0.150}$$

$x = 1.6 \times 10^{-3}$; Assumptions good ($1 \times 10^{-7} << x << 0.150$).

$x = [OH^-] = 1.6 \times 10^{-3}$, pOH = 2.80, pH = 11.20

d. Major species: pyridine (C_5H_5N), H_2O

We will consider pyridine as the major source of OH^-.

	C_5H_5N + H_2O	$\rightleftharpoons$	$C_5H_5NH^+$	+	OH^-
Initial	0.150 *M*		0		0(10^{-7})
	x mol/L pyridine reacts with water				
Change	$-x$	→	$+x$		$+x$
Equil	$0.150 - x$		x		x

$$K_b = \frac{[C_5H_5NH^+][OH^-]}{[C_5H_5N]} = 1.7 \times 10^{-9} = \frac{x^2}{0.150 - x} \approx \frac{x^2}{0.150}$$

$x = 1.6 \times 10^{-5}$; Assumptions good by 5% rule.

$x = [OH^-] = 1.6 \times 10^{-5}$, pOH = 4.80; pH = 9.20

e. Major species: CH_3NH_2, H_2O

The major source of OH^- is CH_3NH_2.

	CH_3NH_2 + H_2O	$\rightleftharpoons$	$CH_3NH_3^+$	+	OH^-
Initial	0.150 *M*		0		0(10^{-7})
	x mol/L CH_3NH_2 reacts with water				
Change	$-x$	→	$+x$		$+x$
Equil	$0.150 - x$		x		x

$$K_b = 4.38 \times 10^{-4} = \frac{[CH_3NH_3^+][OH^-]}{[CH_3NH_2]} = \frac{x^2}{0.150 - x} \approx \frac{x^2}{0.150}$$

$x = 8.11 \times 10^{-3}$; Assumption is borderline (5.4% error).

Using successive approximations:

$$4.38 \times 10^{-4} = \frac{x^2}{0.150 - 0.00811}, \; x = 7.88 \times 10^{-3}$$

$$4.38 \times 10^{-4} = \frac{x^2}{0.150 - 0.00788}, \; x = 7.89 \times 10^{-3} \text{ (consistent answer)}$$

$x = [OH^-] = 7.89 \times 10^{-3}$, pOH = 2.103

pH = 11.897 (We could only measure 11.90 ± 0.01 at best.)

65. $K_b = 3.0 \times 10^{-6}$

	H_2NNH_2 + H_2O	$\rightleftharpoons$	$H_2NNH_3^+$	+	OH^-
Initial	2.0 *M*		0		0(10^{-7})
	x mol/L H_2NNH_2 reacts with H_2O to reach equilibrium				
Change	$-x$	→	$+x$		$+x$
Equil	$2.0 - x$		x		x

$$K_b = 3.0 \times 10^{-6} = \frac{[H_2NNH_3^+][OH^-]}{[H_2NNH_2]} = \frac{x^2}{2.0 - x} \approx \frac{x^2}{2.0}$$

$x = [OH^-] = 2.4 \times 10^{-3}\ M$; pOH = 2.62; pH = 11.38 Assumptions good by 5% rule.

$[H_2NNH_3^+] = [OH^-] = 2.4 \times 10^{-3}\ M$, $[H_2NNH_2] = 2.0 - x = 2.0\ M$

$[H^+] = 10^{-11.38} = 4.2 \times 10^{-12}$

67. a. $C_2H_5NH_2 + H_2O \rightleftharpoons C_2H_5NH_3^+ + OH^-$ $K_b = 5.6 \times 10^{-4}$

	$C_2H_5NH_2$		$C_2H_5NH_3^+$	OH^-
Initial	0.20 M		0	$0(10^{-7})$
	x mol/L $C_2H_5NH_2$ reacts with H_2O to reach equilibrium			
Change	$-x$	→	$+x$	$+x$
Equil	$0.20 - x$		x	x

$$K_b = \frac{[C_2H_5NH_3^+][OH^-]}{[C_2H_5NH_2]} = \frac{x^2}{0.20 - x} \approx \frac{x^2}{0.20}$$

$x = 1.1 \times 10^{-2}$; $0.200 - 0.011 = 0.189$ Assumption borderline (~ 5% error).

Using successive approximations:

$\frac{x^2}{0.189} = 5.6 \times 10^{-4}$, $x = 1.0 \times 10^{-2}\ M$ (consistent answer)

$x = [OH^-] = 1.0 \times 10^{-2}\ M$

$$[H^+] = \frac{K_w}{[OH^-]} = \frac{1.0 \times 10^{-14}}{1.0 \times 10^{-2}} = 1.0 \times 10^{-12}\ M;\ \text{pH} = 12.00$$

b. $Et_2NH + H_2O \rightleftharpoons Et_2NH_2^+ + OH^-$ Et = $-C_2H_5$ $K_b = 1.3 \times 10^{-3}$

	Et_2NH		$Et_2NH_2^+$	OH^-
Initial	0.20 M		0	$0(10^{-7})$
	x mol/L $(C_2H_5)_2NH$ reacts with H_2O			
Change	$-x$	→	$+x$	$+x$
Equil	$0.20 - x$		x	x

$$K_b = 1.3 \times 10^{-3} = \frac{[Et_2NH_2^+][OH^-]}{[Et_2NH]} = \frac{x^2}{0.20 - x} \approx \frac{x^2}{0.20}$$

$x = 1.6 \times 10^{-2}$ Assumption bad (x is 8% of 0.20).

Using successive approximations:

$\frac{x^2}{0.184} = 1.3 \times 10^{-3}$; $x = 1.55 \times 10^{-2}$ (carry extra significant figure)

$\frac{x^2}{0.185} = 1.3 \times 10^{-3}$; $x = 1.55 \times 10^{-2}$

$[OH^-] = x = 1.55 \times 10^{-2}\ M = 1.6 \times 10^{-2}\ M$; $[H^+] = 6.45 \times 10^{-13}\ M = 6.5 \times 10^{-13}\ M$;
pH = 12.19

c. $Et_3N + H_2O \rightleftharpoons Et_3NH^+ + OH^-$ $K_b = 4.0 \times 10^{-4}$

Equil $0.20 - x$ x x

$4.0 \times 10^{-4} = \frac{x^2}{0.20 - x} \approx \frac{x^2}{0.20}$; $x = 8.9 \times 10^{-3}$; Assumptions good

$[OH^-] = 8.9 \times 10^{-3}\ M$; $[H^+] = 1.1 \times 10^{-12}\ M$; pH = 11.96

69. a. $NH_3 + H_2O \rightleftharpoons NH_4^+ + OH^-$ $K_b = 1.8 \times 10^{-5}$

Equil $0.10 - x$ x x

$1.8 \times 10^{-5} = \frac{x^2}{0.10 - x} \approx \frac{x^2}{0.10}$

$x = [NH_4^+] = 1.3 \times 10^{-3}$; % ionized = $\frac{1.3 \times 10^{-3}}{0.10} \times 100 = 1.3\%$ Assumptions good.

b. $NH_3 + H_2O \rightleftharpoons NH_4^+ + OH^-$

Equil $0.010 - x$ x x

$1.8 \times 10^{-5} = \frac{x^2}{0.010 - x} \approx \frac{x^2}{0.010}$

$x = [NH_4^+] = 4.2 \times 10^{-4}$ Assumptions good.

% ionized = $\frac{4.2 \times 10^{-4}}{0.010} \times 100 = 4.2\%$

Note: For the same base, the percent ionized increases as the initial concentration decreases.

71. $\frac{5.0 \text{ mg}}{10.0 \text{ mL}} \times \frac{1 \text{ mmol}}{299.4 \text{ mg}} = 1.7 \times 10^{-3} \frac{\text{mmol}}{\text{mL}} = 1.7 \times 10^{-3}\ M$

Cod + $H_2O \rightleftharpoons CodH^+ + OH^-$ $K_b = 10^{-6.05} = 8.9 \times 10^{-7}$

Initial $1.7 \times 10^{-3}\ M$ 0 $0(10^{-7})$

x mol/L codeine reacts with H_2O to reach equilibrium

Change $-x$ → $+x$ $+x$

Equil $1.7 \times 10^{-3} - x$ x x

$8.9 \times 10^{-7} = \dfrac{x^2}{1.7 \times 10^{-3} - x} \approx \dfrac{x^2}{1.7 \times 10^{-3}}$; $x = 3.9 \times 10^{-5}$ Assumptions good.

$[OH^-] = 3.9 \times 10^{-5}$; $[H^+] = 2.6 \times 10^{-10}$; pH = 9.59

73. PT = p-toluidine

$PT + H_2O \rightleftharpoons PTH^+ + OH^-$; $K_b = \dfrac{[PTH^+][OH^-]}{[PT]}$

From normal setup:

$x = [PTH^+] = [OH^-]$; $[PT] = 0.016 - [OH^-]$

pH = 8.60; pOH = 5.40; $[OH^-] = 4.0 \times 10^{-6}$

$K_b = \dfrac{(4.0 \times 10^{-6})^2}{(0.016 - 4.0 \times 10^{-6})} = 1.0 \times 10^{-9}$

Acid/Base Properties of Salts

75. a. $NaNO_3 \rightarrow Na^+ + NO_3^-$ neutral (Na^+ and NO_3^- have no acidic/basic properties)

b. $NaNO_2 \rightarrow Na^+ + NO_2^-$ basic

NO_2^- is a weak base; it is the conjugate base of the weak acid, HNO_2.

$NO_2^- + H_2O \rightleftharpoons HNO_2 + OH^-$

c. $NH_4NO_3 \rightarrow NH_4^+ + NO_3^-$ acidic

NH_4^+, weak acid: $NH_4^+ \rightleftharpoons H^+ + NH_3$

d. $NH_4NO_2 \rightarrow NH_4^+ + NO_3^-$

NH_4^+ is a weak acid and NO_2^- is a weak base.

$NH_4^+ \rightleftharpoons NH_3 + H^+$ $K_a = \dfrac{K_w}{K_b} = \dfrac{1.0 \times 10^{-14}}{1.8 \times 10^{-5}} = 5.6 \times 10^{-10}$

$NO_2^- + H_2O \rightleftharpoons HNO_2 + OH^-$ $K_b = \dfrac{K_w}{K_a} = \dfrac{1.0 \times 10^{-14}}{4.0 \times 10^{-4}} = 2.5 \times 10^{-11}$

NH_4^+ is stronger as an acid in water than NO_2^- is as a base, $K_a(NH_4^+) > K_b(NO_2^-)$. Therefore, the solution is acidic.

e. $Na_2CO_3 \rightarrow 2\,Na^+ + CO_3^{2-}$, basic

Na^+ has no effect. CO_3^{2-} is a weak base. $CO_3^{2-} + H_2O \rightleftharpoons HCO_3^- + OH^-$

f. $NaF \rightarrow Na^+ + F^-$; F^- weak base; $F^- + H_2O \rightleftharpoons HF + OH^-$ solution is basic.

77. KOH: strong base; KBr: neutral

KCN: CN^- is weak base, $K_b = 1.0 \times 10^{-14}/6.2 \times 10^{-10} = 1.6 \times 10^{-5}$

NH_4Br: NH_4^+ is weak acid, $K_a = 5.6 \times 10^{-10}$

NH_4CN: slightly basic, CN^- is a stronger base compared to NH_4^+ as an acid.

HCN: weak acid, $K_a = 6.2 \times 10^{-10}$

acidic $\rightarrow$ basic: HCN, NH_4Br, KBr, NH_4CN, KCN, KOH

79. a. $CH_3NH_3Cl \rightarrow CH_3NH_3^+ + Cl^-$ Methylammonium ion is a weak acid.

$CH_3NH_3^+ \rightleftharpoons CH_3NH_2 + H^+$ Ignore Cl^-, conjugate base of a strong acid.

$$K_a = \frac{[CH_3NH_2][H^+]}{[CH_3NH_3^+]} = \frac{[CH_3NH_2][H^+][OH^-]}{[CH_3NH_3^+][OH^-]} = \frac{K_w}{K_b} = \frac{1.00 \times 10^{-14}}{4.38 \times 10^{-4}} = 2.28 \times 10^{-11}$$

	$CH_3NH_3^+$	$\rightleftharpoons$	CH_3NH_2	+	H^+
Initial	0.10 *M*		0		$0(10^{-7})$
	x mol/L $CH_3NH_3^+$ dissociates to reach equilibrium				
Change	$-x$	$\rightarrow$	$+x$		$+x$
Equil	$0.10 - x$		x		x

$$2.28 \times 10^{-11} = \frac{x^2}{0.10 - x} \approx \frac{x^2}{0.10}$$

$x = [H^+] = 1.5 \times 10^{-6}\,M$; pH = 5.82 Assumptions good.

b. $NaCN \rightarrow Na^+ + CN^-$ Cyanide ion is a weak base. Ignore Na^+.

$CN^- + H_2O \rightleftharpoons HCN + OH^-$ $\quad K_b = \frac{K_w}{K_a} = \frac{1.0 \times 10^{-14}}{6.2 \times 10^{-10}} = 1.6 \times 10^{-5}$

	$CN^- + H_2O$	$\rightleftharpoons$	HCN	+ OH^-
Initial	0.050 *M*		0	$0(10^{-7})$
	x mol/L CN^- reacts with H_2O			
Change	$-x$	$\rightarrow$	$+x$	$+x$
Equil	$0.050 - x$		x	x

$$K_b = 1.6 \times 10^{-5} = \frac{[HCN][OH^-]}{[CN^-]} = \frac{x^2}{0.050 - x} \approx \frac{x^2}{0.050}$$

$x = [OH^-] = 8.9 \times 10^{-4}$ *M*; pOH = 3.05; pH = 10.95 Assumptions good.

c. $CO_3^{2-} + H_2O \rightarrow HCO_3^- + OH^-$ $K_b = \frac{K_w}{K_a} = \frac{1.0 \times 10^{-14}}{5.6 \times 10^{-11}} = 1.8 \times 10^{-4}$

	CO_3^{2-}		HCO_3^-	OH^-
Initial	0.20 *M*		0	$0(10^{-7})$
	x mol/L CO_3^{2-} reacts with H_2O			
Change	$-x$	→	$+x$	$+x$
Equil	$0.20 - x$		x	x

$$K_b = 1.8 \times 10^{-4} = \frac{[HCO_3^-][OH^-]}{[CO_3^{2-}]} = \frac{x^2}{0.20 - x} \approx \frac{x^2}{0.20}$$

$x = 6.0 \times 10^{-3} = [OH^-]$; pOH = 2.22; pH = 11.78 Assumptions good.

81. $NaN_3 \rightarrow Na^+ + N_3^-$; Azide, N_3^- is a weak base.

$N_3^- + H_2O \rightleftharpoons HN_3 + OH^-$ $K_b = \frac{K_w}{K_a} = \frac{1.0 \times 10^{-14}}{1.9 \times 10^{-5}} = 5.3 \times 10^{-10}$

	N_3^-		HN_3	OH^-
Initial	0.010 *M*		0	$0(10^{-7})$
	x mol/L N_3^- reacts with H_2O to reach equilibrium			
Change	$-x$	→	$+x$	$+x$
Equil	$0.010 - x$		x	x

$$K_b = \frac{[HN_3][OH^-]}{[N_3^-]} = 5.3 \times 10^{-10} = \frac{x^2}{0.010 - x} \approx \frac{x^2}{0.010}$$

$x = [OH^-] = 2.3 \times 10^{-6}$; $[H^+] = \frac{1.0 \times 10^{-14}}{2.3 \times 10^{-6}} = 4.3 \times 10^{-9}$ Assumptions good.

$[HN_3] = [OH^-] = 2.3 \times 10^{-6}$ *M*; $[Na^+] = 0.010$ *M*

$[N_3^-] = 0.010 - 2.3 \times 10^{-6} \approx 0.010$ *M*

83. The stronger an acid, the weaker its conjugate base, since $K_aK_b = K_w$. Acetic acid is a stronger acid than hypochlorous acid. It's conjugate base, $C_2H_3O_2^-$ is a weaker base than the conjugate base of HOCl, OCl^-. Thus, the hypochlorite ion, OCl^-, is a stronger base than the acetate ion, $C_2H_3O_2^-$.

Relationships Between Structure and Strengths of Acids and Bases

85. a. $HBrO < HBrO_2 < HBrO_3$

The greater the number of O-atoms in an oxy-acid, the stronger the acid.

b. $HAsO_4^{2-} < H_2AsO_4^- < H_3AsO_4$

It gets progressively more difficult to remove successive H^+ from a polyprotic acid because the H^+ being removed is attracted to a more negatively charged anion.

87. a. $BrO_3^- < BrO_2^- < BrO^-$

Acids are stronger as more oxygens are present. Thus, the conjugate base of the acid is weaker as more oxygen atoms are present.

b. $H_2PO_4^- < HPO_4^{2-} < PO_4^{3-}$

For the conjugate bases of a polyprotic acid, the more highly negatively charged ions are more strongly attracted to H^+.

Lewis Acids and Bases

89. a. $B(OH)_3$, acid; H_2O, base b. Ag^+, acid; NH_3, base c. BF_3, acid; NH_3, base

91. $Al(OH)_3(s) + 3\ H^+(aq) \longrightarrow Al^{3+}(aq) + 3\ H_2O(l)$

$Al(OH)_3(s) + OH^-(aq) \longrightarrow Al(OH)_4^-(aq)$

93. Fe^{3+} should be the stronger Lewis acid. It is smaller with a greater positive charge and will be more strongly attracted to lone pairs of electrons.

ADDITIONAL EXERCISES

95. Both are strong acids.

$0.050\ mol/L \times 50.0\ mL = 2.5\ mmol\ HCl$

$0.10\ mol/L \times 150.0\ mL = 15\ mmol\ HNO_3$

$$[H^+] = \frac{2.5\ mmol + 15\ mmol}{200.0\ mL} = 0.088\ M,\ [OH^-] = 1.1 \times 10^{-13}$$

$$[Cl^-] = \frac{2.5\ mmol}{200.0\ mL} = 0.0125\ mol/L \approx 0.013\ M$$

$$[NO_3^-] = \frac{15\ mmol}{200.0\ mL} = 0.075\ M$$

97. $NaHSO_4 \rightarrow Na^+ + HSO_4^-$ HSO_4^- is a weak acid with $K_a = 0.012$

H_2SO_4 is a strong acid, so the reaction $HSO_4^- + H_2O \rightarrow H_2SO_4 + OH^-$ does not occur to any appreciable extent.

The solution is acidic and the reaction is:

$$HSO_4^- + H_2O \rightleftharpoons H_3O^+ + SO_4^{2-}$$

CO_3^{2-} is a base, so the reaction:

$$CO_3^{2-} + HSO_4^- \rightleftharpoons HCO_3^- + SO_4^{2-} \text{ can occur.}$$

99.

	HBz	$\rightleftharpoons$	H^+	+	Bz^-
Initial	C		~0		0
	x mol/L HBz dissociates to reach equilibrium				
Change	$-x$	$\rightarrow$	$+x$		$+x$
Equil	$C - x$		x		x

$$K_a = \frac{[H^+][Bz^-]}{[HBz]} = 6.4 \times 10^{-5} = \frac{x^2}{C - x}, \text{ and } x = [H^+]$$

$$6.4 \times 10^{-5} = \frac{[H^+]^2}{C - [H^+]}$$

$pH = 2.8$; $[H^+] = 10^{-2.8} = 1.6 \times 10^{-3}$ (carry 1 extra sig. fig.)

$$C - 1.6 \times 10^{-3} = \frac{(1.6 \times 10^{-3})^2}{6.4 \times 10^{-5}} = 4.0 \times 10^{-2}$$

$$C = 4.0 \times 10^{-2} + 0.16 \times 10^{-2} = 4.16 \times 10^{-2}\, M$$

The molar solubility is 4.16×10^{-2} mol/L $\approx 4 \times 10^{-2}$ mol/L

$$\frac{4 \times 10^{-2} \text{ mol}}{\text{L}} \times \frac{122.1 \text{ g}}{\text{mol}} \times 0.1 \text{ L} = \frac{0.5 \text{ g}}{100 \text{ mL}}$$

101. The N-atom is protonated in each case.

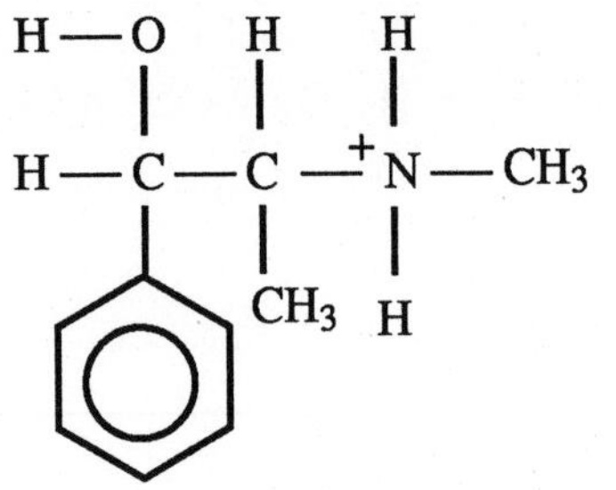

conjugate acid of ephedrine

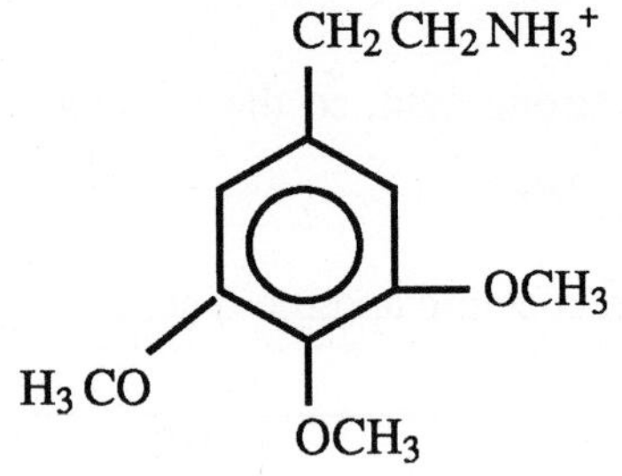

conjugate acid of mescaline

103. a. In the lungs, there is a lot of O_2; the equilibrium favors $Hb(O_2)_4$. In the cells there is a deficiency of O_2; the equilibrium favors HbH_4^{4+}.

b. CO_2 is a weak acid, $CO_2 + H_2O \rightleftharpoons HCO_3^- + H^+$. Removing CO_2 essentially decreases H^+. $Hb(O_2)_4$ is favored and O_2 is not released by hemoglobin in the cells. Breathing into a paper bag increases $[CO_2]$ in the blood.

c. CO_2 builds up in the blood and it becomes too acidic, driving the equilibrium to the left. Hemoglobin can't bind O_2 as strongly in the lungs. Bicarbonate ion acts as a base in water and neutralizes the excess acidity.

CHALLENGE PROBLEMS

105. Weak bonds to H and the presence of atoms with favorable electron affinities.

107. Base strength is enhanced if bonds to hydrogen are strong. The presence of atoms with high electronegativities can decrease base strength.

a. $SeH^- < SH^- < OH^-$ Same order as bond energies.

b. Bond energies N—H (391 kJ/mol), P—H (322 kJ/mol)

stronger bond, stronger base $PH_3 < NH_3$

c. $HONH_2 < NH_3$ Presence of O decreases base strength.

109. a. An acid will generate NH_4^+ in liquid ammonia and a base will generate NH_2^-. NH_4^+ corresponds to H^+ and NH_2^- corresponds to OH^-.

b. $[NH_4^+] = [NH_2^-]$

c. $2\ Na(s) + 2\ NH_3(l) \rightarrow 2\ Na^+ + 2\ NH_2^- + H_2(g)$

d. Ammonia is more basic than water. The acidity of substance is enhanced in liquid ammonia. For example, acetic acid is a strong acid in ammonia.

111. a. $HCO_3^- + HCO_3^- \rightleftharpoons H_2CO_3 + CO_3^{2-}$

$$K_{eq} = \frac{[H_2CO_3][CO_3^{2-}]}{[HCO_3^-][HCO_3^-]} \times \frac{[H^+]}{[H^+]} = \frac{K_{a_2}}{K_{a_1}} = \frac{5.6 \times 10^{-11}}{4.3 \times 10^{-7}} = 1.3 \times 10^{-4}$$

b. $[H_2CO_3] = [CO_3^{2-}]$ if the reaction in (a) is considered to be the only reaction.

c. $H_2CO_3 \rightleftharpoons 2\ H^+ + CO_3^{2-}$; $K_{eq} = \frac{[H^+]^2[CO_3^{2-}]}{[H_2CO_3]} = K_{a_1}K_{a_2}$

Since, $[H_2CO_3] = [CO_3^{2-}]$ from part b, $[H^+]^2 = K_{a_1}K_{a_2}$

$[H^+] = (K_{a_1}K_{a_2})^{1/2}$; $pH = \frac{pK_{a_1} + pK_{a_2}}{2}$

d. $[H^+] = [(4.3 \times 10^{-7})(5.6 \times 10^{-11})]^{1/2}$; $[H^+] = 4.9 \times 10^{-9}$; pH = 8.31

113.

	HBrO	$\rightleftharpoons$	H^+	+	BrO^-	$K_a = 2 \times 10^{-9}$
Initial	$1.0 \times 10^{-6}\ M$		$0(10^{-7})$		0	As usual, assume H^+ contribution from autoionization of water is negligible.
	x mol/L HBrO dissociates					
Change	$-x$	$\rightarrow$	$+x$		$+x$	
Equil	$1.0 \times 10^{-6} - x$		x		x	

$$\frac{x^2}{1.0 \times 10^{-6} - x} \approx \frac{x^2}{1.0 \times 10^{-6}} = 2 \times 10^{-9}$$

$x = [H^+] = 4 \times 10^{-8}$; pH = 7.4 Assumption that we can ignore H^+ contribution from H_2O is bad.

This answer is impossible. We can't add a small amount of a weak acid to a neutral solution and get a basic solution. In the correct solution, we would have to take into account the autoionization of water and solve exactly. This is beyond the scope of this text book.

115. 25.0 mL × 2.00 mmol/mL = 50.0 mmol H^+ from the strong acid HCl. The strong acid is in excess and will neutralize PO_4^{3-} to H_3PO_4 and neutralize OH^- to H_2O.

$$5.0\ \text{mmol}\ PO_4^{3-} \times \frac{3\ \text{mmol}\ H^+}{\text{mmol}\ PO_4^{3-}} + 5.0\ \text{mmol}\ OH^- \times \frac{1\ \text{mmol}\ H^+}{\text{mmol}\ OH^-}$$

= 20. mmol H^+ needed to neutralize bases.

After the neutralization reactions, solution contains 30. mmol H^+ and 5.0 mmol acetic acid. The contribution of H^+ from the weak acid, acetic acid, will be negligible. Ignoring H^+ from acetic acid:

$$[H^+] = \frac{30.\ \text{mmol}}{250.0\ \text{mL}} = 0.12\ \text{mol/L},\ pH = 0.92$$

CHAPTER FIFTEEN: APPLICATIONS OF AQUEOUS EQUILIBRIA

QUESTIONS

1. A common ion is an ion that appears in an equilibrium reaction but came from a source other than that reaction. Addition of a common ion (H^+ or NO_2^-) to the reaction, $HNO_2 \rightleftharpoons H^+ + NO_2^-$, will drive the equilibrium to the left.

3. The capacity of a buffer is a measure of how much strong acid or base the buffer can neutralize. All the buffers listed have the same pH. The 1.0 *M* buffer has the greatest capacity; the 0.01 *M* buffer the least capacity.

5. Between the starting point of the titration and the equivalence point, we are dealing with a buffer solution. Thus,

$$pH = pK_a + \log\frac{[Base]}{[Acid]}$$

Halfway to the equivalence point: [Base] = [Acid] so $pH = pK_a + \log 1 = pK_a$

7. No, since there are three colored forms there must be two proton transfer reactions. Thus, there must be at least two acidic protons in the acid (orange) form of thymol blue.

9. The two forms of an indicator are different colors. To see only one color that form must be in ten fold excess over the other. To go from

$$\frac{[HIn]}{[In^-]} = 10 \text{ to } \frac{[HIn]}{[In^-]} = 0.1 \text{ requires a change of 2 pH units.}$$

EXERCISES

Buffers

11. When strong acid or base is added to an acetic acid/sodium acetate mixture, the strong acid/base is neutralized. The reaction goes to completion. The strong acid/base is replaced with a weak acid/base.

$$H^+ + CH_3CO_2^- \rightarrow CH_3CO_2H;\quad OH^- + CH_3CO_2H \rightarrow CH_3CO_2^- + H_2O$$

13. For $HOAc \rightleftharpoons H^+ + OAc^-$ $\quad K_a = 1.8 \times 10^{-5} \quad pK_a = 4.74$

a.

	HOAc	$\rightleftharpoons$	H^+	+	OAc^-
Initial	0.10 *M*		~0		0.25 *M*
	x mol/L HOAc dissociates to reach equilibrium				
Change	$-x$	$\rightarrow$	$+x$		$+x$
Equil	$0.10 - x$		x		$0.25 + x$

$$1.8 \times 10^{-5} = \frac{x(0.25 + x)}{(0.10 - x)} \approx \frac{x(0.25)}{0.10} \quad \text{(assuming } 0.25 + x \approx 0.25 \text{ and } 0.10 - x \approx 0.10)$$

$x = [H^+] = 7.2 \times 10^{-6}$, pH = 5.14 Assumptions good by the 5% rule.

	OPr^-	+	H^+	→	HOPr	
Before	0.10 *M*		0.020 *M*		0.10 *M*	
Change	-0.020		-0.020	→	+0.020	Reacts completely
After	0.08		0		0.12	

A buffer solution results (weak acid and conjugate base). Using Henderson-Hasselbalch equation:

$$pH = pK_a + \log\frac{[Base]}{[Acid]} = 4.89 + \log\frac{(0.08)}{(0.12)} = 4.7$$

19. a. HOPr, OH^-

0.10 *M* 0.020 *M*

OH^- will react completely with the best acid present, HOPr.

	HOPr	+	OH^-	→	OPr^-	+	H_2O	
Before	0.10 *M*		0.020 *M*		0			
Change	-0.020		-0.020	→	+0.020			Reacts completely
After	0.08		0		0.020			

A buffer solution results after the reaction. Using Henderson-Hasselbalch equation:

$$pH = pK_a + \log\frac{[Base]}{[Acid]} = 4.89 + \log\frac{(0.020)}{(0.08)} = 4.3$$

b.

	OPr^- + H_2O	⇌	HOPr	+	OH^-
Initial	0.10 *M*		0		0.020 *M*
	x mol/L OPr^- reacts with H_2O to reach equilibrium				
Change	$-x$	→	$+x$		$+x$
Equil	$0.10 - x$		x		$0.020 + x$

$[OH^-] = 0.020 + x \approx 0.020$; pOH = 1.70; pH = 12.30 Assumption good.

Note: OH^- contribution from the weak base, OPr^-, was negligible. pH can be determined by only considering the amount of strong base present.

c. $[OH^-] = 0.020$; pOH = 1.70; pH = 12.30

d. HOPr, OPr^- OH^-

0.10 *M* 0.10 *M* 0.020 *M*

OH^- will react completely with HOPr, the best acid present.

	HOPr	+	OH^-	→	OPr^-	+	H_2O	
Before	0.10 *M*		0.020 *M*		0.10 *M*			
Change	-0.020		-0.020	→	+0.020			Reacts completely
After	0.08		0		0.12			

Using the Henderson-Hasselbalch equation to solve for the pH of the resulting buffer solution:

$$pH = pK_a + \log \frac{[Base]}{[Acid]} = 4.89 + \log \frac{(0.12)}{(0.08)} = 5.1$$

21. Consider all of the results to Exercises 15, 17, and 19.

Solution	Initial pH	after added acid	after added base
a	2.94	1.70	4.3
b	8.94	5.5	12.30
c	7.00	1.70	12.30
d	4.89	4.7	5.1

The solution in (d) is a buffer; it contains both weak acid (HOPr) and a weak base (OPr^-). It resists changes in pH when strong acid or base is added.

23. The reaction, $H^+ + NH_3 \rightarrow NH_4^+$, goes to completion. After this reaction occurs there must be both NH_3 and NH_4^+ in solution for it to be a buffer. After the reaction the solutions contain:

a. 0.05 *M* NH_4^+ and 0.05 *M* H^+

b. 0.05 *M* NH_4^+

c. 0.05 *M* NH_4^+ and 0.05 *M* H^+

d. 0.05 *M* NH_4^+ and 0.05 *M* NH_3

Thus, only the combination in (d) results in a buffer.

25. $NH_4^+ \rightleftharpoons H^+ + NH_3 \qquad K_a = \frac{K_w}{K_b} = 5.6 \times 10^{-10} \qquad pK_a = 9.25$

a. $9.00 = 9.25 + \log \frac{[NH_3]}{[NH_4^+]}$

$\log \frac{[NH_3]}{[NH_4^+]} = -0.25$

$\frac{[NH_3]}{[NH_4^+]} = 10^{-0.25} = 0.56$

b. $8.80 = 9.25 + \log \frac{[NH_3]}{[NH_4^+]}$

$\frac{[NH_3]}{[NH_4^+]} = 10^{-0.45} = 0.35$

c. $10.00 = 9.25 + \log \frac{[NH_3]}{[NH_4^+]}$

$\frac{[NH_3]}{[NH_4^+]} = 10^{0.75} = 5.6$

d. $9.60 = 9.25 + \log \frac{[NH_3]}{[NH_4^+]}$

$\frac{[NH_3]}{[NH_4^+]} = 10^{0.35} = 2.2$

27. When OH^- is added, it converts HOAc into OAc^-:

$HOAc + OH^- \rightarrow OAc^- + H_2O$ Reacts completely

The moles of OAc^- produced equals the moles of OH^- added. Since the volume is 1.0 L (assuming no volume change), then:

$[OAc^-] + [HOAc] = 2.0\ M$ and $[OAc^-]$ produced = $[OH^-]$ added

a. $$pH = pK_a + \log \frac{[OAc^-]}{[HOAc]}$$

For $pH = pK_a$, $\log \frac{[OAc^-]}{[HOAc]} = 0$

Therefore, $\frac{[OAc^-]}{[HOAc]} = 1.0$ and $[OAc^-] = [HOAc]$

Since $[OAc^-] + [HOAc] = 2.0$, then $[OAc^-] = [HOAc] = 1.0\ M$

To produce 1.0 L of a 1.0 M OAc^- solution we need to add 1.0 mol of NaOH to 1.0 L of the HOAc solution.

$$1.0 \text{ mol of NaOH} \times \frac{40.00 \text{ g}}{\text{mol NaOH}} = 40. \text{ g NaOH}$$

b. $$4.00 = 4.74 + \log \frac{[OAc^-]}{[HOAc]};\ \frac{[OAc^-]}{[HOAc]} = 10^{-0.74} = 0.18$$

$[OAc^-] = 0.18\ [HOAc]$ or $[HOAc] = 5.6\ [OAc^-]$

$[OAc^-] + [HOAc] = 2.0\ M$; $[OAc^-] + 5.6\ [OAc^-] = 2.0\ M$

$$[OAc^-] = \frac{2.0 \text{ M}}{6.6} = 0.30\ M$$

To produce 0.30 M OAc^- solution we need to add 0.30 mol of NaOH to 1.0 L of the HOAc solution.

$$0.30 \text{ mol} \times \frac{40.00 \text{ g}}{\text{mol}} = 12 \text{ g NaOH}$$

c. $$5.00 = 4.74 + \log \frac{[OAc^-]}{[HOAc]};\ \frac{[OAc^-]}{[HOAc]} = 10^{0.26} = 1.8$$

$1.8\ [HOAc] = [OAc^-]$ or $[HOAc] = 0.56\ [OAc^-]$; $[HOAc] + [OAc^-] = 2.0\ M$

$1.56\ [OAc^-] = 2.0\ M$; $[OAc^-] = 1.3\ M$

We need to add 1.3 mol of NaOH or $1.3 \text{ mol} \times \frac{40.00 \text{ g}}{\text{mol}} = 51 \text{ g NaOH}$

29. $$75.0 \text{ g } CH_3CO_2Na \times \frac{1 \text{ mol}}{82.03 \text{ g}} = 0.914 \text{ mol NaOAc};\ [OAc^-] = \frac{0.914 \text{ mol}}{0.5000 \text{ L}} = 1.83\ M$$

$$pH = pK_a + \log \frac{[OAc^-]}{[HOAc]} = 4.74 + \log\left(\frac{1.83}{0.64}\right) = 5.20$$

31. $[H^+]$ added $= \frac{0.010 \text{ mol}}{0.25 \text{ L}} = \frac{0.040 \text{ mol}}{\text{L}}$ The added H^+ converts NH_3 to NH_4^+.

a.

	NH_4^+	$\rightleftharpoons$	NH_3 +	H^+	
Before	0.15 *M*		0.050 *M*	0.040 *M*	
Change	+0.040	$\leftarrow$	-0.040	-0.040	React completely
After	0.19		0.010	0	

A buffer solution still exists. Using Henderson-Hasselbalch equation:

$$pH = pK_a + \log \frac{[NH_3]}{[NH_4^+]} = 9.25 + \log \left(\frac{0.010}{0.19}\right) = 7.97$$

b.

	NH_4^+	$\rightleftharpoons$	NH_3 +	H^+	
Before	1.5 *M*		0.50 *M*	0.040 *M*	
Change	+0.040	$\leftarrow$	-0.040	-0.040	React completely
After	1.54		0.46	0	(carry extra sig. fig.)

To calculate the pH, plug in the new buffer concentrations into the Henderson-Hasselbalch equation.

$$pH = pK_a + \log \frac{[NH_3]}{[NH_4^+]} = 9.25 + \log \left(\frac{0.46}{1.54}\right) = 8.73$$

The two buffers differ in their capacity. Solution (b) has the greatest capacity, i.e., largest concentrations. Buffers with greater capacities will be able to absorb more acid or base.

Acid-Base Titrations

33.

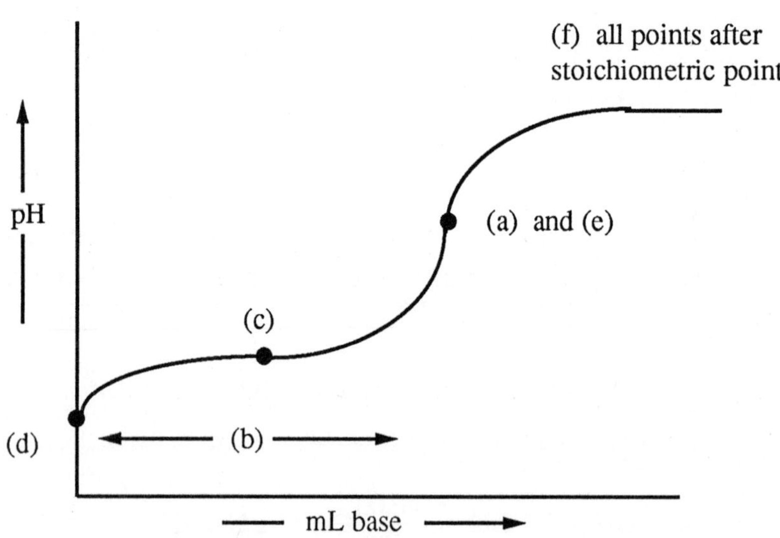

35. At the beginning of the titration, only the weak acid, HLac, present:

	HLac	$\rightleftharpoons$	H^+	+	Lac^-	$K_a = 1.4 \times 10^{-4}$	HLac: $HC_3H_5O_3$
Initial	0.100 *M*		~0		0		Lac^-: $C_3H_5O_3^-$
	x mol/L HLac dissociates too reach equilibrium						
Change	$-x$	$\rightarrow$	$+x$		$+x$		
Equil	$0.100 - x$		x		x		

$$1.4 \times 10^{-4} = \frac{x^2}{0.100 - x} \approx \frac{x^2}{0.100}$$

$x = [H^+] = 3.7 \times 10^{-3}$ *M*; pH = 2.43 Assumptions good.

Up to the stoichiometric point, we calculate the pH using the Henderson-Hasselbalch equation. This is the buffer region. For example, 4.0 mL of NaOH added:

$$\text{initial mmol HLac} = 25.0 \text{ mL} \times \frac{0.100 \text{ mmol}}{\text{mL}} = 2.50 \text{ mmol}$$

$$\text{mmol OH}^- \text{ added} = 4.0 \text{ mL} \times \frac{0.100 \text{ mmol}}{\text{mL}} = 0.40 \text{ mmol OH}^-$$

The base converts 0.40 mmoles HLac to 0.40 mmoles Lac^- according to the equation:

$$HLac + OH^- \rightarrow Lac^- + H_2O$$

mmol HLac remaining = 2.50 - 0.40 = 2.10, mmol Lac^- produced = 0.40

We have a buffer solution. Using the Henderson-Hasselbalch equation:

$$pH = pK_a + \log \frac{[Lac^-]}{[HLac]} = 3.86 + \log \frac{(0.40)}{(2.10)}$$ (total volume cancels, so we can use the ratio of moles)

pH = 3.14

Other points in the buffer region are calculated in a similar fashion. Do a stoichiometry problem first, followed by a buffer problem. The buffer region includes all points up to 24.9 mL OH^- added.

At the stoichiometric point, we have added enough OH^- to convert all of the HLac (2.50 mmol) into its conjugate base, Lac^-. All that is present is a weak base. To determine the pH, we perform a weak base calculation.

$[\text{Lac}^-] = \dfrac{2.50\ \text{mmol}}{50.0\ \text{mL}} = 0.0500\ M$

	$\text{Lac}^- + H_2O \rightleftharpoons$	HLac	$+ \ OH^-$
Initial	$0.0500\ M$	0	~0
	x mol/L Lac^- reacts with H_2O to reach equilibrium		
Change	$-x$ →	$+x$	$+x$
Equil	$0.0500 - x$	x	x

$K_b = \dfrac{1.00 \times 10^{-14}}{1.4 \times 10^{-4}} = 7.1 \times 10^{-11}$

$$\frac{x^2}{0.0500 - x} \approx \frac{x^2}{0.0500} = 7.1 \times 10^{-11}$$

$x = [OH^-] = 1.9 \times 10^{-6}\ M$; pOH = 5.72; pH = 8.28 Assumptions good.

Past the stoichiometric point, we have added more than 2.50 mmol of NaOH. The pH will be determined by the excess OH^- ion present.

At 25.1 mL: OH^- added $= 25.1\ \text{mL} \times \dfrac{0.100\ \text{mmol}}{\text{mL}} = 2.51\ \text{mmol}$

excess $OH^- = 2.51 - 2.50 = 0.01$ mmol

$[OH^-] = \dfrac{0.01\ \text{mmol}}{50.1\ \text{mL}} = 2 \times 10^{-4}\ M$; pOH = 3.7; pH = 10.3

All results are listed in Table 15.1 at the end of the solution to Exercise 15.38.

37. At beginning of titration, only a weak base, NH_3, present:

	$NH_3 + H_2O \rightleftharpoons$	NH_4^+	$+ \ OH^-$
Equil.	$0.100 - x$	x	x

$K_b = 1.8 \times 10^{-5}$

$$\frac{x^2}{0.100 - x} \approx \frac{x^2}{0.100} = 1.8 \times 10^{-5}$$

$x = [OH^-] = 1.3 \times 10^{-3}$; pOH = 2.89; pH = 11.11 Assumptions good.

In buffer region (4 - 24.9 mL):

Use $\text{pH} = \text{p}K_a + \log\dfrac{[\text{Base}]}{[\text{Acid}]}$ $K_a = \dfrac{1.0 \times 10^{-14}}{1.8 \times 10^{-5}} = 5.6 \times 10^{-10}$ $\text{p}K_a = 9.25$

$$\text{pH} = 9.25 + \log\frac{[NH_3]}{[NH_4^+]}$$

For example, after 8.0 mL HCl added:

$$\text{initial mmol } NH_3 = 25.0 \text{ mL} \times \frac{0.100 \text{ mmol}}{\text{mL}} = 2.50 \text{ mmol}$$

$$\text{mmol } H^+ \text{ added} = 8.0 \text{ mL} \times \frac{0.100 \text{ mmol}}{\text{mL}} = 0.80 \text{ mmol}$$

Added H^+ converts: $NH_3 + H^+ \rightarrow NH_4^+$

mmol NH_3 remaining = 2.50 - 0.80 = 1.70 mmol, mmol NH_4^+ produced = 0.80 mmol

$pH = 9.25 + \log \frac{1.70}{0.80} = 9.58$ (mole ratios can be used since total volume cancels)

At stoichiometric point, enough HCl has been added to convert all of the weak base (NH_3) into its conjugate acid, NH_4^+. Perform a weak acid calculation, $[NH_4^+]_o = \frac{2.50 \text{ mmol}}{50.0 \text{ mL}} = 0.0500\ M$:

	NH_4^+	$\rightleftharpoons$	H^+	+	NH_3	$K_a = 5.6 \times 10^{-10}$
Equil	$0.0500 - x$		x		x	

$$5.6 \times 10^{-10} = \frac{x^2}{0.0500 - x} \approx \frac{x^2}{0.0500}$$

$x = [H^+] = 5.3 \times 10^{-6}\ M$; pH = 5.28 Assumptions good.

Beyond the stoichiometric point, the pH is determined by the excess H^+.

At 28 mL: $H^+ \text{ added} = 28 \text{ mL} \times \frac{0.100 \text{ mmol}}{\text{mL}} = 2.8 \text{ mmol}$

Excess H^+ = 2.8 mmol - 2.5 mmol = 0.3 mmol

$$[H^+] = \frac{0.3 \text{ mmol}}{53 \text{ mL}} = 6 \times 10^{-3}\ M;\ pH = 2.2$$

The results are in Table 15.1.

Table 15.1: Summary of results in Exercises 15.35 and 15.37
(Graph below)

mL added titrant	pH Exercise 15.35	pH Exercise 15.37
0.0	2.43	11.11
4.0	3.14	9.97
8.0	3.53	9.58
12.5	3.86	9.25
20.0	4.46	8.65
24.0	5.24	7.87
24.5	5.6	7.6
24.9	6.3	6.9
25.0	8.28	5.28
25.1	10.3	3.7
26.	11.3	2.7
28.	11.8	2.2
30.	12.0	2.0

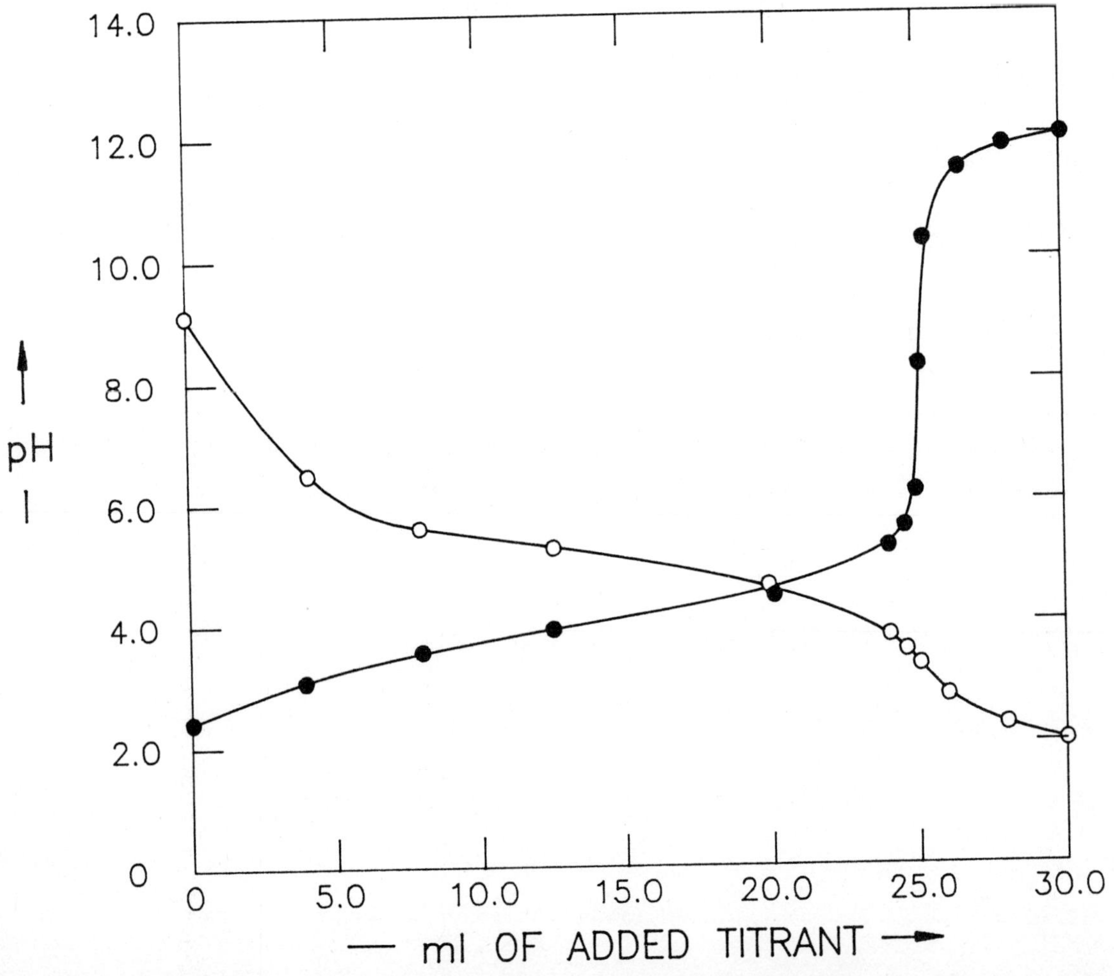

39. a. $0.104 \text{ g } NaC_2H_3O_2 \times \frac{1 \text{ mol } NaC_2H_3O_2}{82.03 \text{ g}} = 1.27 \times 10^{-3} \text{ mol} = 1.27 \text{ mmol } NaC_2H_3O_2$

$1.27 \text{ mmol } NaC_2H_3O_2 \times \frac{1 \text{ mmol HCl}}{\text{mmol } NaC_2H_3O_2} \times \frac{1 \text{ mL}}{0.0996 \text{ mmol HCl}} = 12.7 \text{ mL HCl}$

Add H^+ reacts with the acetate ion to completion:

$$H^+ + C_2H_3O_2^- \longrightarrow HC_2H_3O_2$$

The titration converts the acetate ion into acetic acid. At the equivalence point, there is a solution of only acetic acid with a concentration of:

$\frac{1.27 \text{ mmol}}{25 \text{ mL} + 12.7 \text{ mL}} = 3.4 \times 10^{-2} \, M$

The acetic acid dissociates, making the equivalence point acidic:

	HOAc	$\rightleftharpoons$	H^+	+	OAc^-
Initial	0.034 M		~0		0
	x mol/L HOAc dissociates to reach equilibrium				
Change	$-x$	$\longrightarrow$	$+x$		$+x$
Equil	$0.034 - x$		x		x

$K_a = 1.8 \times 10^{-5} = \frac{[H^+][OAc^-]}{[HOAc]} = \frac{x^2}{0.034 - x} \approx \frac{x^2}{0.034}$

$x = [H^+] = 7.8 \times 10^{-4} \, M$; pH = 3.11 Assumptions good.

b. $50.00 \text{ mL} \times \frac{0.0426 \text{ mmol HOCl}}{\text{mL}} = 2.13 \text{ mmol HOCl}$

$2.13 \text{ mmol HOCl} \times \frac{1 \text{ mmol NaOH}}{\text{mmol HOCl}} \times \frac{1 \text{ mL}}{0.1028 \text{ mol NaOH}} = 20.7 \text{ mL of NaOH soln.}$

Added OH^- reacts with HOCl to completion:

$$OH^- + HOCl \longrightarrow OCl^-$$

The titration converts HOCl into OCl^- ion. At the equivalence point, there are 2.13 mmol of OCl^- in 70.7 mL of solution (50.0 mL + 20.7 mL). The concentration of OCl^- is:

$\frac{2.13 \text{ mmol}}{70.7 \text{ mL}} = 3.01 \times 10^{-2} \text{ mol/L}$

The OCl^- reacts with H_2O, to produce a basic solution.

	OCl^-	+	H_2O	$\rightleftharpoons$	HOCl	+	OH^-
Initial	0.0301 M				0		~0
	x mol/L OCl^- reacts with H_2O to reach equilibrium						
Change	$-x$			$\longrightarrow$	$+x$		$+x$
Equil	$0.0301 - x$				x		x

$$K_b = \frac{K_w}{K_a} = \frac{1.0 \times 10^{-14}}{3.5 \times 10^{-8}} = 2.9 \times 10^{-7} = \frac{[HOCl]\,[OH^-]}{[OCl^-]} = \frac{x^2}{0.0301 - x}$$

$2.9 \times 10^{-7} \approx \frac{x^2}{0.0301}$; $x = [OH^-] = 9.3 \times 10^{-5}$; pOH = 4.03; pH = 9.97

Assumptions good.

41. The pH will be less than about 0.5.

43. a. yellow b. orange c. blue d. bluish green

45.

Exercise	pH at stoich. pt.	indicator
15.35	8.28	phenolphthalein
15.37	5.28	bromcresol green
15.39a	3.11	erythrosin B or 2,4-dinitrophenol
15.39b	9.97	thymolphthalein

{ titration in 15.39a not feasible, pH break too small (refers to 15.37 and 15.39a rows)

Solubility Equilibria

47. a. $AgC_2H_3O_2(s) \rightleftharpoons Ag^+(aq) + C_2H_3O_2^-(aq)$; $K_{sp} = [Ag^+]\,[C_2H_3O_2^-]$

b. $MnS(s) \rightleftharpoons Mn^{2+}(aq) + S^{2-}(aq)$; $K_{sp} = [Mn^{2+}]\,[S^{2-}]$

c. $Al(OH)_3(s) \rightleftharpoons Al^{3+}(aq) + 3\ OH^-(aq)$; $K_{sp} = [Al^{3+}]\,[OH^-]^3$

d. $Ca_3(PO_4)_2(s) \rightleftharpoons 3\ Ca^{2+}(aq) + 2\ PO_4^{3-}(aq)$; $K_{sp} = [Ca^{2+}]^3\,[PO_4^{3-}]^2$

49. a. $CaC_2O_4(s) \rightleftharpoons Ca^{2+}(aq) + C_2O_4^{2-}(aq)$

$[Ca^{2+}] = [C_2O_4^{2-}] = 4.8 \times 10^{-5}$ mol/L

$K_{sp} = [Ca^{2+}]\,[C_2O_4^{2-}] = (4.8 \times 10^{-5})^2 = 2.3 \times 10^{-9}$

b. $PbBr_2(s) \rightleftharpoons Pb^{2+}(aq) + 2\ Br^-(aq)$

s = solubility in mol/L → s 2s s = 2.14×10^{-2} *M*

$K_{sp} = [Pb^{2+}]\,[Br^-]^2 = (2.14 \times 10^{-2})\,(4.28 \times 10^{-2})^2 = 3.92 \times 10^{-5}$

c. $BiI_3(s) \rightleftharpoons Bi^{3+}(aq) + 3\,I^-(aq)$

s = solubility in mol/L → s 3s s = 1.32×10^{-5} *M*

$[I^-] = 3s = 3.96 \times 10^{-5}$

$K_{sp} = [Bi^{3+}]\,[I^-]^3 = (1.32 \times 10^{-5})\,(3.96 \times 10^{-5})^3 = 8.20 \times 10^{-19}$

51. s = solubility in mol/L of solid

a.

	$Al(OH)_3(s)$	$\rightleftharpoons$	$Al^{3+}(aq)$	+	$3\,OH^-(aq)$
Initial			0		10^{-7} from water
	s mol/L $Al(OH)_3(s)$ dissolves to reach equilibrium				
Change	-s	→	+s		+3s
Equil			s		10^{-7} + 3s

$K_{sp} = 2 \times 10^{-32} = [Al^{3+}]\,[OH^-]^3 = (s)\,(10^{-7} + 3s)^3 = s(10^{-7})^3$ Assuming $10^{-7} + 3s \approx 10^{-7}$

$s = \frac{2 \times 10^{-32}}{10^{-21}} = 2 \times 10^{-11}$ mol/L Assumption good.

$\frac{2 \times 10^{-11}\ \text{mol}}{\text{L}} \times \frac{78.00\ \text{g}}{\text{mol}} = 2 \times 10^{-9}$ g/L

b.

	$Be(IO_4)_2$	$\rightleftharpoons$	Be^{2+}	+	$2\,IO_4^-$
Initial			0		0
	s mol/L dissolves to reach equilibrium				
Change	-s	→	+s		+2s
Equil			s		2s

$K_{sp} = 1.57 \times 10^{-9} = [Be^{2+}]\,[IO_4^-]^2 = (s)\,(2s)^2 = 4s^3,\ s = \left(\frac{1.57 \times 10^{-9}}{4}\right)^{\frac{1}{3}}$

$s = 7.32 \times 10^{-4}$ mol/L; $\frac{7.32 \times 10^{-4}\ \text{mol}}{\text{L}} \times \frac{390.8\ \text{g}}{\text{mol}} = 0.286$ g/L

c.

	$CaSO_4$	$\rightleftharpoons$	Ca^{2+}	+	SO_4^{2-}
Initial			0		0
	s mol/L dissolves to reach equilibrium				
Change	-s	→	+s		+s
Equil			s		s

$K_{sp} = 6.1 \times 10^{-5} = [Ca^{2+}]\,[SO_4^{2-}] = s^2,\ s = (6.1 \times 10^{-5})^{\frac{1}{2}}$

$s = 7.8 \times 10^{-3}$ mol/L; $\frac{7.8 \times 10^{-3}\ \text{mol}}{\text{L}} \times \frac{136.2\ \text{g}}{\text{mol}} = 1.1$ g/L

d.

	$CaCO_3$	$\rightleftharpoons$	Ca^{2+}	+	CO_3^{2-}
Initial			0		0
	s mol/L dissolves to reach equilibrium				
Change	-s	$\rightarrow$	+s		+s
Equil			s		s

$K_{sp} = 8.7 \times 10^{-9} = [Ca^{2+}]\,[CO_3^{2-}] = s^2,\ s = (8.7 \times 10^{-9})^{\frac{1}{2}}$

$s = 9.3 \times 10^{-5}$ mol/L; $\dfrac{9.3 \times 10^{-5}\ \text{mol}}{\text{L}} \times \dfrac{100.1\ \text{g}}{\text{mol}} = 9.3 \times 10^{-3}$ g/L

Note: This ignores any reaction between CO_3^{2-} and H_2O to form HCO_3^-.

e.

	$Mg(NH_4)PO_4$	$\rightleftharpoons$	Mg^{2+}	+	NH_4^+	+	PO_4^{3-}
Initial			0		0		0
	s mol/L dissolves to reach equilibrium						
Change	-s	$\rightarrow$	+s		+s		+s
Equil			s		s		s

$K_{sp} = 3 \times 10^{-13} = [Mg^{2+}]\,[NH_4^+]\,[PO_4^{3-}] = s^3,\ s = (3 \times 10^{-13})^{\frac{1}{3}}$

$s = 7 \times 10^{-5}$ mol/L; $\dfrac{7 \times 10^{-5}\ \text{mol}}{\text{L}} \times \dfrac{137.3\ \text{g}}{\text{mol}} = 1 \times 10^{-2}$ g/L

Note: We are ignoring any reaction between NH_4^+ and PO_4^{3-}.

53. If the anion in the salt can act as a base in water, the solubility will increase as the solution becomes more acidic.

This is true for $Al(OH)_3$, $CaCO_3$, $Mg(NH_4)PO_4$, Ag_2CO_3, and Ag_2CrO_4.

$Al(OH)_3$, $Al(OH)_3 + 3\ H^+ \rightarrow Al^{3+} + 3\ H_2O$

$CaCO_3 + H^+ \rightarrow Ca^{2+} + HCO_3^- \xrightarrow{H^+} Ca^{2+} + H_2O + CO_2(g)$

$Mg(NH_4)PO_4 + H^+ \rightarrow Mg^{2+} + NH_4^+ + HPO_4^{2-}$

$HPO_4^{2-} \xrightarrow{H^+} H_2PO_4^- \xrightarrow{H^+} H_3PO_4$

$Ag_2CO_3 + H^+ \rightarrow 2\ Ag^+ + HCO_3^- \xrightarrow{H^+} 2\ Ag^+ + H_2O + CO_2(g)$

$2Ag_2CrO_4 + 2\ H^+ \rightarrow 4\ Ag^+ + Cr_2O_7^{2-} + H_2O$

$CaSO_4$ and $SrSO_4$ will be slightly more soluble in acid. Since,

$MSO_4 + H^+ \rightarrow M^{2+} + HSO_4^-$

occurs. However, K_b for SO_4^{2-} is only about 10^{-12}, and the effect of acid on these salts is not as great as on the others.

55. s = solubility in mol/L

a.

	$Fe(OH)_3(s)$	$\rightleftharpoons$	$Fe^{3+}(aq)$	+	$3\ OH^-(aq)$
Initial			0		$1 \times 10^{-7}\ M$
	s mol/L dissolves to reach equilibrium				
Change	-s	→	+s		+3s
Equil			s		$1 \times 10^{-7} + 3s$

$K_{sp} = 4 \times 10^{-38} = [Fe^{3+}]\,[OH^-]^3 = (s)\,(1 \times 10^{-7} + 3s)^3 \approx s(1 \times 10^{-7})^3$

$s = 4 \times 10^{-17}$ mol/L Assumption good ($s << 1 \times 10^{-7}$).

b.

	$Fe(OH)_3$	$\rightleftharpoons$	Fe^{3+}	+	$3\ OH^-$	pH = 5.0, pOH = 9.0, $[OH^-] = 1 \times 10^{-9}$
Initial			0		$1 \times 10^{-9}\ M$	(buffered)
	s mol/L dissolves to reach equilibrium					
Change	-s	→	+s		-----	(assume no pH change in buffer)
Equil			s		1×10^{-9}	

$K_{sp} = 4 \times 10^{-38} = [Fe^{3+}]\,[OH^-]^3 = (s)\,(1 \times 10^{-9})^3$

$s = 4 \times 10^{-11}$ mol/L

c. pH = 11.0; pOH = 3.0, $[OH^-] = 1 \times 10^{-3}\ M$

	$Fe(OH)_3$	$\rightleftharpoons$	Fe^{3+}	+	$3\ OH^-$	
Initial			0		$0.001\ M$	(buffered)
	s mol/L dissolves to reach equilibrium					
Change	-s	→	+s		-----	(assume no pH change)
Equil			s		0.001	

$K_{sp} = 4 \times 10^{-38} = [Fe^{3+}]\,[OH^-]^3 = (s)\,(0.001)^3$

$s = 4 \times 10^{-29}$ mol/L

57. After mixing $[Pb^{2+}]_o = [Cl^-]_o = 0.010\ M$

$[Pb^{2+}]_o\,[Cl^-]_o^{\,2} = (0.010)\,(0.010)^2 = 1.0 \times 10^{-6} < K_{sp}\ (1.6 \times 10^{-5})$

No precipitate will form. Thus,

$[Pb^{2+}] = [Cl^-] = 0.010\ M$, $[NO_3^-] = 2(0.010) = 0.020\ M$

$[Na^+] = [Cl^-] = 0.010\ M$

Complex Ion Equilibria

59. a.

$$Co^{2+} + NH_3 \rightleftharpoons CoNH_3^{2+} \qquad K_1$$
$$CoNH_3^{2+} + NH_3 \rightleftharpoons Co(NH_3)_2^{2+} \qquad K_2$$
$$Co(NH_3)_2^{2+} + NH_3 \rightleftharpoons Co(NH_3)_3^{2+} \qquad K_3$$
$$Co(NH_3)_3^{2+} + NH_3 \rightleftharpoons Co(NH_3)_4^{2+} \qquad K_4$$
$$Co(NH_3)_4^{2+} + NH_3 \rightleftharpoons Co(NH_3)_5^{2+} \qquad K_5$$
$$Co(NH_3)_5^{2+} + NH_3 \rightleftharpoons Co(NH_3)_6^{2+} \qquad K_6$$

$$Co^{2+} + 6\,NH_3 \rightleftharpoons Co(NH_3)_6^{2+} \qquad K_f = K_1K_2K_3K_4K_5K_6$$

b.

$$Fe^{3+} + SCN^- \rightleftharpoons FeSCN^{2+} \qquad K_1$$
$$FeSCN^{2+} + SCN^- \rightleftharpoons Fe(SCN)_2^+ \qquad K_2$$
$$Fe(SCN)_2^+ + SCN^- \rightleftharpoons Fe(SCN)_3 \qquad K_3$$
$$Fe(SCN)_3 + SCN^- \rightleftharpoons Fe(SCN)_4^- \qquad K_4$$
$$Fe(SCN)_4^- + SCN^- \rightleftharpoons Fe(SCN)_5^{2-} \qquad K_5$$
$$Fe(SCN)_5^{2-} + SCN^- \rightleftharpoons Fe(SCN)_6^{3-} \qquad K_6$$

$$Fe^{3+} + 6\,SCN^- \rightleftharpoons Fe(SCN)_6^{3-} \qquad K_f = K_1K_2K_3K_4K_5K_6$$

c.

$$Ag^+ + NH_3 \rightleftharpoons AgNH_3^+ \qquad K_1$$
$$AgNH_3^+ + NH_3 \rightleftharpoons Ag(NH_3)_2^+ \qquad K_2$$

$$Ag^+ + 2\,NH_3 \rightleftharpoons Ag(NH_3)_2^+ \qquad K_f = K_1K_2$$

61. An ammonia solution is basic. Initially the reaction that occurs is:

$$Cu^{2+}(aq) + 2\,OH^-(aq) \rightarrow Cu(OH)_2(s) \text{ (white ppt.)}$$

As the concentration of NH_3 increases the complex ion, $Cu(NH_3)_4^{2+}$, forms:

$$Cu(OH)_2(s) + 4\,NH_3(aq) \rightleftharpoons Cu(NH_3)_4^{2+}(aq) + 2\,OH^-(aq)$$

63. $\frac{65\text{ g KI}}{0.500\text{ L}} \times \frac{1\text{ mol}}{166.0\text{ g}} = \frac{0.78\text{ mol}}{\text{L}}$

	Hg^{2+} +	$4\,I^-$	$\rightleftharpoons$	HgI_4^{2-}	$K_f = 1.0 \times 10^{30}$
Before	0.010 *M*	0.78 *M*		0	
Change	-0.010	-0.040	→	+0.010	React completely (K large)
After (New Initial)	0	0.74		0.010	

x mol/L HgI_4^{2-} dissociates to reach equilibrium

Change	+*x*	+4*x*	←	-*x*
Equil	*x*	0.74 + 4*x*		0.010 - *x*

$$K_f = 1.0 \times 10^{30} = \frac{[HgI_4^{2-}]}{[Hg^{2+}][I^-]^4} = \frac{(0.010 - x)}{(x)(0.74 + 4x)^4}$$

$$1.0 \times 10^{30} \approx \frac{(0.010)}{(x)(0.74)^4};\quad x = [Hg^{2+}] = 3.3 \times 10^{-32}\ \text{mol/L}$$ Assumptions good.

Note: 3.3×10^{-32} mol/L corresponds to one Hg^{2+} ion per 5×10^7 L. It is very reasonable to approach the equilibrium in two steps. The reaction really does go to completion.

ADDITIONAL EXERCISES

65. $NH_3 + H_2O \rightleftharpoons NH_4^+ + OH^-$ $\quad K_b = \frac{[NH_4^+][OH^-]}{[NH_3]}$

$$-\log K_b = -\log [OH^-] - \log \frac{[NH_4^+]}{[NH_3]}$$

$$-\log [OH^-] = -\log K_b + \log \frac{[NH_4^+]}{[NH_3]}$$

$$pOH = pK_b + \log \frac{[NH_4^+]}{[NH_3]} \text{ or } pOH = pK_b + \log \frac{[Acid]}{[Base]}$$

67. a. The optimum pH for a buffer is when pH = pK_a. At this pH a buffer will have equal capacity for both added acid and base.

$K_b = 1.19 \times 10^{-6}$; $K_a = K_w/K_b = 8.40 \times 10^{-9}$; $pH = pK_a = 8.076 = 8.08$

b. $$pH = pK_a + \log \frac{[TRIS]}{[TRIS\text{-}H^+]}$$

$$7.0 = 8.08 + \log \frac{[TRIS]}{[TRIS\text{-}H^+]}$$

$$\frac{[TRIS]}{[TRIS\text{-}H^+]} = 10^{-1.08} = 0.083 \approx 0.08 \text{ at pH} = 7.0$$

$$9.0 = 8.08 + \log \frac{[TRIS]}{[TRIS\text{-}H^+]}$$

$$\frac{[TRIS]}{[TRIS\text{-}H^+]} = 10^{0.92} = 8.3 \approx 8 \text{ at pH} = 9.0$$

c. $$\frac{50.0\text{ g TRIS}}{2.0\text{ L}} \times \frac{1\text{ mol}}{121.1\text{ g}} = 0.206\ M = 0.21\ M = [TRIS]$$

$$\frac{65.0\text{ g TRIS-HCl}}{2.0\text{ L}} \times \frac{1\text{ mol}}{157.6\text{ g}} = 0.206\ M = 0.21\ M = [TRIS\text{-}HCl] = [TRIS\text{-}H^+]$$

$$pH = pK_a + \log \frac{[TRIS]}{[TRIS\text{-}H^+]} = 8.08 + \log \frac{(0.21)}{(0.21)} = 8.08$$

Amount of H^+ added from HCl is:

$$0.50 \times 10^{-3}\,L \times \frac{12\ mol}{L} = 6.0 \times 10^{-3}\ mol$$

The H^+ from HCl will convert TRIS to TRIS-H^+. The reaction is:

	TRIS	+ H^+	→	TRIS-H^+	
Before	0.21 *M*	$\frac{6.0 \times 10^{-3}}{0.2005} = 0.030\ M$		0.21 *M*	
Change	-0.030	-0.030	→	+0.030	Reacts completely
After	0.18	0		0.24	

Now use the Henderson-Hasselbalch equation to solve the buffer problem.

$$pH = 8.08 + \log\frac{0.18}{0.24} = 7.96$$

69.

	pK_a =
cacodylic acid	6.19
TRIS-HCl	8.08
benzoic acid	4.19
acetic acid	4.74
HF	3.14
NH_4Cl	9.25

a. potassium fluoride + HCl

b. benzoic acid + NaOH

c. sodium acetate + acetic acid

d. $(CH_3)_2AsO_2Na$ + HCl:

This is the best choice to produce a conjugate acid/base pair. Actually the best choice is an equimolar mixture ammonium chloride and sodium acetate. NH_4^+ is a weak acid ($K_a = 5.6 \times 10^{-10}$). $C_2H_3O_2^-$ is a weak base ($K_b = 5.6 \times 10^{-10}$). A mixture of the two will give a buffer at pH = 7 since the weak acid and weak base are the same strengths. $NH_4C_2H_3O_2$ is commercially available and its solutions are used as pH = 7 buffers.

e. ammonium chloride + NaOH

71. a. $CH_3CO_2H + OH^- \rightleftharpoons CH_3CO_2^- + H_2O$

$$K_{eq} = \frac{[OAc^-]}{[HOAc]\,[OH^-]} \times \frac{[H^+]}{[H^+]} = \frac{K_a}{K_w} = \frac{1.8 \times 10^{-5}}{1.0 \times 10^{-14}} = 1.8 \times 10^{9}$$

b. $CH_3CO_2^- + H^+ \rightleftharpoons CH_3CO_2H$

$$K_{eq} = \frac{[HOAc]}{[H^+]\,[OAc^-]} = \frac{1}{K_a} = 5.6 \times 10^{4}$$

c. $CH_3CO_2H + NH_3 \rightleftharpoons CH_3CO_2^- + NH_4^+$

$$K_{eq} = \frac{[OAc^-][NH_4^+]}{[HOAc][NH_3]} \times \frac{[H^+]}{[H^+]} = \frac{K_a\,(HOAc)}{K_a\,(NH_4^+)}$$

$$= \frac{K_a(HOAc)\,K_b(NH_3)}{K_w} = 3.2 \times 10^4$$

d. $HCl(aq) + NaOH\,(aq) \rightarrow NaCl(aq) + H_2O$

Net ionic equation is: $H^+ + OH^- \rightleftharpoons H_2O$

$$K_{eq} = \frac{1}{K_w} = 1.0 \times 10^{14}$$

Although all four reactions have large equilibrium constants, only (a), (b), and (d) can be useful in a titration. The solution in (c) results in a buffer; it still contains both a weak acid and a weak base. For a titration to be useful there must be a large change in pH for a small amount of added titrant near the equivalence point. This condition is precisely what a buffer is used to prevent.

73. At equivalence point, $C_6H_4(CO_2)_2^{2-} = P^{2-}$ is the major species. It is a weak base in water.

	P^{2-} + H_2O	$\rightleftharpoons$	HP^-	+ OH^-	
Initial	$\frac{0.5\text{ g}}{0.1\text{ L}} \times \frac{1\text{ mol}}{204.2\text{ g}} = 0.025\,M$		0	~0	(carry extra sig. fig.)
	x mol/L P^{2-} reacts with H_2O to reach equilibrium				
Change	$-x$	→	$+x$	$+x$	
Equil	$0.025 - x$		x	x	

$$K_b = \frac{[HP^-][OH^-]}{[P^{2-}]} = \frac{K_w}{K_{a_2}} = \frac{1.0 \times 10^{-14}}{10^{-5.51}} = 3.2 \times 10^{-9} = \frac{x^2}{0.025 - x} \approx \frac{x^2}{0.025}$$

$x = [OH^-] = 8.9 \times 10^{-6}$; pH = 8.9 Assumptions good.

Phenolphthalein would be the best indicator for this titration.

Note: Although HP^- is also a weak base, it is a much weaker base than P^{2-}. We can ignore the OH^- contribution from HP^-.

75. $Ca_5(PO_4)_3(OH)(s) \rightleftharpoons 5\,Ca^{2+} + 3\,PO_4^{3-} + OH^-$

s = solubility (mol/L) → 5s 3s $s + 1.0 \times 10^{-7} \approx s$

$K_{sp} = 6.8 \times 10^{-37} = [Ca^{2+}]^5\,[PO_4^{3-}]^3\,[OH^-] = (5s)^5(3s)^3(s)$

$6.8 \times 10^{-37} = (3125)(27)s^9$; $s = 2.7 \times 10^{-5}$ mol/L

The solubility of hydroxyapatite will increase as a solution gets more acidic, since both phosphate and hydroxide can react with H^+.

$Ca_5(PO_4)_3(F)$	$\rightleftharpoons$	$5\ Ca^{2+}$	+	$3\ PO_4^{3-}$	+	F^-
s = solubility (mol/L)	$\rightarrow$	5s		3s		s

$1 \times 10^{-60} = (5s)^5(3s)^3(s) = (3125)(27)s^9$; $s = 6 \times 10^{-8}$ mol/L

The hydroxyapatite in the tooth enamel is converted to the less soluble fluorapatite by fluoride treated water. The less soluble fluorapatite will then be more difficult to remove making teeth less susceptible to decay.

77. $Fe(OH)_3 \rightleftharpoons Fe^{3+} + 3\ OH^-$ $K_{sp} = [Fe^{3+}][OH^-]^3 = 4 \times 10^{-38}$

pH = 7.41; pOH = 6.59; $[OH^-] = 2.6 \times 10^{-7}$

$$[Fe^{3+}] = \frac{4 \times 10^{-38}}{(2.6 \times 10^{-7})^3} = 2 \times 10^{-18}\ M$$

The lowest level of iron in serum:

$$\frac{60 \times 10^{-6}\ g}{0.1\ L} \times \frac{1\ mol}{55.85\ g} = 1 \times 10^{-5}\ M$$

The actual concentration of iron is much greater than expected when considering only the solubility of $Fe(OH)_3$. Thus, there must be complexing agents present to increase the solubility.

79.

	Cr^{3+}	+ H_2EDTA^{2-}	$\rightleftharpoons$	$CrEDTA^-$	+ $2\ H^+$	
Before	0.0010 M	0.050 M		0	$1 \times 10^{-6}\ M$	(buffer pH = 6.0)
Change	-0.0010	-0.0010	$\rightarrow$	+0.0010	buffered (no change)	(React completely)
After (New Initial)	0	0.049		0.0010	1×10^{-6} (buffer)	
	x mol/L $CrEDTA^-$ dissociates to reach equilibrium					
Change	$+x$	$+x$	$\leftarrow$	$-x$	-----	
Equil	x	$0.049 + x$		$0.0010 - x$	1×10^{-6} (buffer)	

$$K_f = 1.0 \times 10^{23} = \frac{[CrEDTA^-][H^+]^2}{[Cr^{3+}][H_2EDTA^{2-}]} = \frac{(0.0010 - x)(1 \times 10^{-6})^2}{(x)(0.049 + x)}$$

$$1.0 \times 10^{23} \approx \frac{(0.0010)(1 \times 10^{-12})}{(x)(0.049)}$$

$x = [Cr^{3+}] = 2 \times 10^{-37}\ M$ Assumptions good.

CHALLENGE PROBLEMS

81. Phenolphthalein will change color at pH ~ 9. Phenolphthalein will mark the second end point. Therefore, at the phenolphthalein end point we will have titrated both protons on malonic acid.

$$H_2Mal + 2\ OH^- \longrightarrow 2\ H_2O + Mal^{2-}$$

$$31.50\ mL \times \frac{0.0984\ mmol\ NaOH}{mL} \times \frac{1\ mmol\ H_2Mal}{2\ mmol\ NaOH} = 1.55\ mmol\ H_2Mal$$

$$[H_2Mal] = \frac{1.55\ mmol}{25.00\ mL} = 0.0620\ M$$

83. It will be more soluble in base. Addition of OH^- will effectively remove H^+:
$H^+ + OH^- \longrightarrow H_2O$. This drives the equilibrium:

$$H_4SiO_4 \rightleftharpoons H_3SiO_4^- + H^+$$

to the right. This in turn, drives the solubility reaction:

$$SiO_2(s) + 2\ H_2O \rightleftharpoons H_4SiO_4$$

to the right, increasing the solubility of the silica.

85. $K_{sp} = 6.4 \times 10^{-9} = [Mg^{2+}]\ [F^-]^2$

$6.4 \times 10^{-9} = (0.00375 - y)\ (0.0625 - 2y)^2$

This is a cubic equation. No simplifying assumptions can be made since y is relatively large. There is a formula for solving a cubic equation (I've heard), but I've never seen it. We could use a computer or a programmable calculator to solve the equation by numerical methods. By the time we've done all that, we could have solved the problem several times over using the approximations based on our "chemical common sense."

87.

$$Cu(OH)_2 \rightleftharpoons Cu^{2+} + 2\ OH^- \qquad K_{sp} = 1.6 \times 10^{-19}$$

$$Cu^{2+} + 4\ NH_3 \rightleftharpoons Cu(NH_3)_4^{2+} \qquad K_f = 1.0 \times 10^{13}$$

$$Cu(OH)_2 + 4\ NH_3 \rightleftharpoons Cu(NH_3)_4^{2+} + 2\ OH^- \qquad K = K_{sp}K_f = 1.6 \times 10^{-6}$$

89. $1.0\ mL \times \frac{1.0\ mmol}{mL} = 1.0\ mmol\ Cd^{2+}$ added to the ammonia solution.

Thus, $[Cd^{2+}]_o = 1.0 \times 10^{-3}$ mol/L. We will treat each equilibrium separately, considering the one with the largest K first.

	Cd^{2+}	+	$4\ NH_3$	$\rightleftharpoons$	$Cd(NH_3)_4^{2+}$	
Before	$1.0 \times 10^{-3}\ M$		$5.0\ M$		0	
Change	-1.0×10^{-3}		-4.0×10^{-3}	$\rightarrow$	$+1.0 \times 10^{-3}$	React completely
After (New Initial)	0		$4.996 \approx 5.0$		1.0×10^{-3}	

x mol/L $Cd(NH_3)_4^{2+}$ dissociates to reach equilibrium

Change	$+x$	$+4x$	$\leftarrow$	$-x$
Equil	x	$5.0 + 4x$		$0.0010 - x$

$$K_f = 1.0 \times 10^7 = \frac{(0.0010 - x)}{(x)\ (5.0 + 4x)^4} \approx \frac{(0.0010)}{(x)\ (5.0)^4}$$

$x = [Cd^{2+}] = 1.6 \times 10^{-13}\ M$. Assumptions good. This is the maximum $[Cd^{2+}]$ possible. If a precipitate forms, $[Cd^{2+}]$ will be less than this value. Calculate the pH of a 5.0 M NH_3 solution.

	NH_3 + H_2O	$\rightleftharpoons$	NH_4^+	+	OH^-	$K_b = 1.8 \times 10^{-5}$
Initial	$5.0\ M$		0		~0	

y mol/L NH_3 reacts with H_2O to reach equilibrium

Change	$-y$	$\rightarrow$	$+y$	$+y$
Equil	$5.0 - y$		y	y

$$K_b = 1.8 \times 10^{-5} = \frac{[NH_4^+]\ [OH^-]}{[NH_3]} = \frac{y^2}{5.0 - y} \approx \frac{y^2}{5.0}$$

$y = [OH^-] = 9.5 \times 10^{-3}\ M$ Assumptions good.

We now calculate the value of the solubility quotient.

$[Cd^{2+}]\ [OH^-]^2 = (1.6 \times 10^{-13})\ (9.5 \times 10^{-3})^2 = 1.4 \times 10^{-7}$

$1.4 \times 10^{-17} < K_{sp}\ (5.9 \times 10^{-15})$ Therefore, no precipitate forms.

CHAPTER SIXTEEN: SPONTANEITY, ENTROPY AND FREE ENERGY

QUESTIONS

1. A spontaneous process is one that occurs without any outside intervention.

3. a. Entropy increases; there is a greater volume accessible to the randomly moving gas molecules. Positional disorder will increase.

   b. The positional entropy doesn't change. There is no change in volume and thus, no change in the numbers of position of the molecules. The total entropy increases because the increase in temperature increases the energy disorder.

   c. Entropy decreases since volume decreases.

5. Living organisms need an external energy source to carry out these processes. Green plants use the energy from sunlight to produce glucose from carbon dioxide and water by photosynthesis. The energy released from the metabolism of glucose (food) helps drive the synthesis of proteins. A living cell is not a closed system, i.e. a system that has no external energy source.

7. The more widely something is dispersed, the greater the disorder. We must do work to overcome this disorder. In terms of the 2nd law it would be more advantageous to prevent contamination of the environment rather than clean it up later.

EXERCISES

Spontaneity and Entropy

9. a and c. Note: "a" requires external energy because it is an endothermic process. It is spontaneous, however, since ΔS is favorable and provides the driving force.

11. We draw all of the possible arrangements of the two particles in the three levels.

2kJ	___	___	X	___	X	XX
1 kJ	___	X	___	XX	X	___
0 kJ	XX	X	X	___	___	___
Total E =	0kJ	1kJ	2kJ	2kJ	3kJ	4kJ

The most likely total energy is 2kJ.

13.

2kJ	___	___	X	___	X	___
1kJ	___	X	___	XX	X	XXX
0kJ	XXX	XX	XX	X	X	___
E_T (kJ) =	0	1	2	2	3	3

2kJ	xx	x	xx	xxx
1kJ	___	xx	x	___
0kJ	x	___	___	___
E_T (kJ) =	4	4	5	6

A total energy of 2, 3, or 4 kJ is equally probable.

15. Of the three phases (solid, liquid, and gas), solids are most ordered and gases are most disordered. Thus, a, b, and c involve an increase in entropy.

17. a. N_2O: It is a more complex molecule, larger positional disorder.

b. H_2 at 100°C and 0.5 atm: Higher temperature and lower pressure means greater volume and hence, greater positional entropy.

c. N_2 at STP has the greater volume. d. H_2O (l)

Entropy and the Second Law of Thermodynamics: Free Energy

19. a. $C_{12}H_{22}O_{11}$ larger molecule, more complex, larger positional disorder.

b. H_2O (0°C) higher temp., S = 0 at 0 K.

c. $H_2S(g)$ A gas has greater disorder than a liquid.

21. a. Decrease in disorder; ΔS (-) b. Increase in disorder; ΔS (+)

c. Decrease in disorder; ΔS (-) The number of moles of gas decreases.

23. a. $H_2(g) + 1/2\ O_2(g) \rightarrow H_2O(g)$

$\Delta S° = S°(H_2O(g)) - [S°(H_2) + 1/2\ S°(O_2)]$

$\Delta S° = 1$ mol (189 J/K•mol) - [1 mol (131 J/K•mol) + 1/2 mol (205 J/K•mol)]

$\Delta S° = 189$ J/K - 234 J/K = -45 J/K

b. $3\ O_2(g) \rightarrow 2\ O_3(g)$

$\Delta S° = 2$ mol(239 J/mol•K) - 3 mol (205 J/mol•K) = -137 J/K

c. $N_2(g) + O_2(g) \rightarrow 2\ NO(g)$

$\Delta S° = 2(211) - (192 + 205) = +25$ J/K

25. $CS_2(g) + 3\ O_2(g) \rightarrow CO_2(g) + 2\ SO_2(g)$

$\Delta S° = S°(CO_2) + 2\ S°(SO_2) - [3\ S°(O_2) + S°(CS_2)]$

-143 J/K = 214 J/K + 2(248 J/K) - 3(205 J/K) - (1 mol) S°

S° = 238 J/K•mol

27. a. $\Delta G = \Delta H - T\Delta S$; $\Delta G = 25 \times 10^3$ J - (300 K) (5 J/K)

Note: We must be consistent on units. Typically enthalpy values are in kJ and entropy in J/K. We must use the same energy units for both if we are using the equation $\Delta G = \Delta H - T\Delta S$.

$\Delta G = +$ 23,500 J $\approx$ 24,000 K Not spontaneous

b. $\Delta G =$ 25,000 J - (300 K) (100 J/K) = -5000 J Spontaneous

c. Without calculating ΔG, we know this reaction will be spontaneous at all temperatures. ΔH is negative and ΔS is positive ($-T\Delta S < 0$). ΔG will always be less than zero.

d. $\Delta G =$ -10,000 J - (200 K) (-40 J/K) = -2000 J Spontaneous

29. At the boiling point $\Delta G = 0$, so $\Delta H = T\Delta S$

$$\Delta S = \frac{\Delta H}{T} = \frac{31.4 \text{ kJ/mol}}{(273.2 + 61.7) \text{ K}} = 9.38 \times 10^{-2} \text{ kJ/mol•K}$$

or 93.8 J/K•mol

Free Energy and Chemical Reactions

31. a. $CH_4(g) + 2\,O_2(g) \longrightarrow CO_2(g) + 2\,H_2O\,(g)$

	$CH_4(g)$	$2\,O_2(g)$	$CO_2(g)$	$2\,H_2O\,(g)$	
ΔH_f°	-76 kJ/mol	0	-393.5	-242	
ΔG_f°	-51 kJ/mol	0	-394	-229	Data from Appendix 4
S°	186 J/K•mol	205	214	189	

$\Delta H^\circ =$ 2 mol (-242 kJ/mol) + 1 mol (-393.5 kJ/mol) - [1 mol (-75 kJ/mol)] = -803 kJ

$\Delta S^\circ =$ 2 mol (189 J/K•mol) + 1 mol (214 J/K•mol)

- [1 mol (186 J/K•mol) + 2 mol (205 J/K•mol)] = -4 J/K

There are two ways to get ΔG°:

$\Delta G^\circ = \Delta H^\circ - T\Delta S^\circ = -803 \times 10^3$ J - 298 K (-4 J/K)

$= -8.018 \times 10^5$ J = -802 kJ

or from ΔG_f°, we get

$\Delta G^\circ =$ 2 mol (-229 kJ mol) + 1 mol (-394 kJ/mol) - [1 mol (-51 kJ/mol)]

$\Delta G^\circ =$ -801 kJ (Answers are the same within roundoff error)

b. $6\ CO_2(g) + 6\ H_2O(l) \rightarrow C_6H_{12}O_6(s) + 6\ O_2(g)$

	$6\ CO_2(g)$	$6\ H_2O(l)$	$C_6H_{12}O_6(s)$	$6\ O_2(g)$
ΔH_f°	-393.5 kJ/mol	-286	-1275	0
S°	214 J/K•mol	70	212	205

$\Delta H^\circ = -1275 - [6(-286) + 6(-393.5)] = 2802$ kJ

$\Delta S^\circ = 6(205) + 212 - [6(214) + 6(70)] = -262$ J/K

$\Delta G^\circ = 2802$ kJ - 298 K (-0.262 kJ/K) = 2880. kJ

c. $P_4O_{10}(s) + 6\ H_2O(l) \rightarrow 4\ H_3PO_4(s)$

	$P_4O_{10}(s)$	$6\ H_2O(l)$	$4\ H_3PO_4(s)$
ΔH_f° (kJ/mol)	-2984	-286	-1279
S° (J/K•mol)	229	70	110

$\Delta H^\circ = 4$ mol (-1279 kJ/mol) - [1 mol (-2984 kJ/mol) + 6 mol (-286 kJ/mol)]

$\Delta H^\circ = -416$ kJ

$\Delta S^\circ = 4(110) - [229 + 6(70)] = -209$ J/K

$\Delta G^\circ = \Delta H^\circ - T\Delta S^\circ = -416$ kJ - 298 K(-0.209 kJ/K) = -354 kJ

d. $HCl(g) + NH_3(g) \rightarrow NH_4Cl(s)$

	$HCl(g)$	$NH_3(g)$	$NH_4Cl(s)$
ΔH_f° (kJ/mol)	-92	-46	-314
S° (J/K•mol)	187	193	96

$\Delta H^\circ = -314 - [-92 - 46] = -176$ kJ

$\Delta S^\circ = 96 - [187 + 193] = -284$ J/K

$\Delta G^\circ = \Delta H^\circ - T\Delta S^\circ = -176$ kJ - (298 K)(-0.284 kJ/K) = -91 kJ

33. $SF_4(g) + F_2(g) \rightarrow SF_6(g)$

-374 kJ = -1105 kJ - $\Delta G_f^\circ(SF_4)$

$\Delta G_f^\circ = -731$ kJ/mol

35. $P_4\ (s,\alpha) \longrightarrow P_4\ (s,\beta)$

a. At T < -76.9°C, this reaction is spontaneous. Thus, the sign of ΔG is (-). At 76.9°C, $\Delta G = 0$ and above 76.9 °C, the sign of ΔG is (+). This is consistent with ΔH (-) and ΔS (-).

b. Since the sign of ΔS is negative, then the β form has the more ordered structure.

37. $H_2(g) \longrightarrow 2\ H(g)$

a. A bond is broken; ΔH (+); Increase in disorder; ΔS (+)

b. $\Delta G = \Delta H - T\Delta S$

For the reaction to be spontaneous, the entropy term must be dominate. The reaction will be spontaneous at higher temperatures.

39. $NO(g) + O_3(g) \rightleftharpoons NO_2(g) + O_2(g)$

$\Delta G° = \Sigma\ \Delta G_f^\circ$ (Products - $\Sigma\ \Delta G_f^\circ$ (Reactants)

$= 1$ mol (52 kJ/mol) - [1 mol (87 kJ/mol) + 1 mol (163 kJ/mol)]

$\Delta G° = -198$ kJ and $\Delta G° = -RT \ln K$

$$K = \exp \frac{-\Delta G°}{RT} = \exp\left(\frac{+\ 1.98 \times 10^5\ J}{8.3145\ J/K{\cdot}mol\ (298\ K)}\right) = e^{79.912} = 5.07 \times 10^{34}$$

41. a. $\Delta G° = -RT \ln K$

$\Delta G° = -(8.3145\ J/K{\cdot}mol)\ (298\ K) \ln (1.00 \times 10^{-14}) = 7.99 \times 10^4\ J/mol = 79.9\ kJ/mol$

b. $\Delta G° = -RT \ln K = -(8.3145\ J/K{\cdot}mol)\ (313\ K) \ln (2.92 \times 10^{-14})$

$= 8.11 \times 10^4\ J/mol = 81.1\ kJ/mol$

Free Energy and Pressure

43. From Exercise 16.39, we get $\Delta G° = -198$ kJ

$$\Delta G = \Delta G° + RT \ln \frac{P_{NO_2} P_{O_2}}{P_{NO} P_{O_3}}$$

$$\Delta G = -198\ kJ + \frac{8.3145\ J/K{\cdot}mol}{1000\ J/kJ} (298\ K) \ln \frac{(1.00 \times 10^{-7})\ (1.00 \times 10^{-3})}{(1.00 \times 10^{-6})\ (2.00 \times 10^{-6})}$$

$= -198$ kJ + 9.69 kJ = -188 kJ

45. $N_2(g) + 3\ H_2(g) \rightleftharpoons 2\ NH_3(g)$

$\Delta H° = 2\ \Delta H_f° (NH_3) = 2(-46) = -92$ kJ

$\Delta G° = 2\ \Delta G_f° (NH_3) = 2(-17) = -34$ kJ

$\Delta S° = 2(193) - [192 + 3(131)]$

$\Delta S° = -199$ J/K

$$K = \exp \frac{-\Delta G°}{RT} = \exp\left(\frac{34{,}000 \text{ J}}{(8.3145 \text{ J/K•mol})\ (298 \text{ K})}\right)$$

$K = e^{13.72} = 9.1 \times 10^5$

a. $$\Delta G = \Delta G° + RT \ln \frac{P^2_{NH_3}}{P_{N_2} P^3_{H_2}}$$

$$\Delta G = -34 \text{ kJ} + \frac{(8.3145 \text{ J/K•mol})\ (298 \text{ K})}{1000 \text{ J/kJ}} \ln \frac{(50.)^2}{(200.)\ (200.)^3}$$

$\Delta G = -34$ kJ $- 33$ kJ $= -67$ kJ

b. $$\Delta G = -34 \text{ kJ} + \frac{(8.3145 \text{ J/K•mol})\ (298 \text{ K})}{1000 \text{ J/kJ}} \ln \frac{(200.)^2}{(200.)\ (600.)^3}$$

$\Delta G = -34$ kJ $- 34.4$ kJ $= -68$ kJ

c. $\Delta G°_{100} = \Delta H° - T\Delta S°$, $\Delta G°$ depends on T. Assume $\Delta H°$ and $\Delta S°$ are T independent.

$\Delta G°_{100} = -92$ kJ $-$ (100. K) (-0.199 kJ/K)

$\Delta G°_{100} = -72$ kJ

$\Delta G = \Delta G° + RT \ln Q$

$$\Delta G = -72 \text{ kJ} + \frac{(8.3145 \text{ J/K•mol})\ (100. \text{ K})}{1000 \text{ J/kJ}} \ln \frac{(10.)^2}{(50.)\ (200.)^3} = -72 - 13 = -85 \text{ kJ}$$

d. $\Delta G°_{700} = -92$ kJ $-$ (700. K) (-0.199 kJ/K) $= 47$ kJ

$$\Delta G = 47 \text{ kJ} + \frac{(8.3145 \text{ J/K•mol})\ (700. \text{ K})}{1000 \text{ J/kJ}} \ln \frac{(10.)^2}{(50.)\ (200.)^3}$$

$\Delta G = 47$ kJ $- 88$ kJ $= -41$ kJ

47. a.

$$C\equiv O(g) + 2\,H{-}H(g) \longrightarrow H{-}\underset{\displaystyle H}{\overset{\displaystyle H}{\underset{|}{\overset{|}{C}}}}{-}O{-}H(g)$$

$\Delta S° \approx 100(-2) = -200$ J/K

Bonds broken:		Bonds made:	
$C\equiv O$	1072 kJ/mol	3 C—H	413 kJ/mol
2 H—H	432 kJ/mol	1 C—O	358 kJ/mol
		1 O—H	467 kJ/mol

$\Delta H° = 1072 + 2(432) - [3(413) + 358 + 467] = -128$ kJ

$\Delta G° = \Delta H° - T\Delta S° = -128\text{ kJ} - (298\text{ K})(-0.2\text{ kJ/K}) = -68\text{ kJ} \approx -70\text{ kJ}$

b. $\Delta H° = -201 - [-110.5] = -91$ kJ

$\Delta S° = 240 - [198 + 2(131)] = -220.$ J/K

$\Delta G° = -91\text{ kJ} - (298\text{ K})(-0.220\text{ kJ/K}) = -25$ kJ

The values are reasonably close for $\Delta H°$ and $\Delta S°$ (within 30%). $\Delta G°$ is the right order of magnitude. Even a factor of 3 difference (as in $\Delta G°$) translates into less than 1 power of 10 in the equilibrium constant. The approximations give good ball park estimates.

c. Since ΔS is unfavorable and ΔH is favorable, we would run the reaction at relatively low temperatures.

$\Delta G = 0$; $\Delta H - T\Delta S = 0$

$$T = \frac{\Delta H}{\Delta S} = \frac{91 \times 10^3\text{ J}}{220\text{ J/K}} = 410\text{ K}$$

Reaction would be spontaneous at $T < 410$ K. Note: We may need high temperatures to get the reaction to occur at a reasonable rate.

ADDITIONAL EXERCISES

49. There are more ways to roll a seven. We can consider all of the possible throws by constructing a table.

one die	1	2	3	4	5	6	
1	2	3	4	5	6	7	
2	3	4	5	6	7	8	
3	4	5	6	7	8	9	sum of the two dice
4	5	6	7	8	9	10	
5	6	7	8	9	10	11	
6	7	8	9	10	11	12	

There are six ways to get a seven, more than any other number. The seven is not favored by energy; rather it is favored by probability. To change the probability we would have to expend energy (do work).

51. $\Delta G = 0$ so $\Delta H = T\Delta S$

$$\Delta S = \frac{\Delta H}{T} = \frac{35.23 \times 10^3 \text{ J/mol}}{3650 \text{ K}} = 9.65 \text{ J/K•mol}$$

53. It appears that the sum of the two processes is no net change. This is not so, ΔS_{univ} has increased even though it looks as if we have gone through a cyclic process.

55. $CH_4(g) + CO_2(g) \longrightarrow CH_3CO_2H\ (l)$

$\Delta H° = -484 - [-75 + (-393.5)] = -16$ kJ

$\Delta S° = 160 - [186 + 214] = -240.$ J/K

$\Delta G° = \Delta H° - T\Delta S° = -16$ kJ - (298 K) (-0.240 kJ/K) = +56 kJ

This reaction is spontaneous only at a temperature below $T = \frac{\Delta H°}{\Delta S°} = 67$ K.

This is not practical. Substances will be in condensed phases and rates will be very slow.

$CH_3OH(g) + CO(g) \longrightarrow CH_3CO_2H(l)$

$\Delta H° = -484 - [-110.5 + (-201)] = -173$ kJ

$\Delta S° = 160 - [198 + 240] = -278$ J/K

$\Delta G° = -173$ kJ - (298 K) (-0.278 kJ/K) = -90. kJ

This reaction is spontaneous at a temperature below $T = \frac{\Delta H°}{\Delta S°} = 622$ K.

Thus, the reaction of CH_3OH and CO will be preferred. It is spontaneous at high enough temperatures that the rates of reaction should be reasonable.

57. K^+ (blood) $\rightleftharpoons$ K^+ (muscle) $\Delta G° = 0$, so $\Delta G = RT \ln\left(\frac{[K^+]_m}{[K^+]_b}\right)$

$$\Delta G = \frac{8.3145 \text{ J}}{\text{K mol}}(310.\text{ K}) \ln\left(\frac{0.15}{0.0050}\right); \ \Delta G = 8.8 \times 10^3 \text{ J/mol} = 8.8 \text{ kJ/mol} = \text{work}$$

$$\frac{8.8 \text{ kJ}}{\text{mol } K^+} \times \frac{1 \text{ mol ATP}}{30.5 \text{ kJ}} = 0.29 \text{ mol ATP}$$

Other ions will have to be transported in order to maintain electroneutrality. Either anions must be transported into the cells, or cations (Na^+) in the cell must be transported to the blood. The latter is what happens: $[Na^+]$ in blood is greater than $[Na^+]$ in cells as a result of this pumping.

59. As ΔS_{univ} continually increases, there will be less energy available for work. When no work can be done, the world ends.

CHALLENGE PROBLEMS

61. Arrangement I:

$S = k \ln W$
$W = 1$
$S = k \ln 1 = 0$

Arrangement II.

$W = 4$
$S = k \ln 4 = 1.38 \times 10^{-23}$ J/K ln 4
$S = 1.91 \times 10^{-23}$ J/K

Arrangement III:

$W = 6$
$S = k \ln 6 = 2.47 \times 10^{-23}$ J/K

63. We shall graph ln K vs 1/T

Temp (°C)	T (K)	1000/T	K	ln K
0	273	3.66	1.14×10^{-15}	-34.41
25	298	3.36	1.00×10^{-14}	-32.24
35	308	3.25	2.09×10^{-14}	-31.50
40	313	3.19	2.92×10^{-14}	-31.17
50	323	3.10	5.47×10^{-14}	-30.54

The graph was drawn previously for Exercise 14.29. The equation of the straight line is:

$$\ln K = -6.91 \times 10^3 \left(\frac{1}{T}\right) - 9.09$$

$$\text{Slope} = -6.91 \times 10^3 = \frac{-\Delta H^\circ}{R}; \qquad \Delta H^\circ = 5.74 \times 10^4 \text{ J} = 57.4 \text{ kJ}$$

$$\text{Intercept} = -9.09 = \frac{\Delta S^\circ}{R}; \qquad \Delta S^\circ = -75.6 \text{ J/K}$$

65. a. $\Delta G^\circ = -RT \ln K$

$K = \exp(-\Delta G^\circ/RT)$

$$K = \exp\left(\frac{30{,}500 \text{ J/mol}}{8.3145 \text{ J/K}\cdot\text{mol} \times 298 \text{ K}}\right) = 2.22 \times 10^5$$

b. $C_6H_{12}O_6(s) + 6\, O_2(g) \longrightarrow 6\, CO_2(g) + 6\, H_2O(l)$

$\Delta G^\circ = 6 \text{ mol } (-394 \text{ kJ/mol}) + 6 \text{ mol } (-237 \text{ kJ/mol}) - 1 \text{ mol } (-911 \text{ kJ/mol})$

$\Delta G^\circ = -2875 \text{ kJ}$

$$\frac{2875 \text{ kJ}}{\text{mol glucose}} \times \frac{1 \text{ mol ATP}}{30.5 \text{ kJ}} = 94.3 \text{ mol ATP}$$

This is an overstatement. The assumption that all of the free energy goes into this reaction is false. Actually only 38 moles of ATP are produced by metabolism of one mole of glucose.

67. The hydration of the dissociated ions results in a more ordered "structure" in solution. The smallest ion (F^-) has the largest charge density and should show the most ordering.

69. At equilibrium:

$$P_{H_2} = \frac{nRT}{V} = \frac{\left(\frac{1.10 \times 10^{13} \text{ molecules}}{6.022 \times 10^{23} \text{ molecules/mol}}\right)\left(\frac{0.08206 \text{ L atm}}{\text{mol K}}\right)(298 \text{ K})}{1.0 \text{ L}}$$

$P_{H_2} = 4.5 \times 10^{-10}$ atm

Essentially all of the H_2 and Br_2 have reacted, therefore $P_{HBr} = 2.0$ atm.

Since we began with equal moles of H_2 and Br_2, we will have equal moles of H_2 and Br_2 at equilibrium. Therefore, $P_{H_2} = P_{Br_2} = 4.5 \times 10^{-10}$ atm

$$K = \frac{P_{HBr}^2}{P_{H_2}P_{Br_2}} = \frac{(2.0)^2}{(4.5 \times 10^{-10})^2} = 2.0 \times 10^{19}$$

$\Delta G^\circ = -RT \ln K = -(8.3145 \text{ J/K}\cdot\text{mol})(298 \text{ K}) \ln(2.0 \times 10^{19})$

$\Delta G^\circ = -1.1 \times 10^5 \text{ J/mol} = -110 \text{ kJ/mol}$

$\Delta G^\circ = \Delta H^\circ - T\Delta S^\circ$

$$\Delta S^\circ = \frac{\Delta H^\circ - \Delta G^\circ}{T} = \frac{-103{,}800 \text{ J/mol} - (-110{,}000 \text{ J/mol})}{298 \text{ K}}$$

$\Delta S^\circ = 21 \text{ J/K}\cdot\text{mol} = 20 \text{ J/K}\cdot\text{mol}$

CHAPTER SEVENTEEN: ELECTROCHEMISTRY

REVIEW OF OXIDATION - REDUCTION REACTIONS

1. Oxidation: increase in oxidation number
   loss of electrons

   Reduction: decrease in oxidation number
   gain of electrons

2. a. H (+1), 0 (-2), N (+5) b. Cl (-1), Cu (+2)

   c. zero d. H (+1), 0 (-1)

   e. Mg (+2), 0 (-2), S (+6) f. zero

   g. Pb (+2), 0 (-2), S (+6) h. O (-2), Pb (+4)

   i. Na (+1), 0 (-2), C (+3) j. O (-2), C (+4)

   k. $(NH_4)_2Ce(SO_4)_3$ contains NH_4^+ ions and SO_4^{2-} ions. Thu, cerium exists as the Ce^{4+} ion.
   H (+1), N (-3), Ce (+4), S (+6), O (-2)

   l. O (-2), Cr (+3)

3.

	Redox?	Ox. Agent	Red. Agent	Substance Oxidized	Substance Reduced
a.	Yes	H_2O	CH_4	CH_4 (C)	H_2O (H)
b.	Yes	$AgNO_3$	Cu	Cu	$AgNO_3$ (Ag)
c.	No	-----	-----	-----	-----
d.	Yes	HCl	Zn	Zn	HCl (H)
e.	No	-----	-----	-----	-----
f.	Yes	HCl	Fe	Fe	HCl (H)

4. a. $Cr \rightarrow Cr^{3+} + 3\,e^-$

$$NO_3^- \rightarrow NO$$

$$4\,H^+ + NO_3^- \rightarrow NO + 2\,H_2O$$

$$3\,e^- + 4\,H^+ + NO_3^- \rightarrow NO + 2\,H_2O$$

$$Cr \rightarrow Cr^{3+} + 3\,e^-$$

$$3\,e^- + 4\,H^+ + NO_3^- \rightarrow NO + 2\,H_2O$$

$$4\,H^+(aq) + NO_3^-(aq) + Cr(s) \rightarrow Cr^{3+}(aq) + NO(g) + 2\,H_2O(l)$$

b. $(Al \rightarrow Al^{3+} + 3\ e^-) \times 5$

$$MnO_4^- \rightarrow Mn^{2+}$$
$$8\ H^+ + MnO_4^- \rightarrow Mn^{2+} + 4\ H_2O$$
$$(5\ e^- + 8\ H^+ + MnO_4^- \rightarrow Mn^{2+} + 4\ H_2O) \times 3$$

$$5\ Al \rightarrow 5\ Al^{3+} + 15\ e^-$$
$$15\ e^- + 24\ H^+ + 3\ MnO_4^- \rightarrow 3\ Mn^{2+} + 12\ H_2O$$

$$24\ H^+(aq) + 3\ MnO_4^-(aq) + 5\ Al(s) \rightarrow 5\ Al^{3+}(aq) + 3\ Mn^{2+}(aq) + 12\ H_2O(l)$$

c. $(Ce^{4+} + e^- \rightarrow Ce^{3+}) \times 6$

$$CH_3OH \rightarrow CO_2$$
$$H_2O + CH_3OH \rightarrow CO_2 + 6\ H^+$$
$$H_2O + CH_3OH \rightarrow CO_2 + 6\ H^+ + 6\ e^-$$

$$6\ Ce^{4+} + 6\ c^- \rightarrow 6\ Cc^{3+}$$
$$H_2O + CH_3OH \rightarrow CO_2 + 6\ H^+ + 6\ e^-$$

$$H_2O(l) + CH_3OH(aq) + 6\ Ce^{4+}(aq) \rightarrow 6\ Ce^{3+}(aq) + CO_2(g) + 6\ H^+(aq)$$

d.

$$SO_3^{2-} \rightarrow SO_4^{2-}$$
$$H_2O + SO_3^{2-} \rightarrow SO_4^{2-} + 2\ H^+$$
$$(H_2O + SO_3^{2-} \rightarrow SO_4^{2-} + 2\ H^+ + 2\ e^-) \times 5$$

$$(5\ e^- + 8\ H^+ + MnO_4^- \rightarrow Mn^{2+} + 4\ H_2O) \times 2$$
See 17.4 b

$$5\ H_2O + 5\ SO_3^{2-} \rightarrow 5\ SO_4^{2-} + 10\ H^+ + 10\ e^-$$
$$10\ e^- + 16\ H^+ + 2\ MnO_4^- \rightarrow 2\ Mn^{2+} + 8\ H_2O$$

$$6\ H^+(aq) + 5\ SO_3^{2-}\ (aq) + 2\ MnO_4^-\ (aq) \rightarrow 5\ SO_4^{2-}\ (aq) + 2\ Mn^{2+}(aq) + 3\ H_2O(l)$$

e.

$$PO_3^{3-} \rightarrow PO_4^{3-}$$
$$H_2O + PO_3^{3-} \rightarrow PO_4^{3-} + 2\ H^+ + 2\ e^-$$
$$2\ OH^- + H_2O + PO_3^{3-} \rightarrow PO_4^{3-} + 2\ H^+ + 2\ OH^- + 2\ e^-$$
$$(2\ OH^- + PO_3^{3-} \rightarrow PO_4^{3-} + H_2O + 2\ e^-) \times 3$$

$$MnO_4^- \rightarrow MnO_2$$
$$3\ e^- + 4\ H^+ + MnO_4^- \rightarrow MnO_2 + 2\ H_2O$$
$$3\ e^- + 4\ H^+ + 4\ OH^- + MnO_4 \rightarrow MnO_2 + 2\ H_2O + 4\ OH^-$$
$$(3\ e^- + 2\ H_2O + MnO_4^- \rightarrow MnO_2 + 4\ OH^-) \times 2$$

$$6\ OH^- + 3\ PO_3^{3-} \rightarrow 3\ PO_4^{3-} + 3\ H_2O + 6\ e^-$$
$$6\ e^- + 4\ H_2O + 2\ MnO_4^- \rightarrow 2\ MnO_2 + 8\ OH^-$$

$$H_2O\ (l) + 3\ PO_3^{3-}(aq) + 2\ MnO_4^-(aq) \rightarrow 3\ PO_4^{3-}(aq) + 2\ MnO_2(s) + 2\ OH^-(aq)$$

f.

$$Mg \rightarrow Mg(OH)_2$$
$$2\ OH^- + Mg \rightarrow Mg(OH)_2 + 2\ e^-$$

$$OCl^- \rightarrow Cl^-$$
$$2\ e^- + 2\ H^+ + OCl^- \rightarrow Cl^- + H_2O$$
$$2\ e^- + 2\ OH^- + 2\ H^+ + OCl^- \rightarrow Cl^- + H_2O + 2\ OH^-$$
$$2\ e^- + H_2O + OCl^- \rightarrow Cl^- + 2\ OH^-$$

$$2\ OH^- + Mg \rightarrow Mg(OH)_2 + 2\ e^-$$
$$2\ e^- + H_2O + OCl^- \rightarrow Cl^- + 2\ OH^-$$

$$H_2O(l) + Mg(s) + OCl^-(aq) \rightarrow Mg(OH)_2(s) + Cl^-(aq)$$

g.

$$Ce^{4+} \rightarrow Ce(OH)_3$$
$$(e^- + 3\ OH^- + Ce^{4+} \rightarrow Ce(OH)_3) \times 40$$

$$Ni(CN)_4^{2-} \rightarrow Ni(OH)_2 + CO_3^{2-} + NO_3^-$$
$$2\ OH^- + Ni(CN)_4^{2-} \rightarrow Ni(OH)_2 + 4\ CO_3^{2-} + 4\ NO_3^-$$
$$24\ H_2O + 2\ OH^- + Ni(CN)_4^{2-} \rightarrow Ni(OH)_2 + 4\ CO_3^{2-} + 4\ NO_3^- + 48\ H^+ + 40\ e^-$$
$$24\ H_2O + 48\ OH^- + 2\ OH^- + Ni(CN)_4^{2-} \rightarrow Ni(OH)_2 + 4\ CO_3^{2-} + 4\ NO_3^{2-} + 48\ H^+ + 48\ OH^- + 40\ e^-$$
$$50\ OH^- + Ni(CN)_4^{2-} \rightarrow Ni(OH)_2 + 4\ CO_3^{2-} + 4\ NO_3^- + 24\ H_2O + 40\ e^-$$

$$40\ e^- + 120\ OH^- + 40\ Ce^{4+} \rightarrow 40\ Ce(OH)_3$$
$$50\ OH^- + Ni(CN)_4^{2-} \rightarrow Ni(OH)_2 + 4\ CO_3^{2-} + 4\ NO_3^- + 24\ H_2O + 40\ e^-$$

$$170\ OH^-(aq) + 40\ Ce^{4+}(aq) + Ni(CN)_4^{2-}(aq) \rightarrow Ni(OH)_2(s) + 40\ Ce(OH)_3(s) + 4\ CO_3^{2-}(aq) + 4\ NO_3^-(aq) + 24\ H_2O(l)$$

h.

$$H_2CO \rightarrow HCO_3^-$$
$$2\,H_2O + H_2CO \rightarrow HCO_3^- + 5\,H^+ + 4\,e^-$$
$$5\,OH^- + 2\,H_2O + H_2CO \rightarrow HCO_3^- + 5\,H^+ + 5\,OH^- + 4\,e^-$$
$$5\,OH^- + H_2CO \rightarrow HCO_3^- + 3\,H_2O + 4\,e^-$$

$$Ag(NH_3)_2^+ \rightarrow Ag + 2\,NH_3$$
$$(e^- + Ag(NH_3)_2^+ \rightarrow Ag + 2\,NH_3) \times 4$$

$$5\,OH^- + H_2CO \rightarrow HCO_3^- + 3\,H_2O + 4\,e^-$$
$$4\,e^- + 4\,Ag(NH_3)_2^+ \rightarrow 4\,Ag + 8\,NH_3$$

$$5\,OH^-(aq) + H_2CO(aq) + 4\,Ag(NH_3)_2^+(aq) \rightarrow 4\,Ag(s) + 8\,NH_3(aq) + HCO_3^-(aq) + 3\,H_2O(l)$$

QUESTIONS:

5. In a galvanic cell a spontaneous reaction occurs, producing an electric current. In an electrolytic cell electricity is used to force a reaction to occur that is not spontaneous.

7. a. cathode: electrode at which reduction occurs.

   b. anode: electrode at which oxidation occurs.

   c. oxidation half-reaction: half reaction in which e^- are products.

   d. reduction half-reaction: half reaction in which e^- are reactants.

9. 1. protective coating 2. alloying 3. cathodic protection

EXERCISES:

Galvanic Cells

11.

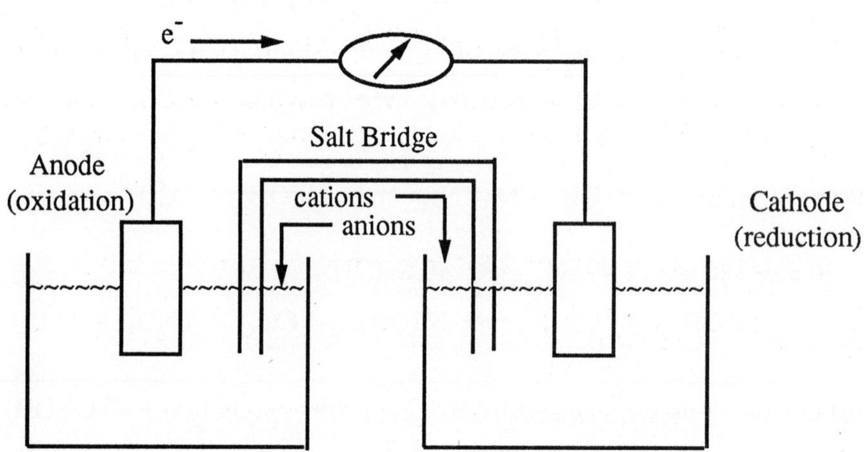

The diagram for all cells will look like this. The contents of each half cell will be identified for each reaction.

a.

$$7\,H_2O + 2\,Cr^{3+} \rightarrow Cr_2O_7^{2-} + 14\,H^+ + 6\,e^-$$
$$(Cl_2 + 2\,e^- \rightarrow 2\,Cl^-) \times 3$$

$$7\,H_2O\,(l) + 2\,Cr^{3+}(aq) + 3\,Cl_2(g) \rightarrow Cr_2O_7^{2-}(aq) + 6\,Cl^-(aq) + 14\,H^+(aq)$$

Cathode: Pt electrode; $Cl_2(g)$ bubbled into solution, Cl^- in solution

Anode: Pt electrode; Cr^{3+}, H^+, and $Cr_2O_7^{2-}$ in solution

We need a nonreactive metal to use as the electrode in each case, since all of the reactants and products are in solution. Pt is the most common choice. Other possibilities are gold and graphite.

b.

$$Cu^{2+} + 2\,e^- \rightarrow Cu$$ Cathode: Cu electrode; Cu^{2+} in solution
$$Mg \rightarrow Mg^{2+} + 2\,e^-$$ Anode: Mg electrode; Mg^{2+} in solution

$$Cu^{2+}(aq) + Mg(s) \rightarrow Cu(s) + Mg^{2+}(aq)$$

13. The diagram will look like that for Exercise 17.11. We get the reaction by remembering that the reduction occurs at the cathode in a galvanic cell.

a.

$$Cl_2 + 2\,e^- \rightarrow 2\,Cl^-$$
$$2\,Br^- \rightarrow Br_2 + 2\,e^-$$

$$Cl_2(g) + 2\,Br^-(aq) \rightarrow Br_2(aq) + 2\,Cl^-(aq)$$

Cathode: Pt electrode; $Cl_2(g)$ bubbled in, Cl^- in solution

Anode: Pt electrode; Br_2 and Br^- in solution

b.

$$(2\,e^- + 2\,H^+ + IO_4^- \rightarrow IO_3^- + H_2O) \times 5$$
$$(4\,H_2O + Mn^{2+} \rightarrow MnO_4^- + 8\,H^+ + 5\,e^-) \times 2$$

$$10\,H^+ + 5\,IO_4^- + 8\,H_2O + 2\,Mn^{2+} \rightarrow 5\,IO_3^- + 5\,H_2O + 2\,MnO_4^- + 16\,H^+$$

This simplifies to:

$$3\,H_2O\,(l) + 5\,IO_4^-\,(aq) + 2\,Mn^{2+}(aq) \rightarrow 5\,IO_3^-(aq) + 2\,MnO_4^-(aq) + 6\,H^+(aq)$$

Cathode: Pt electrode; IO_4^-, IO_3^- and H^+ in solution

Anode: Pt electrode; Mn^{2+}, MnO_4^- and H^+ in solution

15. 11a. $Pt \mid Cr^{3+}$ (1 M), $Cr_2O_7^{2-}$(1 M), H^+ (1 M) || Cl_2 (1 atm), Cl^- (1 M) | Pt

11b. Mg | Mg^{2+} (1 M) || Cu^{2+} (1 M) | Cu

13a. Pt | Br^- (1 M), Br_2 (1 M) || Cl_2 (1 atm), Cl^- (1 M) | Pt

13b. Pt | Mn^{2+} (1 M), MnO_4^- (1 M), H^+ (1 M) || H^+ (1 M), IO_4^- (1 M), IO_3^- (1 M) | Pt

Cell Potential, Standard Reduction Potentials, and Free Energy

17. 11a.

$3\,Cl_2 + 6\,e^- \rightarrow 6\,Cl^-$ $E° = 1.36$ V

$2\,Cr^{3+} + 7\,H_2O \rightarrow 14\,H^+ + Cr_2O_7^{2-} + 6\,e^-$ $-E° = -1.33$ V

$2\,Cr^{3+} + 7\,H_2O + 3\,Cl_2 \rightarrow 6\,Cl^- + 14\,H^+ + Cr_2O_7^{2-}$ $E°_{cell} = 0.03$ V

11b. $Cu^{2+} + 2\,e^- \rightarrow Cu$ $E° = 0.34$ V

$Mg \rightarrow Mg^{2+} + 2\,e^-$ $-E° = -(-2.37$ V)

$Cu^{2+} + Mg \rightarrow Mg^{2+} + Cu$ $E°_{cell} = 2.71$ V

13a. $Cl_2 + 2\,e^- \rightarrow 2\,Cl^-$ $E° = 1.36$ V

$2\,Br^- \rightarrow Br_2 + 2\,e^-$ $-E° = -1.09$ V

$Cl_2 + 2\,Br^- \rightarrow Br_2 + 2\,Cl^-$ $E°_{cell} = 0.27$ V

13b. $10\,e^- + 10\,H^+ + 5\,IO_4^- \rightarrow 5\,IO_3^- + 5\,H_2O$ $E° = 1.60$ V

$8\,H_2O + 2\,Mn^{2+} \rightarrow 2\,MnO_4^- + 16\,H^+ + 10\,e^-$ $-E° = -1.51$ V

$3\,H_2O + 5\,IO_4^- + 2\,Mn^{2+} \rightarrow 5\,IO_3^- + 2\,MnO_4^- + 6\,H^+$ $E°_{cell} = 0.09$ V

19. a. $2\,Ag^+ + 2\,e^- \rightarrow 2\,Ag$ $E° = 0.80$ V

$Cu \rightarrow Cu^{2+} + 2\,e^-$ $-E° = -0.34$ V

$2\,Ag^+ + Cu \rightarrow Cu^{2+} + 2\,Ag$ $E°_{cell} = 0.46$ V Spontaneous

b. $Zn^{2+} + 2\,e^- \rightarrow Zn$ $E° = -0.76$ V

$Ni \rightarrow Ni^{2+} + 2\,e^-$ $-E° = -(-0.23$ V)

$Zn^{2+} + Ni \rightarrow Zn + Ni^{2+}$ $E°_{cell} = -0.53$ V Not spontaneous

21. a. $2\,H^+ + 2\,e^- \rightarrow H_2$ $E° = 0.0$ V

$Cu \rightarrow Cu^{2+} + 2\,e^-$ $-E° = -0.34$ V

$E°_{cell} = -0.34$ V; No, H^+(aq) cannot oxidize Cu to Cu^{2+}at standard conditions.

b. $2\,H^+ + 2\,e^- \rightarrow H_2$ $E° = 0.0$ V

$Mg \rightarrow Mg^{2+} + 2\,e^-$ $-E° = -(-2.37$ V)

$E°_{cell} = 2.37$ V; Yes

c. $Fe^{3+} + e^- \rightarrow Fe^{2+}$ $E° = 0.77$ V

$2\,I^- \rightarrow I_2 + 2\,e^-$ $-E° = -0.54$ V

$E°_{cell} = 0.23$ V; Yes

d. $Fe^{3+} + e^- \rightarrow Fe^{2+}$ $E° = 0.77$ V

$Fe^{3+} + 3\,e^- \rightarrow Fe$ $E° = -0.036$ V

$Br_2 + 2\,e^- \rightarrow 2\,Br^-$ $E° = 1.09$ V

No, for $2\,Fe^{3+} + 2\,Br^- \rightarrow Br_2 + 2\,Fe^{2+}$ $E°_{cell} = -0.32$ V

The reaction is not spontaneous ($E°_{cell} < 0$). Fe^{3+} cannot oxidize Br^- at standard conditions.

23. The general form of a reduction half reaction is $Ox + ne^- \rightarrow Red$. Oxidizing agents (Ox) are on the left hand side. We look for the largest E° values to correspond to the best oxidizing agents.

	MnO_4^-	>	Cl_2	>	$Cr_2O_7^{2-}$	>	Fe^{3+}	>	Fe^{2+}	>	Mg^{2+}
E°	1.51		1.36		1.33		0.77		-0.44		-2.37

25. $Br_2 + 2\,e^- \rightarrow 2\,Br^-$ $E° = 1.09$ V

$2\,H^+ + 2\,e^- \rightarrow H_2$ $E° = 0.00$ V

$Cd^{2+} + 2\,e^- \rightarrow Cd$ $E° = -0.40$ V

$La^{3+} + 3\,e^- \rightarrow La$ $E° = -2.37$ V

$Ca^{2+} + 2\,e^- \rightarrow Ca$ $E° = -2.76$ V

a. Oxidizing agents are on the left side of the half reaction. Br_2 is the best oxidizing agent.

b. Reducing agents are on the right side of the half reaction. Ca is the best reducing agent.

c. $MnO_4^- + 8\,H^+ + 5\,e^- \rightarrow Mn^{2+} + 4\,H_2O$

$E° = 1.51$ V Permanganate can oxidize Br^-, H_2, Ca, and Cd.

d. $Zn^{2+} + 2\,e^- \rightarrow Zn$, $E° = -0.76$ V So for $Zn \rightarrow Zn^{2+} + 2\,e^-$, $-E° = 0.76$ V.

Thus, zinc can reduce Br_2 and H^+.

27. a. $E^\circ_{red} > 0.80$ V oxidize Hg; $E^\circ_{red} < 0.91$ V not oxidize Hg_2^{2+}

No half reaction in this table fits this requirement. However, by changing concentrations, Ag^+ may be able to work.

b. $E^\circ_{red} > 1.09$ V, oxidize Br^-

$E^\circ_{red} < 1.36$ V, not oxidize Cl^-

$Cr_2O_7^{2-}$, O_2, MnO_2, and IO_3^- are all possible.

c. $Ni^{2+} + 2\,e^- \rightarrow Ni \qquad E^\circ = -0.23$ V

$Mn^{2+} + 2\,e^- \rightarrow Mn \qquad E^\circ = -1.18$ V

Any oxidizing agent with $-0.23 \text{ V} > E^\circ_{red} > -1.18$ V will work. $PbSO_4$, Cd^{2+}, Fe^{2+}, Cr^{3+}, Zn^{2+} and H_2O will be able to do this.

29. Reduce I_2 to I^-, $E^\circ_{red} < 0.54$ V

Not reduce Cu^{2+} to Cu, $E^\circ_{red} > 0.34$ V

Yes, $0.34 < E^\circ_{red} < 0.54$

31. a.

$$(ClO_2^- \rightarrow ClO_2 + e^-) \times 2 \qquad -E^\circ = -0.954 \text{ V}$$
$$Cl_2 + 2\,e^- \rightarrow 2\,Cl^- \qquad E^\circ = 1.36 \text{ V}$$

$$2\,ClO_2^- + Cl_2 \rightarrow 2\,ClO_2 + 2\,Cl^- \qquad E^\circ_{cell} = 0.41 \text{ V}$$

$\Delta G^\circ = -nFE^\circ$

$\Delta G^\circ = -(2 \text{ mol } e^-)(96{,}485 \text{ C/mol } e^-)(0.41 \text{ J/C}) = -7.91 \times 10^4 \text{ J} = -79 \text{ kJ}$

$\Delta G^\circ = -RT \ln K$; so $K = \exp(-\Delta G^\circ/RT)$

$K = \exp[(7.9 \times 10^4 \text{ J}) / (8.3145 \text{ J/K}\cdot\text{mol})(298 \text{ K}) = 7.0 \times 10^{13}$

or $E^\circ = \dfrac{0.0592}{n} \log K^\circ$; $\log K = \dfrac{nE^\circ}{0.0592} = \dfrac{2(0.41)}{0.0592} = 13.85$

$K = 10^{13.85} = 7.1 \times 10^{13}$

b.

$$(H_2O + ClO_2 \rightarrow ClO_3^- + 2\,H^+ + e^-) \times 5$$
$$5\,e^- + 4\,H^+ + ClO_2 \rightarrow Cl^- + 2\,H_2O$$

$$5\,H_2O + 5\,ClO_2 + 4\,H^+ + ClO_2 \rightarrow 5\,ClO_3^- + 10\,H^+ + Cl^- + 2\,H_2O$$

$$3\,H_2O\,(l) + 6\,ClO_2(g) \rightarrow 5\,ClO_3^-(aq) + Cl^-(aq) + 6\,H^+(aq)$$

33. 11a. $E° = +0.03\ V,\ \Delta G° = -nFE°$

$\Delta G° = -(6\ mol\ e^-)\ (96{,}485\ C/mol\ e^-)\ (0.03\ J/C) = -1.7 \times 10^4\ J = -20\ kJ$

$\log K = \frac{nE°}{0.0592} = \frac{6(0.03)}{0.0592} = 3.04;\ K = 10^{3.04} = 1 \times 10^3$

11b. $E° = 2.71\ V$

$\Delta G° = -(2\ mol\ e^-)\ (96{,}485\ C/mol\ e^-)\ (2.71\ J/C) = -5.23 \times 10^5\ J = -523\ kJ$

$\log K = \frac{2(2.71)}{0.0592} = 91.55;\ K = 3.55 \times 10^{91}$

13a. $E° = 0.27\ V$

$\Delta G° = -(2\ mol\ e^-)\ (96{,}485\ C/mol\ e^-)\ (0.27\ J/C) = -5.21 \times 10^4\ J = -52\ kJ$

$\log K = \frac{2(0.27)}{0.0592} = 9.12;\ K = 1.3 \times 10^9$

13b. $E° = 0.09\ V$

$\Delta G° = -(10\ mol\ e^-)\ (96{,}485\ C/mol\ e^-)\ (0.09\ J/C) = -8.7 \times 10^4\ J = -90\ kJ$

$\log K = \frac{10(0.09)}{0.0592} = 15.20;\ K = 2 \times 10^{15}$

35. $\Delta G° = -nFE° = -(1\ mol\ e^-)\ (96{,}485\ C/mol\ e^-)\ (0.80\ V)$

$\Delta G° = -77{,}200\ J = -77\ kJ$

$-77\ kJ = \Delta G_f°(Ag) - \Delta G_f°\ (Ag^+)$, $\Delta G_f°$ for e^- equals zero.

$-77\ kJ = 0 - \Delta G_f°(Ag^+);\ \Delta G_f° = 77\ kJ/mol$

37. $2\ H_2O + 2\ e^- \longrightarrow H_2 + 2\ OH^-$

$\Delta G° = 2(-157) - 2(-237) = +160.\ kJ$

$\Delta G° = -nFE°$

$$E° = \frac{-\Delta G°}{nF} = \frac{-1.60 \times 10^5\ J}{(2\ mol\ e^-)\ (96{,}485\ C/mol\ e^-)} = -0.829\ V$$

The two values agree. (-0.83 V in Table 17.1)

39. $PbSO_4(s) + 2\ e^- \longrightarrow Pb + SO_4^{2-}$ $E° = -0.35\ V$

$Pb \longrightarrow Pb^{2+} + 2\ e^-$ $-E° = -(-0.13\ V)$

$PbSO_4(s) \longrightarrow Pb^{2+}(aq) + SO_4^{2-}(aq)$ $E°_{cell} = -0.22\ V$

$K = K_{sp}$ and $E^\circ_{cell} = \frac{0.0592}{n} \log K_{sp}$

$\log K_{sp} = \frac{nE^\circ}{0.0592} = \frac{2(-0.22)}{0.0592} = -7.43$

$K_{sp} = 10^{-7.43} = 3.7 \times 10^{-8}$

The Nernst Equation

41. For the half cell on the left, $E_l = E^\circ = 0.80$ V since $[Ag^+] = 1.0$ *M*. Let's calculate E for the half cell on the right, and thus figure out what reaction occurs. The half cell with the smallest reduction potential is the anode.

a. $E_r = E^\circ = 0.80$ V

Since $E_l = E_r$, then $E_{cell} = 0$, No reaction occurs. Concentration cells only produce a voltage when the ion concentrations are not equal.

b. For $Ag^+ + e^- \longrightarrow Ag$

$E_r = E^\circ - \frac{0.0592}{1} \log \frac{1}{[Ag^+]}$

$E_r = 0.80 \text{ V} + 0.0592 \log [Ag^+]$

$E_r = 0.80 \text{ V} + 0.0592 \log 2.0 = 0.80 + 0.018 = 0.82 \text{ V}$

The half cell with 2 *M* Ag^+ is the cathode. Electrons flow from anode to cathode, to the right in the diagram.

$E_{cell} = E_r - E_l = 0.82 - 0.80 = 0.02 \text{ V}$

Note: The driving force for the reaction is to equalize the concentration of Ag^+ in the two half cells. The half cell with the largest Ag^+ concentrations will always be the cathode.

c. $E_r = 0.80 + 0.0592 \log 0.10 = 0.80 - 0.059 = 0.74 \text{ V}$

Half cell with 0.10 *M* Ag^+ is the anode. Electrons move to the cathode (to the left in the diagram).

$E_{cell} = E_l - E_r = 0.80 - 0.74 = 0.06 \text{ V}$

d. $E_r = 0.80 + 0.0592 \log (4.0 \times 10^{-5}) = 0.80 - 0.26 = 0.54 \text{ V}$

$E_{cell} = E_l - E_r = 0.80 - 0.54 = 0.26 \text{ V}$

Half cell with 4.0×10^{-5} *M* Ag^+ is the anode. Electrons move from anode to cathode (to the left in the diagram).

e. If the concentrations are the same in each half cell, $E_l = E_r$ and no reaction occurs. $E_{cell} = 0$.

43. $5\ e^- + 8\ H^+ + MnO_4^- \rightarrow Mn^{2+} + 4\ H_2O$ $E° = 1.51$ V

$(Fe^{2+} \rightarrow Fe^{3+} + e^-) \times 5$ $-E° = -0.77$ V

$8\ H^+ + MnO_4^- + 5\ Fe^{2+} \rightarrow 5\ Fe^{3+} + Mn^{2+} + H_2O$ $E°_{cell} = 0.74$ V

$$E = E° - \frac{0.0592}{n} \log Q$$

$$E = 0.74\ V - \frac{0.0592}{5} \log \frac{[Fe^{3+}]^5\ [Mn^{2+}]}{[Fe^{2+}]^5\ [MnO_4^-]\ [H^+]^8}$$

$$E = 0.74 - \frac{0.0592}{5} \log \frac{(1 \times 10^{-6})^5\ (1 \times 10^{-6})}{(1 \times 10^{-3})^5\ (1 \times 10^{-2})\ (1 \times 10^{-4})^8}$$

$$E = 0.74 - \frac{0.0592}{5} \log 1 \times 10^{13} = 0.74 - 0.15 = 0.59\ V = 0.6\ V$$

Yes, $E > 0$ so reaction will occur as written.

45. $Cu^{2+} + H_2 \rightarrow 2\ H^+ + Cu$ $E° = 0.34\ V - 0.0\ V = 0.34\ V$

a. $E = E° - \frac{0.0592}{2} \log \frac{1}{[Cu^{2+}]}$ since $P_{H_2} = 1$ atm and $[H^+] = 1$ mol/L

$$E = E° + \frac{0.0592}{2} \log [Cu^{2+}]$$

$$E = 0.34 + \frac{0.0592}{2} \log (2.5 \times 10^{-4}) = 0.23\ V$$

b. $E = 0.34 + \frac{0.0592}{2} \log [Cu^{2+}]$

$Cu(OH)_2$	$\rightleftharpoons$	Cu^{2+}	+	$2\ OH^-$	$K_{sp} = 1.6 \times 10^{-19}$
s mol/L dissolves	$\rightarrow$	s		$0.10 + 2\ s$	

$$1.6 \times 10^{-19} = (s)\ (0.10 + 2s)^2 \approx s\ (0.10)^2$$

$s = [Cu^{2+}] = 1.6 \times 10^{-17}$ Assumption good.

$$E = E° + \frac{0.0592}{2} \log [Cu^{2+}] = 0.34 + \frac{0.0592}{2} \log (1.6 \times 10^{-17})$$

$E = 0.34 - 0.50 = -0.16$ V, not spontaneous, reverse reaction occurs. $E_{cell} = 0.16$ V, Cu electrode becomes the anode.

c. $0.195 = 0.34 + \frac{0.0592}{2}\log[Cu^{2+}]$, $\log[Cu^{2+}] = -4.899$

$[Cu^{2+}] = 10^{-4.899} = 1.3 \times 10^{-5}\ M$

Note: If cell is reversed as in (b), $0.195 = -0.34 - \frac{0.0592}{2}\log[Cu^{2+}]$ and $[Cu^{2+}] = 8.4 \times 10^{-19}$

d. $E = E° + \frac{0.0592}{2}\log[Cu^{2+}]$

Graph E vs. $\log[Cu^{2+}]$. The slope will be 0.0296 V or 29.6 mV.

47. The potential oxidizing agents are NO_3^- and H^+. Hydrogen ion cannot oxidize Pt under any condition. Nitrate cannot oxidize Pt unless there is Cl^- in the solution. Aqua regia has both Cl^- and NO_3^-: The nitrate oxidizes Pt and it is complexed by the Cl^-.

$$12\ Cl^- + 3\ Pt \rightarrow 3\ PtCl_4^{2-} + 6\ e^- \qquad -E° = -0.76$$

$$2\ NO_3^- + 8\ H^+ + 6\ e^- \rightarrow 2\ NO + 4\ H_2O \qquad E° = 0.96$$

$$12\ Cl^- + 3\ Pt + 2\ NO_3^- + 8\ H^+ \rightarrow 3\ PtCl_4^{2-} + 2\ NO + 4\ H_2O \qquad E°_{cell} = +0.20\ V$$

Electrolysis

49. a. $Al^{3+} + 3\ e^- \rightarrow Al$

$$1.0 \times 10^3\ g\ Al \times \frac{1\ mol\ Al}{26.98\ g\ Al} \times \frac{3\ mol\ e^-}{mol\ Al} \times \frac{96,485\ C}{mol\ e^-} \times \frac{1\ s}{100.0\ C}$$

$$= 1.07 \times 10^5\ s = 3.0 \times 10^1\ hours$$

b. $$1.0\ g\ Ni \times \frac{1\ mol}{58.69\ g} \times \frac{2\ mol\ e^-}{mol\ Ni} \times \frac{96,485\ C}{mol\ e^-} \times \frac{1\ s}{100.0\ C} = 33\ s$$

c. $$5.0\ mol\ Ag \times \frac{1\ mol\ e^-}{mol\ Ag} \times \frac{96,485\ C}{mol\ e^-} \times \frac{1\ s}{100.0\ C} = 4.8 \times 10^3\ s = 1.3\ hr$$

51. a. Cathode: reduction: $K^+ + e^- \rightarrow K$ $E° = -2.92\ V$

Anode: oxidation: $2F^- \rightarrow F_2 + 2\ e^-$ $-E° = -2.87\ V$

b. Cathode: easier to reduce H_2O than K^+

$$2\ H_2O + 2\ e^- \rightarrow H_2 + 2\ OH^- \qquad E° = -0.83\ V$$

Anode: easier to oxidize H_2O than F^-

$$2\ H_2O \rightarrow 4\ H^+ + O_2 + 4\ e^- \qquad -E° = -1.23\ V$$

c. Species present that can be reduced: H_2O_2, SO_4^{2-}, H^+ and H_2O

Possible cathode reactions:

$$2\,H_2O + 2\,e^- \rightarrow H_2 + 2\,OH^- \qquad E^\circ = -0.83\text{ V}$$

$$2\,H^+ + 2\,e^- \rightarrow H_2 \qquad E^\circ = 0.00\text{ V}$$

$$SO_4^{2-} + 4\,H^+ + 2\,e^- \rightarrow H_2SO_3 + H_2O \qquad E^\circ = 0.20\text{ V}$$

$$H_2O_2 + 2\,H^+ + 2\,e^- \rightarrow 2\,H_2O \qquad E^\circ = 1.78\text{ V}$$

Reduction of H_2O_2 will occur (most positive E°) at the cathode.

Possible anode reactions:

$$2\,H_2O \rightarrow O_2 + 4\,H^+ + 4\,e^- \qquad -E^\circ = -1.23\text{ V}$$

$$H_2O_2 \rightarrow O_2 + 2\,H^+ + 2\,e^- \qquad -E^\circ = -0.68\text{ V}$$

Easier to oxidize H_2O_2 than H_2O, so

$H_2O_2 \rightarrow O_2 + 2\,H^+ + 2\,e^-$ occurs at the anode.

53. $600.\text{ s} \times \frac{5.0\text{ C}}{\text{s}} \times \frac{1\text{ mol e}^-}{96{,}485\text{ C}} \times \frac{1\text{ mol M}^{3+}}{3\text{ mol e}^-} = 1.036 \times 10^{-2}\text{ mol}$ (carry extra significant figures)

$$\text{Atomic mass} = \frac{1.18\text{ g}}{1.036 \times 10^{-2}\text{ mol}} = \frac{114\text{ g}}{\text{mol}} \approx \frac{110\text{ g}}{\text{mol}}$$

The element is most likely indium, In. Indium forms 3+ ions, Cd, Ag and Pd do not.

55. F_2 is produced at the anode.

$$2\,F^- \rightarrow F_2 + 2\,e^-$$

$$2.00\text{ hr} \times \frac{60\text{ min}}{\text{hr}} \times \frac{60\text{ s}}{\text{min}} \times \frac{10.0\text{ C}}{\text{s}} \times \frac{1\text{ mol e}^-}{96{,}485\text{ C}} = 0.746\text{ mol e}^-$$

$$0.746\text{ mol e}^- \times \frac{1\text{ mol F}_2}{2\text{ mol e}^-} = 0.373\text{ mol F}_2$$

$$V = \frac{nRT}{P} = \frac{(0.373\text{ mol})\left(\frac{0.08206\text{ L atm}}{\text{mol K}}\right)(298\text{ K})}{1.00\text{ atm}} = 9.12\text{ L}$$

K is produced at the cathode.

$K^+ + e^- \longrightarrow K$

$$0.746 \text{ mol } e^- \times \frac{1 \text{ mol K}}{\text{mol } e^-} \times \frac{39.10 \text{ g K}}{\text{mol K}} = 29.2 \text{ g K}$$

57. $Au(CN)_4^- + 3\,e^- \longrightarrow Au + 4\,CN^-$

Water is oxidized: $2\,H_2O \longrightarrow O_2 + 4\,H^+ + 4\,e^-$

Actually, this process is done in basic solution. You don't want $H^+ + CN^- \longrightarrow HCN(g)$ occurring. So oxidation of water in base is:

$$4\,OH^- \longrightarrow O_2 + 2\,H_2O + 4\,e^-$$

Overall reaction:

$$4\,Au(CN)_4^-(aq) + 12\,OH^-(aq) \longrightarrow 4\,Au(s) + 16\,CN^-(aq) + 3\,O_2(g) + 6\,H_2O(l)$$

$$1.00 \times 10^3 \text{ g Au} \times \frac{1 \text{ mol Au}}{197.0 \text{ g}} \times \frac{3 \text{ mol } O_2}{4 \text{ mol Au}} = 3.81 \text{ mol } O_2$$

$PV = nRT$

$$V = \frac{nRT}{P} = \frac{3.81 \text{ mol}\left(\frac{0.08206 \text{ L atm}}{\text{mol K}}\right)(298 \text{ K})}{740 \text{ torr}} \times \frac{760 \text{ torr}}{\text{atm}} = 96 \text{ L}$$

59. $$\frac{1.0 \times 10^6 \text{ g}}{\text{hr}} \times \frac{1 \text{ hr}}{60 \text{ min}} \times \frac{1 \text{ min}}{60 \text{ s}} \times \frac{1 \text{ mol } C_6H_8N_2}{108.1 \text{ g } C_6H_8N_2} \times \frac{2 \text{ mol } e^-}{\text{mol } C_6H_8N_2} \times \frac{96{,}485 \text{ C}}{\text{mol } e^-}$$

$$= 5.0 \times 10^5 \text{ C/s or current of } 5.0 \times 10^5 \text{ A.}$$

61. $$2.30 \text{ min} \times \frac{60 \text{ s}}{\text{min}} = 138 \text{ s}$$

$$138 \text{ s} \times \frac{2.00 \text{ C}}{\text{s}} \times \frac{1 \text{ mol } e^-}{96{,}485 \text{ C}} \times \frac{1 \text{ mol Ag}}{\text{mol } e^-} = 2.86 \times 10^{-3} \text{ mol Ag}$$

$$\text{Conc.} = \frac{2.86 \times 10^{-3} \text{ mol}}{0.250 \text{ L}} = 1.14 \times 10^{-2} \text{ mol/L}$$

ADDITIONAL EXERCISES

63. The half reaction for the SCE is:

$$Hg_2Cl_2(s) + 2\,e^- \longrightarrow 2\,Hg + 2\,Cl^- \qquad E_{SCE} = +0.242 \text{ V}$$

a. $Cu^{2+} + 2\,e^- \longrightarrow Cu$ $\qquad E° = 0.34$ V

a. $Cu^{2+} + 2\ e^- \rightarrow Cu$ $E° = 0.34\ V$

$E_{cell} = 0.10\ V$, SCE is anode

b. $Fe^{3+} + e^- \rightarrow Fe^{2+}$ $E° = 0.77\ V$

$E_{cell} = 0.53\ V$, SCE is anode

c. $AgCl + e^- \rightarrow Ag + Cl^-$ $E° = 0.22\ V$

$E_{cell} = 0.02\ V$, SCE is cathode

d. $Al^{3+} + 3\ e^- \rightarrow Al$ $E° = -1.66\ V$

$E_{cell} = 1.90\ V$, SCE is cathode

e. $Ni^{2+} + 2\ e^- \rightarrow Ni$ $E° = -0.23\ V$

$E_{cell} = 0.47\ V$, SCE is cathode

65. $2\ Cu^+ \rightarrow Cu + Cu^{2+}$ $E^\circ_{cell} = 0.52\ V - 0.16\ V = 0.36\ V$

$\Delta G° = -nFE° = -(1\ mol\ e^-)\ (96{,}485\ C/mol\ e^-)\ (0.36\ V)$

$\Delta G° = -34{,}700\ CV = -34.7\ kJ = -35\ kJ$

$\log K = \frac{nE°}{0.0592} = \frac{0.36}{0.0592} = 6.08;\ K = 1.2 \times 10^6$

67. $H_2O_2 + 2\ H^+ + 2\ e^- \rightarrow 2\ H_2O$ $E° = 1.78\ V$

$O_2 + 2\ H^+ + 2\ e^- \rightarrow H_2O_2$ $E° = 0.68\ V$

$H_2O_2 + 2\ H^+ + 2\ e^- \rightarrow 2\ H_2O$ $E° = 1.78\ V$ H_2O_2 as an oxidizing agent

$H_2O_2 \rightarrow O_2 + 2\ H^+ + 2\ e^-$ $-E° = -0.68\ V$ H_2O_2 as a reducing agent

---

$2\ H_2O_2 \rightarrow 2\ H_2O + O_2$ $E^\circ_{cell} = 1.10\ V$

69. 68b, d, and e are spontaneous.

b. $Cl_2 + 2\ I^- \rightarrow I_2 + 2\ Cl^-;\ E° = +0.82\ V$

$\Delta G° = -nFE° = -(2\ mol\ e^-)\ (96{,}485\ C/mol\ e^-)\ (0.82\ J/C)$

$\Delta G° = -1.58 \times 10^5\ J = -160\ kJ$

$\Delta G° = -RT \ln K$

$\log K = \frac{nE°}{0.0592} = \frac{2(0.82)}{0.0592} = 27.70$

$K = 10^{27.70} = 5.0 \times 10^{27}$

d. $Pb + Cu^{2+} \rightarrow Cu + Pb^{2+}$ $E° = 0.47$ V Ignore $PbCl_2(s)$ formation.

$\Delta G° = -nFE° = -(2 \text{ mol } e^-)(96{,}485 \text{ C/mol } e^-)(0.47 \text{ J/C})$

$\Delta G° = -9.07 \times 10^4 \text{ J} = -91 \text{ kJ}$

$\log K = \frac{2(0.47)}{0.0592} = 15.88$

$K = 7.6 \times 10^{15}$

e. $(Fe^{2+} \rightarrow Fe^{3+} + e^-) \times 4$ $-E° = -0.77$ V

$4 H^+ + O_2 + 4 e^- \rightarrow 2 H_2O$ $E° = 1.23$ V

$4 H^+ + O_2 + 4 Fe^{2+} \rightarrow 4 Fe^{3+} + 2 H_2O$ $E°_{cell} = 0.46$ V

$\Delta G° = -nFE° = -(4 \text{ mol } e^-)(96{,}485 \text{ C/mol } e^-)(0.46 \text{ J/C}) = -180 \text{ kJ}$

$\log K = \frac{4(0.46)}{0.0592} = 31.08;\ K = 1.2 \times 10^{31}$

71. $Zn^{2+} + 2 e^- \rightarrow Zn$ $E° = -0.76$ V

$Fe^{2+} + 2 e^- \rightarrow Fe$ $E° = -0.44$ V

It is easier to oxidize Zn than Fe, so the Zn will be oxidized protecting the iron of the *Monitor's* hull.

73. $2 H^+ + 2 e^- \rightarrow H_2$ $E° = 0.0000$ V

$D_2 \rightarrow 2 D^+ + 2 e^-$ $-E° = 0.0034$ V

$2 H^+ + D_2 \rightarrow 2 D^+ + H_2$ $E°_{cell} = 0.0034$ V

$\Delta G° = -(2 \text{ mol } e^-)(96{,}485 \text{ C/mol } e^-)(0.0034 \text{ J/C}) = -660 \text{ J}$

$\log K = \frac{2(0.0034)}{0.0592} = 0.115;\ K = 1.3$

75. a. $Hg_2Cl_2 + 2 e^- \rightarrow 2 Hg + 2 Cl^-$ $E_{SCE} = 0.242$ V

$H_2 \rightarrow 2 H^+ + 2 e^-$ $-E° = 0.000$ V

$HgCl_2 + H_2 \rightarrow 2 H^+ + 2 Cl^- + 2 Hg$ $E_{cell} = 0.242$ V

b. The hydrogen half cell is the oxidation half reaction; thus, the hydrogen electrode is the anode.

c. Ignoring concentrations from the SCE (they are constant).

$$E = 0.242\ V - \frac{0.0592}{2}\log\frac{[H^+]^2}{P_{H_2}}$$

If we keep $P_{H_2} = 1$ atm:

$$E = 0.242\ V - \frac{0.0592}{2}\log[H^+]^2$$

Since, $\log[H^+]^2 = 2\log[H^+] = -2\ pH$

Then $E = 0.242 - 0.0592\log[H^+]$ or $E = 0.242\ V + 0.0592\ pH$

d. i) $E = 0.242 - 0.0592\log[H^+] = 0.242 - 0.0592\log(1.0 \times 10^{-3})$

$E = 0.42\ V$

ii) $E = 0.242 - 0.0592\log 2.5 = 0.242 - 0.024 = 0.218\ V$

iii) $E = 0.242 - 0.0592\log(1.0 \times 10^{-9}) = 0.242 + 0.53 = 0.77\ V$

e. $E = 0.242\ V + 0.0592\ pH$

$0.285 = 0.242 + 0.0592\ pH$; $pH = 0.726$, $[H^+] = 0.188\ M$

f. Primarily the reason is convenience. It is inconvenient to deal with the H_2 gas and particularly to keep P_{H_2} constant. Gas cylinders are bulky; and H_2 presents a fire and explosion hazard.

77. $Fe^{2+} + 2e^- \longrightarrow Fe$ $\qquad E = E° = -0.44\ V$

$Ag^+ + e^- \longrightarrow Ag$ $\qquad E = E° - \frac{0.0592}{1}\log\frac{1}{[Ag^+]}$

$E = 0.80 + 0.0592\log[Ag^+] = 0.80 + 0.0592\log(0.010) = 0.68\ V$

It is still easier to plate out Ag.

79. $Pt^{2+} + 2e^- \longrightarrow Pt$ $\qquad E° = 1.2\ V$

$Pd^{2+} + 2e^- \longrightarrow Pd$ $\qquad E° = 0.99\ V$

$Ni^{2+} + 2e^- \longrightarrow Ni$ $\qquad E° = -0.23\ V$

It looks like the electrolysis should work. The only problem might be that some Pd will begin to plate before Pt^{2+} is gone. To check this, let's calculate the potential of a Pt half cell when 99.9% of the Pt is gone.

$$E = E° - \frac{0.0592}{2}\log\frac{1}{[Pt^{2+}]} = 1.2\ V - \frac{0.0592}{2}\log\left(\frac{1}{10^{-3}}\right)$$

$E = 1.2\ V - 0.089\ V = 1.1\ V$

So no Pd^{2+} will begin to plate. Therefore, the technique is feasible.

81. For the Hall process it requires 15 kWh of energy.

$$15\ kWh = \frac{15000\ J}{s} \times 1\ hr \times \frac{60\ min}{hr} \times \frac{60\ s}{min} = 5.4 \times 10^7\ J \text{ or } 5.4 \times 10^4\ kJ$$

To melt Al it requires:

$$1.0 \times 10^3\ g\ Al \times \frac{1\ mol}{26.98\ g} \times \frac{10.7\ kJ}{mol\ Al} = 4.0 \times 10^2\ kJ$$

It is feasible to recycle Al because it takes less than 1% of the energy required to produce the same amount of Al by the Hall process.

83. $1/2\ O_2 + 2\ H^+ + 2\ e^- \longrightarrow H_2O \qquad E° = 1.23\ V$

$H_2 \longrightarrow 2\ H^+ + 2\ e^- \qquad -E° = 0.00\ V$

$H_2 + 1/2\ O_2 \longrightarrow H_2O \qquad E°_{cell} = 1.23\ V$

$\Delta G° = -nFE° = -(2\ mol\ e^-)\ (96{,}485\ C/mol\ e^-)\ (1.23\ V)$

$\Delta G° = -237{,}000\ J = -237\ kJ$

$$1.00 \times 10^3\ g\ H_2O \times \frac{1\ mol\ H_2O}{18.02\ g} \times \frac{-237\ kJ}{mol\ H_2O} = -13{,}200\ kJ = w_{max}$$

The work done can be no larger than the free energy change. The best that could happen is that all of the free energy released goes into doing work, but this does not occur in any real process.

Fuel cells are more efficient in converting chemical energy to electrical energy; they are also less massive. The major disadvantage is that they are expensive.

CHALLENGE PROBLEMS

85. a. $(2\ H^+ + 2\ e^- \longrightarrow H_2) \times 2 \qquad E° = 0.0\ V$

$2\ H_2O \longrightarrow O_2 + 4\ H^+ + 4\ e^- \qquad -E° = -1.23\ V$

$2\ H_2O \longrightarrow 2\ H_2 + O_2 \qquad E°_{cell} = -1.23\ V$

$\Delta G° = -nFE° = -(4\ mol\ e^-)\ (96{,}485\ C/mol\ e^-)\ (-1.23\ V)$

$\Delta G° = 4.75 \times 10^5\ J = 475\ kJ$

b. $\Delta H° = -2\Delta H_f°\ (H_2O) = -2\ mol\ (-286\ kJ/mol) = 572\ kJ$

$\Delta S° = 2\ mol\ (131\ J/K{\cdot}mol) + 1\ mol\ (205\ J/K{\cdot}mol) - 2\ mol\ (70\ J/K{\cdot}mol)$

$\Delta S° = 327\ J/K$

c. at 90°C, T = 363 K (carry extra significant figure in T)

$\Delta G° = 572\text{ kJ} - (363\text{ K})(0.327\text{ kJ/K}) = 453\text{ kJ} = 450\text{ kJ}$

$\Delta G° = -nFE°$

$$E° = \frac{-\Delta G°}{nF} = \frac{-4.5 \times 10^5\text{ J}}{(4\text{ mol e}^-)(96{,}485\text{ C/mol e}^-)} = -1.17\text{ V} = -1.2\text{ V}$$

at 0°C, $\Delta G° = 572\text{ kJ} - (273\text{ K})(0.327\text{ kJ/K}) = 483\text{ kJ}$

$$E° = \frac{-\Delta G°}{nF} = \frac{-4.83 \times 10^5\text{ J}}{(4\text{ mol e}^-)(96{,}485\text{ C/mol e}^-)} = -1.25\text{ V}$$

87. $Ag^+ + e^- \longrightarrow Ag$ $E° = 0.80\text{ V}$

$Ag + 2\, S_2O_3^{2-} \longrightarrow Ag(S_2O_3)_2^{3-} + e^-$ $-E° = -0.017\text{ V}$

$Ag^+ + 2\, S_2O_3^{2-} \longrightarrow Ag(S_2O_3)_2^{3-}$ $E°_{cell} = 0.78\text{ V}$

$$\log K = \frac{nE°}{0.0592} = \frac{(1)(0.78)}{0.0592} = 13.18$$

$K = 10^{13.18} = 1.5 \times 10^{13}$

89. a. $$E = E° - \frac{0.0592}{n} \log \frac{[OH^-]^5}{[CrO_4^{2-}]}$$

pH = 7.40 and pOH = 6.60; $[OH^-] = 10^{-6.60} = 2.5 \times 10^{-7}\ M$

$$E = -0.13 - \frac{0.0592}{3} \log \frac{(2.5 \times 10^{-7})^5}{1.0 \times 10^{-6}}$$

$$E = -0.13 - \frac{0.0592}{3} \log (9.8 \times 10^{-28}) = -0.13 + 0.53 = +0.40\text{ V}$$

b. $$E = E° - \frac{0.0592}{n} \log \frac{[Cr^{3+}]^2}{[Cr_2O_7^{2-}][H^+]^{14}}$$

$$E = 1.33\text{ V} - \frac{0.0592}{6} \log \frac{(1.0 \times 10^{-6})^2}{(1.0 \times 10^{-6})(1.0 \times 10^{-2})^{14}}$$

$$E = 1.33 - \frac{0.0592}{6} \log (1 \times 10^{22}) = 1.33 - 0.22 = 1.11\text{ V}$$

91. a. $E_{meas} = E_{ref} + 0.05916\text{ pH}$

$0.480\text{ V} = 0.250\text{ V} + 0.05916\text{ pH}$

$$pH = \frac{0.480 - 0.250}{0.05916} = 3.888$$

Uncertainty = ± 1 mV = ± 0.001 V

$$pH_{max} = \frac{0.481 - 0.250}{0.05916} = 3.905; \quad 3.905 - 3.888 = 0.017$$

So if the uncertainty in potential is ± 0.001 V, the uncertainty in pH is ± 0.017 or about ± 0.02 pH units.

For this measurement, $[H^+] = 1.29 \times 10^{-4}$

For an error of + 1 mV, $[H^+] = 1.25 \times 10^{-4}$

For an error of -1 mV, $[H^+] = 1.35 \times 10^{-4}$

So the uncertainty in $[H^+]$ is $\pm 0.06 \times 10^{-4}$

b. From the previous example, we will be within ± 0.02 pH units if we measure the potential to the nearest ± 0.001 V (1 mV).

CHAPTER EIGHTEEN: THE REPRESENTATIVE ELEMENTS - GROUP 1A THROUGH 4A

QUESTIONS

1. The gravity of the earth is not strong enough to keep H_2 in the atmosphere.

3. Ionic, covalent, and metallic (or insterstitial)
The ionic and covalent hydrides are true compounds obeying the law of Definite Proportions and differ from each other in the type of bonding. The interstitial hydrides are more like solid solutions of hydrogen and a transition metal and do not obey the Laws of Definite Proportions.

5. Hydrogen forms many compounds in which the oxidation state is +1, as the Group 1A elements. For example H_2SO_4 and HCl compared to Na_2SO_4 and NaCl. On the other hand hydrogen forms diatomic H_2 molecules and is a nonmetal, while the Group 1A elements are metals. Hydrogen also forms hydride anion, H^-, which the Group 1A metals do not.

7. The metals are all easily oxidized. They must be produced in the absence of materials (H_2O, O_2) that are capable of oxidizing them.

9. Planes of carbon atoms slide easily along each other. Graphite is not volatile. The lubricant will not be lost when used in a high vacuum environment.

11. The bonds in SnX_4 compounds have a large covalent character. SnX_4 acts as discrete molecules held together by dispersion forces. SnX_2 compounds are ionic and held in the solid state by strong ionic forces.

13. Size decreases from left to right and increases going down. So going one element right and one element down would result in a similar size for the two elements diagonal to each other. The ionization energies will also be similar for the diagonal elements. Electron affinities are harder to predict, but the similar size and ionization energies would lead to similar properties.

EXERCISES

Group 1A Elements

15. a. $\Delta H° = -110.5 - [-242 - 75] = 207$ kJ

$\Delta S° = 3(131) + 198 - [186 + 189] = 216$ J/K

b. $\Delta G° = \Delta H° - T\Delta S°$ and when $\Delta G = 0$,

$$T = \frac{\Delta H°}{\Delta S°} = \frac{207 \times 10^3 \text{ J}}{216 \text{ J/K}} = 958 \text{ K}$$

Reaction is spontaneous at T > 958 K (when all partial pressures = 1 atm). Pressure won't affect equilibrium if the volume of the reaction vessel is constant.

17. sodium oxide: Na_2O, sodium superoxide: NaO_2, sodium peroxide: Na_2O_2

19. a. $Li_3N(s) + 3\ HCl(aq) \rightarrow 3\ LiCl(aq) + NH_3(aq)$

b. $Rb_2O(s) + H_2O(l) \rightarrow 2\ RbOH(aq)$

c. $Cs_2O_2(s) + 2\ H_2O(l) \rightarrow 2\ CsOH(aq) + H_2O_2(aq)$

d. $NaH(s) + H_2O(l) \rightarrow NaOH(aq) + H_2(g)$

21. $2\ Li(s) + 2\ C_2H_2(g) \rightarrow 2\ LiC_2H(s) + H_2(g)$. It is an oxidation-reduction reaction.

Group 2A Elements

23. barium oxide: BaO, barium peroxide: BaO_2

25. $Mg_3N_2(s) + 6\ H_2O(l) \rightarrow 2\ NH_3(g) + 3\ Mg^{2+}(aq) + 6\ OH^-(aq)$

$Mg_3P_2(s) + 6\ H_2O(l) \rightarrow 2\ PH_3(g) + 3\ Mg^{2+}(aq) + 6\ OH^-(aq)$

27.

H H H
N
Be
:Cl: :Cl:

Trigonal planar

Be uses sp^2 hybrid orbitals.
N uses sp^3 hybrid orbitals.
$BeCl_2$ is a Lewis acid.

29. $Be + H_2O \rightarrow Be(OH)_4^{2-} + H_2$

$Be + 4\ OH^- \rightarrow Be(OH)_4^{2-} + 2\ e^-$

$2\ H_2O + 2\ e^- \rightarrow H_2 + 2\ OH^-$

$Be(s) + 2\ H_2O(l) + 2\ OH^-(aq) \rightarrow Be(OH)_4^{2-}(aq) + H_2(g)$

Be is the reducing agent. H_2O is the oxidizing agent.

Group 3A Elements

31. a. Thallium (I) hydroxide b. Indium(III) sulfide c. Gallium(III) oxide

33. $B_2H_6 + O_2 \rightarrow B(OH)_3$

$B_2H_6 + O_2 \rightarrow 2\ B(OH)_3$ (6 extra O atoms)

$B_2H_6 + 3\ O_2 \rightarrow 2\ B(OH)_3$

35. $In_2O_3(s) + 6\ H^+(aq) \longrightarrow 2\ In^{3+}(aq) + 3\ H_2O(l)$

$In_2O_3(s) + OH^-(aq) \longrightarrow$ No Reaction.

$Ga_2O_3(s) + 6\ H^+(aq) \longrightarrow 2\ Ga^{3+}(aq) + 3\ H_2O(l)$

$Ga_2O_3(s) + 2\ OH^-(aq) + 3\ H_2O(l) \longrightarrow 2\ Ga(OH)_4^-(aq)$

37. $2\ In(s) + 3\ F_2(g) \longrightarrow 2\ InF_3(s)$

$2\ In(s) + 3\ Cl_2(g) \longrightarrow 2\ InCl_3(s)$

$4\ In(s) + 3\ O_2(g) \longrightarrow 2\ In_2O_3(s)$

$2\ In(s) + 6\ HCl(aq) \longrightarrow 3\ H_2(g) + 2\ InCl_3(aq)$

or $2\ In(s) + 6\ HCl(g) \longrightarrow 3\ H_2(g) + 2\ InCl_3(s)$

Group 4A Elements

39. a. linear b. sp

41. a. $K_2SiF_6 + K \longrightarrow KF + Si$

$K_2SiF_6(s) + 4\ K(l) \longrightarrow 6\ KF(s) + Si(s)$

b. K_2SiF_6 is an ionic compound, composed of K^+ cations and SiF_6^{2-} anions. The SiF_6^{2-} anion is held together by covalent bonds. The structure is:

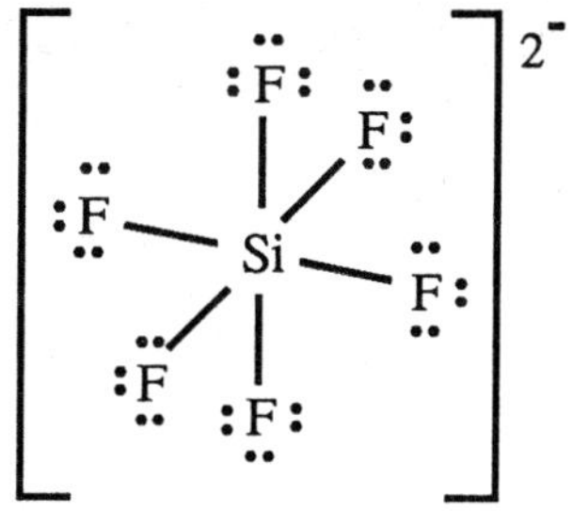

$[SiF_6]^{2-}$ (Si bonded to six F atoms, each F with three lone pairs)

The anion is octahedral.

43. First, we must balance the equation:

$Pb \longrightarrow Pb(OH)_2$

$(Pb + 2\ OH^- \longrightarrow Pb(OH)_2 + 2\ e^-) \times 2$

$H_2O + O_2 \longrightarrow OH^-$

$2\ H_2O + O_2 \longrightarrow 4\ OH^-$

$4\ e^- + 2\ H_2O + O_2 \longrightarrow 4\ OH^-$

$$2\ Pb + 4\ OH^- \longrightarrow 2\ Pb(OH)_2 + 4\ e^- \qquad -E° = +0.57\ V$$

$$4\ e^- + 2\ H_2O + O_2 \longrightarrow 4\ OH^- \qquad E° = +0.40\ V$$

$$2\ Pb + 2\ H_2O + O_2 \longrightarrow 2\ Pb(OH)_2 \qquad E°_{cell} = +0.97\ V$$

Fe pipes corrode more easily than Pb pipes. However, the corrosion of Pb pipes is still spontaneous and Pb(II) is very toxic. Pb pipes were extensively used by the Romans and it has been proposed that chronic lead poisoning is one of the factors leading to the decline and fall of the Roman Empire.

45. Tin (II) fluoride

47. The π electrons are free to move in graphite, thus giving it a greater conductivity (lower resistance). The electrons have the greatest mobility within sheets of carbon atoms. Electrons in diamond are not mobile (high resistance). The structure of diamond is uniform in all directions; thus, there is no directional dependence of the resistivity.

ADDITIONAL EXERCISES

49. a. $2\ Na + 2\ NH_3 \longrightarrow 2\ NaNH_2 + H_2$

b. $$\frac{251.4\ g}{1000.\ g + 251.4\ g} \times 100 = 20.09\%\ \text{by mass}$$

$$\text{mol Na} = 251.4\ g \times \frac{1\ mol}{22.99\ g} = 10.94\ \text{mol Na}$$

$$\text{mol } NH_3 = 1000.\ g \times \frac{1\ mol}{17.03\ g} = 58.72\ \text{mol } NH_3$$

$$\chi_{Na} = \frac{10.94}{10.94 + 58.72} = 0.1570$$

$$\text{Molality} = \frac{251.4\ g\ Na}{kg} \times \frac{1\ mol\ Na}{22.99\ g\ Na} = 10.94\ mol/kg$$

51.

	Li	Na	K	Rb	Cs	Fr
Atomic number	3	11	19	37	55	87
MP	180	98	63	39	29	(≈22)

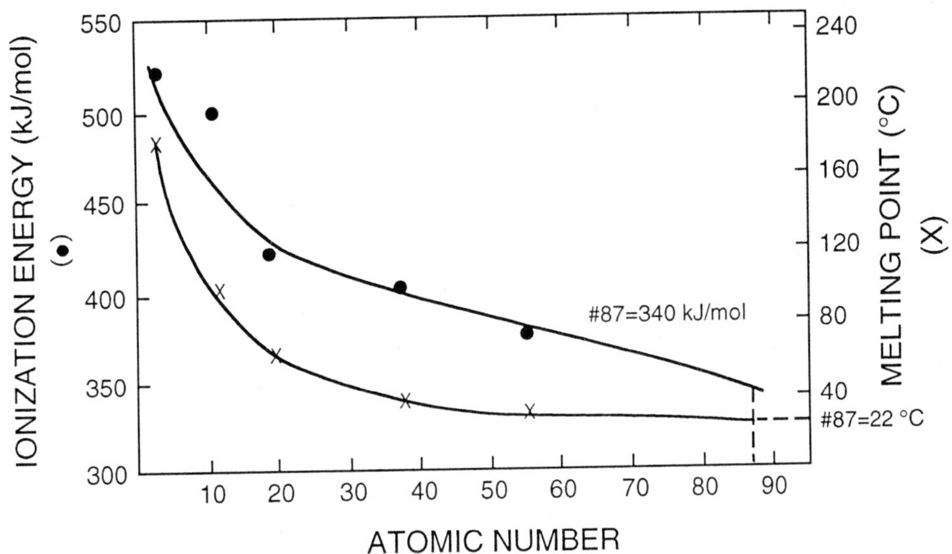

From the graph, we would estimate the melting point of Fr to be around 20°C. Thus, it would be a liquid at room temperature.

53. $1.00 \times 10^3 \text{ kg} \times \dfrac{1000 \text{ g}}{\text{kg}} \times \dfrac{1 \text{ mol Ca}}{40.08 \text{ g}} \times \dfrac{2 \text{ mol e}^-}{\text{mol Ca}} \times \dfrac{96{,}485 \text{ C}}{\text{mol e}^-} = 4.81 \times 10^9 \text{ C}$

$i = \dfrac{4.81 \times 10^9 \text{ C}}{8.00 \text{ hr}} \times \dfrac{1 \text{ hr}}{60 \text{ min}} \times \dfrac{1 \text{ min}}{60 \text{ s}} = \dfrac{1.67 \times 10^5 \text{ C}}{\text{s}} = 1.67 \times 10^5 \text{ A}$

$1.00 \times 10^3 \text{ kg Ca} \times \dfrac{70.90 \text{ g } Cl_2}{40.08 \text{ g Ca}} = 1.77 \times 10^3 \text{ kg of } Cl_2$

55. A single $AlCl_3$ has the Lewis structure:

```
 ..        ..
:Cl — Al — Cl:
 ..   |    ..
      |
     :Cl:
      ..
```

The Al has room for a pair of electrons. It can act as a Lewis acid. The only lone pairs are on Cl, so the structure is:

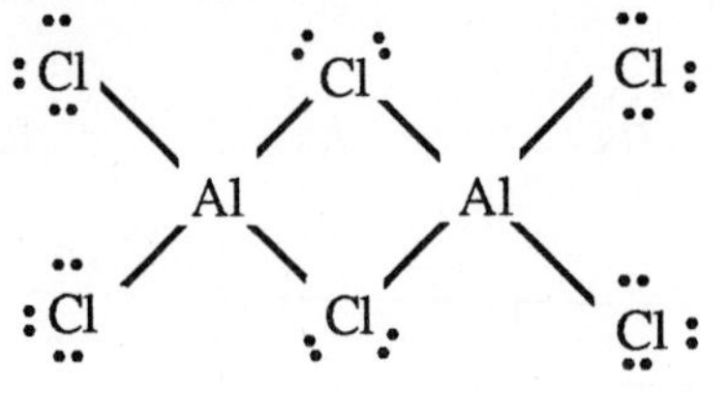

57. $LiAlH_4$ H -1 (metal hydride), Li +1

Al +3; $0 = +1 + x + 4(-1)$; $x = +3$

59. Ga(I) $[Ar]3d^{10}4s^2$ no unpaired e^-

Ga(III) $[Ar]3d^{10}$ no unpaired e^-

Ga(II) $[Ar]3d^{10}4s^1$ 1 unpaired e^-

If the compound contained Ga(II) it would be paramagnetic. This can easily be determined by measuring the mass of a sample in the presence and absence of a magnetic field. Paramagnetic compounds will have an apparent greater mass in a magnetic field.

61.

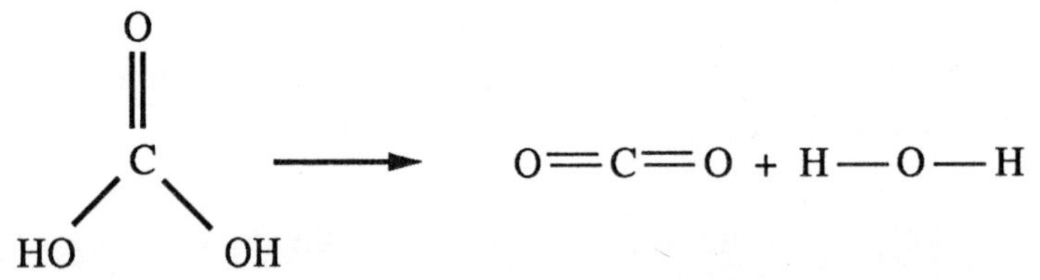

$$\text{(HO)}_2\text{C=O} \longrightarrow \text{O=C=O} + \text{H—O—H}$$

Bonds broken:

2 C—O 358 kJ/mol

Bonds made:

C=O 799 kJ/mol

$\Delta H = 716 - 799 = -83$ kJ

ΔH is favorable for the decomposition of H_2CO_3 to CO_2 and H_2O. ΔS is also favorable for the decomposition as there is an increase in disorder. Hence, H_2CO_3 will spontaneously decompose to CO_2 and H_2O. Carbonic acid should be written as $CO_2(aq)$.

63. Sn and Pb can reduce H^+ to H_2

$Sn(s) + 2 H^+(aq) \longrightarrow Sn^{2+}(aq) + H_2(g)$

$Pb(s) + 2 H^+(aq) \longrightarrow Pb^{2+}(aq) + H_2(g)$

CHALLENGE PROBLEMS

65. a. Na^+ can oxidize Na^- to Na.

$$Na^- \rightarrow Na + e^- \qquad \Delta H = +52.9 \text{ kJ}$$

$$Na^+ + e^- \rightarrow Na \qquad \Delta H = -495 \text{ kJ}$$

$$Na^+ + Na^- \rightarrow 2\ Na \qquad \Delta H = -442 \text{ kJ}$$

The purpose of the cryptand is to encapsulate the Na^+ ion so that it does not come in contact with the Na^- ion and oxidize sodium to sodium metal.

67. White tin is stable at normal temperatures. Gray tin is stable at temperatures below 13.2°C. Thus for the phase change:

$$Sn(gray) \rightarrow Sn(white)$$

ΔG is (-) at T > 13.2°C and ΔG is (+) at T < 13.2°C

This is only possible if ΔH is (+) and ΔS is (+). Thus, gray tin has the more ordered structure.

69. Carbon is much smaller than Si and cannot form a fifth bond in the transition state since carbon has no low energy d orbitals available to expand the octet.

CHAPTER NINETEEN: THE REPRESENTATIVE ELEMENTS - GROUP 5A THROUGH 8A

QUESTIONS

1. The reaction $N_2(g) + 3\ H_2(g) \rightarrow 2\ NH_3(g)$ is exothermic. Thus, K_p decreases as the temperature increases. Lower temperatures are favored for maximum yield of ammonia. However, at lower temperatures the rate is slow; without a catalyst the rate is too slow for the process to be feasible. The discovery of a catalyst increased the rate of reaction at a lower temperature favored by thermodynamics.

3. The pollution provides nitrogen and phosphorous nutrients so the algae can grow. The algae consume oxygen, causing fish to die.

5. Plastic sulfur consists of long S_n chains of sulfur atoms. As plastic sulfur becomes brittle the long chains break down into S_8 rings.

7. Fluorine is the most reactive of the halogens because it is the most electronegative and the bond in the F_2 molecule is very weak.

9. Helium is unreactive and doesn't combine with any other elements. It is a very light gas and would easily escape the earth's gravitational pull as the planet was formed.

EXERCISES

Group 5A Elements

11.

NO_4^{3-} $\quad \left[\ \text{N bonded to four O by single bonds, each O with three lone pairs}\ \right]^{3-}$

N is small. There is probably not enough room for all 4 oxygen atoms. P is larger, thus, PO_4^{3-} is stable.

PO_3^- $\quad \left[\ \text{P double-bonded to one O and single-bonded to two O}\ \right]^{-}$

$P{=}O$ bonds not particularly stable. $N{=}O$ bonds are. Thus, NO_3^- is stable.

13. a. $8\ H^+(aq) + 2\ NO_3^-(aq) + 3\ Cu(s) \rightarrow 3\ Cu^{2+}(aq) + 4\ H_2O(l) + 2\ NO(g)$

b. $NH_4NO_3(s) \xrightarrow{heat} N_2O(g) + 2\ H_2O(g)$

c. $NO(g) + NO_2(g) + 2\ KOH(aq) \rightarrow 2\ KNO_2(aq) + H_2O(l)$

15. NH_3 sp^3; N_2H_4 sp^3; NH_2OH sp^3; N_2 sp

N_2O central N sp; NO sp^2; N_2O_3 both N's are sp^2; NO_2 sp^2; HNO_3 sp^2

17. $\Delta H° = (90 \text{ kJ}) \times 2 = 180.\text{ kJ}$ and $\Delta G° = (87 \text{ kJ}) \times 2 = 174 \text{ kJ}$

$\Delta S° = 2(211 \text{ J/K}) - [192 + 205] = 25 \text{ J/K}$

At high temperature the reaction $N_2 + O_2 \rightarrow 2\ NO$ becomes spontaneous. In the atmosphere, even though $2\ NO \rightarrow N_2 + O_2$ is spontaneous, it doesn't occur because the rate is slow.

19. a. $H_3PO_4 > H_3PO_3$ More O - atoms, stronger acid

b. $H_3PO_4 > H_2PO_4^- > HPO_4^{2-}$

21. Production of antimony:

$$2\ Sb_2S_3(s) + 9\ O_2(g) \rightarrow 2\ Sb_2O_3(s) + 6\ SO_2(g)$$

$$2\ Sb_2O_3(s) + 3\ C(s) \rightarrow 4\ Sb(s) + 3\ CO_2(g)$$

Production of bismuth:

$$2\ Bi_2S_3(s) + 9\ O_2(g) \rightarrow 2\ Bi_2O_3(s) + 6\ SO_2(g)$$

$$2\ Bi_2O_3(s) + 3\ C(s) \rightarrow 4\ Bi(s) + 3\ CO_2(g)$$

Group 6A Elements

23. $O{=}O{-}O \rightarrow O{=}O + O$

Break $O{-}O$

$$\Delta H = +146 \text{ kJ/mol} \times \frac{1 \text{ mol}}{6.022 \times 10^{23}} = 2.42 \times 10^{-22} \text{ kJ} = 2.42 \times 10^{-19} \text{ J}$$

A photon of light must contain at least 2.42×10^{-19} J.

$$E_{photon} = \frac{hc}{\lambda},\ \lambda = \frac{hc}{E} = \frac{(6.626 \times 10^{-34} \text{ J s})(2.998 \times 10^{8} \text{ m/s})}{2.42 \times 10^{-19} \text{ J}}$$

$$\lambda = 8.21 \times 10^{-7} \text{ m} = 821 \text{ nm}$$

25. a. oxidation - reduction reaction b. NO, see (c)

c.

$$(S^{2-} \rightarrow S + 2\ e^-) \times 3$$

$$(3\ e^- + 3\ H^+ + HNO_3 \rightarrow NO + 2\ H_2O) \times 2$$

$$6\ H^+ + 2\ HNO_3 + 3\ S^{2-} \rightarrow 3\ S + 2\ NO + 4\ H_2O$$

27. SF_5^- has 6 + 5(7) + 1 = 42 valence electrons.

square pyramid

Group 7A Elements

29. a. ClF_5, 7 + 5(7) = 42 e^-

Square pyramid

b. IF_3, 7 + 3(7) = 28 e^-

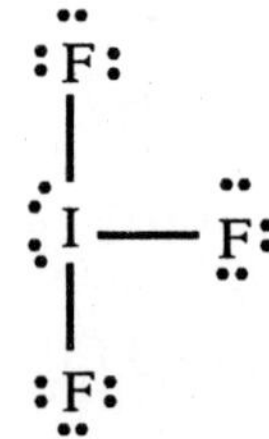

T-shaped

c. Cl_2O_7, 2(7) + 7(6) = 56 e^-

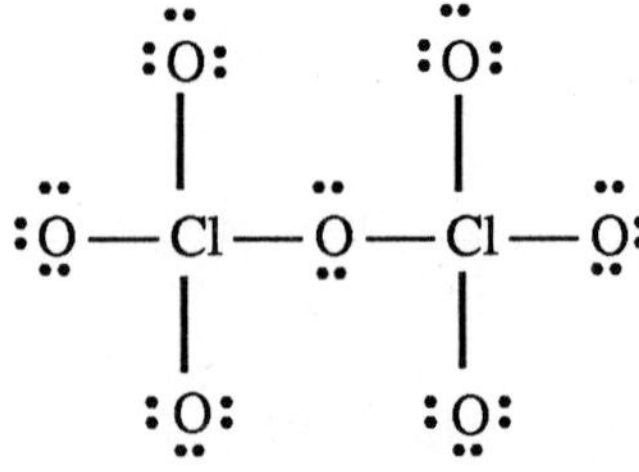

The four O atoms are tetrahedrally arranged about each Cl. The Cl—O—Cl bond angle is close to the tetrahedral bond angle.

d. $FBrO_2$, 7 + 7 + 2(6) = 26 e^-

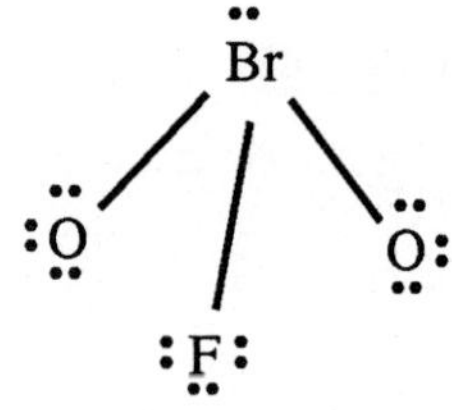

Trigonal pyramidal

31.

$$BrO_3^- + XeF_2 \rightarrow BrO_4^- + Xe + HF$$

$$H_2O + BrO_3^- \rightarrow BrO_4^- + 2\,H^+ + 2\,e^-$$

$$2\,e^- + 2\,H^+ + XeF_2 \rightarrow Xe + 2\,HF$$

$$H_2O(l) + BrO_3^-(aq) + XeF_2(aq) \rightarrow BrO_4^-(aq) + Xe(g) + 2\,HF(aq)$$

33. a.

$$2\,e^- + Cl_2 \rightarrow 2\,Cl^- \qquad E° = +1.36\ V$$

$$2\,H_2O + Cl_2 \rightarrow 2\,OCl^- + 4\,H^+ + 2\,e^- \qquad -E° = -1.63\ V$$

$$2\,H_2O + 2\,Cl_2 \rightarrow 2\,Cl^- + 2\,OCl^- + 4\,H^+ \qquad E° = -0.27\ V$$

b. E° < 0, Reaction is not spontaneous.

c. Reaction is favored as solution becomes more basic.

Group 8A Elements: The Noble Gases

35. Xe has one more valence electron than I. Thus, the isoelectric species will have I and one extra electron substituted for Xe, giving a species with a net minus one charge.

a. IO_4^- b. IO_3^- c. IF_2^- d. IF_4^- e. IF_6^- f. IOF_3^-

37. XeF_2 can react with oxygen to produce explosive xenon oxides and oxyfluorides.

ADDITIONAL EXERCISES

39. As the halogen atoms get larger, it becomes more difficult to fit three halogen atoms around the small N, and the NX_3 molecule becomes less stable.

41. $3\,NO(g) \rightleftharpoons N_2O(g) + NO_2(g)$

$\Delta H° = 82 + 34 - 3(90) = -154$ kJ

$\Delta S° = 220 + 240 - 3(211) = -173$ J/K

$\Delta G° = \Delta H° - T\Delta S° = -154 - 298(-0.173) = -102$ kJ

$$\Delta G° = 0 \text{ when } T = \frac{\Delta H°}{\Delta S°} = \frac{-154{,}000\ J}{-173\ J/K} = 890.\ K$$

The reaction is spontaneous at temperatures below 890. K (all partial pressures = 1 atm).

43. OCN^- has 6 + 4 + 5 + 1 = 16 valence electrons.

$$[\ddot{O}=C=\ddot{N}]^- \longleftrightarrow [:\ddot{O}-C\equiv N:]^- \longleftrightarrow [:O\equiv C-\ddot{N}:]^-$$

Formal Charge	0	0	-1	-1	0	0	+1	0	-2

Only the first two resonance structures should be important. The third places a positive formal charge on the most electronegative atom in the ion and a -2 charge on N.

CNO^-

$[\,\dot{\underset{\cdot\cdot}{}}C{=}N{=}\ddot{O}\colon]^- \longleftrightarrow [\colon C{\equiv}N{-}\ddot{\underset{\cdot\cdot}{O}}\colon]^- \longleftrightarrow [\colon\underset{\cdot\cdot}{C}{-}N{\equiv}O\colon]^-$

Formal Charge	-2	+1	0	-1	+1	-1	-3	+1	+1

All of the resonance structures for fulminate involve greater formal charges than in cyanate, making fulminate more reactive (less stable).

45.

		Bond order	# unpaired e^-
M.O.	NO	2.5	1
	NO^+	3	0
	NO^-	2	2

Lewis NO^+ $[\colon N{\equiv}O\colon]^+$

NO $\dot{N}{=}\ddot{O} \longleftrightarrow \dot{N}{=}\ddot{O} \longleftrightarrow \dot{N}{=}\ddot{O}$

NO^- $[\ddot{N}{=}\ddot{O}]^-$

Lewis structures are not adequate for NO and NO^-. M.O. model gives correct results for all three species. For NO, Lewis structures fail for odd electron species. For NO^-, Lewis structures fail to predict that NO^- is paramagnetic.

47. a.

$$Mn^{2+} + NaBiO_3 \rightarrow MnO_4^- + BiO_3^{3-}$$

$$(4\,H_2O + Mn^{2+} \rightarrow MnO_4^- + 8\,H^+ + 5\,e^-) \times 2$$

$$(2\,e^- + NaBiO_3 \rightarrow BiO_3^{3-} + Na^+) \times 5$$

$$8\,H_2O(l) + 2\,Mn^{2+}(aq) + 5\,NaBiO_3(s) \rightarrow 2\,MnO_4^-(aq) + 16\,H^+(aq) + 5\,BiO_3^{3-}(aq) + 5\,Na^+(aq)$$

b. Bismuthate exists as a covalent network solid : $(BiO_3^-)_x$

49. $$2.42\text{ eV} \times \frac{96.5\text{ kJ/mol}}{\text{eV}} \times \frac{1\text{ mol photons}}{6.022 \times 10^{23}\text{ photons}} \times \frac{1000\text{ J}}{\text{kJ}} = \frac{3.88 \times 10^{-19}\text{ J}}{\text{photon}}$$

$$E = \frac{hc}{\lambda}$$

$$\lambda = \frac{hc}{E} = \frac{(6.626 \times 10^{-34}\text{ J s})(2.998 \times 10^8\text{ m/s})}{3.88 \times 10^{-19}\text{ J}} = 5.12 \times 10^{-7}\text{ m} = 512\text{ nm},$$ Green light

51. a.

$$ClO_3^- + 2\,H_2O \longrightarrow ClO_4^- + 2\,H^+ + 2\,e^- \qquad -E° = -1.19\ V$$

$$2\,H^+ + 2\,e^- \longrightarrow H_2 \qquad E° = 0.0\ V$$

$$ClO_3^- + H_2O \longrightarrow ClO_4^- + H_2 \qquad E°_{cell} = -1.19\ V$$

Therefore, a potential of 1.19 V must be applied (assuming standard conditions).

b. $3\,Al + 3\,NH_4ClO_4 \longrightarrow Al_2O_3 + AlCl_3 + 3\,NO + 6\,H_2O(g)$

$\Delta H° = 3(90) + (-704) + (-1676) + 6\,(-242) - 3(-295)$

$\Delta H° = -2677\ kJ$

$$7 \times 10^5\ kg \times \frac{1000\ g}{kg} \times \frac{1\ mol}{117.49\ g} \times \frac{2677\ kJ}{3\ mol\ NH_4ClO_4} = 5 \times 10^9\ kJ$$

CHALLENGE PROBLEMS

53. For $NCl_3 \longrightarrow NCl_2 + Cl$, only the N—Cl bond is being broken. For O=N—Cl $\longrightarrow$ NO + Cl, when the N—Cl bond is broken, the NO bond gets stronger (bond order increases from 2.0 to 2.5). This makes ΔH for the reaction smaller than just the energy necessary to break the N—Cl bond.

55. a. As we go down the family, K_a increases. This is consistent with the bond to hydrogen getting weaker.

b. Po is below Te, so K_a should be larger. The K_a for H_2Po should be on the order of 10^{-2} or 10^{-1}.

CHAPTER TWENTY: TRANSITION METALS AND COORDINATION CHEMISTRY

QUESTIONS

1. a. ligand: Species that donates a pair of electrons to form a covalent bond to a metal ion (a Lewis base).

   b. chelate: Ligand that can form more than one bond.

   c. bidentate: Ligand that can form two bonds.

   d. complex ion: Metal ion plus ligands.

3. Both electrons in the bond originally came from the same atom.

5. a. isomers: Species with the same formulas but different properties. See text for examples of the following types of isomers.

   b. structural isomers: Isomers that have one or more bonds that are different.

   c. stereoisomers: Isomers that contain the same bonds but differ in how the atoms are arranged in space.

   d. coordination isomers: Isomers that differ in what atoms are found in the coordination sphere.

   e. linkage isomers: Isomers that differ in how one or more ligands are attached to the transition metal.

   f. geometric isomers: (cis-trans isomerism) Isomers that differ in the position of atoms with respect to a rigid ring, bond, or each other.

   g. optical isomerism: Isomers that differ by being non-superimposable mirror images of each other; that is, they are different in the same way our left and right hands differ.

7. Cu^{2+}: $[Ar]3d^9$, Cu^+: $[Ar]3d^{10}$

   Cu(II) is d^9; Cu(I) is d^{10}. Color is a result of the electron transfer between split d orbitals. This cannot occur for the filled d orbitals in Cu(I).

9. Sc^{3+} has no unpaired electrons. V^{3+} and Ti^{3+} have unpaired d-electrons present. Color of transition metal compounds results from the presence of unpaired d electrons.

11. $Fe_2O_3(s) + 6\ H_2C_2O_4(aq) \longrightarrow 2\ Fe(C_2O_4)_3^{3-}(aq) + 3\ H_2O(l) + 6\ H^+(aq)$
    The oxalate anion forms a soluble complex ion with the iron in rust.

13. There is a steady decrease in the atomic radii of the lanthanide elements, going from left to right. As a result of the lanthanide contraction the properties of the 4d and 5d elements in each group are very similar because of the similar size of atoms and ions (See Exercise 7.123).

15. Advantages: cheap energy cost

less pollution

Disadvantages: chemicals used in hydrometallurgy are expensive.

EXERCISES

Transition Metals

17. a. Ni: $[Ar]4s^2 3d^8$ c. Zr: $[Kr]5s^2 4d^2$

b. Cd: $[Kr]5s^2 4d^{10}$ d. Nd: $[Xe]6s^2 5d^1 4f^3$ or $[Xe]6s^2 4f^4$

19. a. Ni^{2+}: $[Ar]3d^8$ c. Zr^{3+}: $[Kr]4d^1$; Zr^{4+}: [Kr]

b. Cd^{2+}: $[Kr]4d^{10}$ d. Nd^{3+}: $[Xe]4f^3$

21. a. Co: $[Ar]4s^2 3d^7$ b. Pt: $[Xe]6s^1 4f^{14} 5d^9$ c. Fe: $[Ar]4s^2 3d^6$

Co^{2+}: $[Ar]3d^7$ Pt^{2+}: $[Xe]4f^{14} 5d^8$ Fe^{2+}: $[Ar]3d^6$

Co^{3+}: $[Ar]3d^6$ Pt^{4+}: $[Xe]4f^{14} 5d^6$ Fe^{3+}: $[Ar]3d^5$

23. $1.00 \times 10^6 \text{ g FeTiO}_3 \times \frac{47.88 \text{ g Ti}}{151.7 \text{ g FeTiO}_3} = 3.16 \times 10^5$ g or 316 kg

25. a. Molybdenum(IV) sulfide and Molybdenum(VI) oxide

b. MoS_2 +4 MoO_3 +6

$(NH_4)_2Mo_2O_7$ +6 $(NH_4)_6Mo_7O_{24}\cdot 4\,H_2O$ +6

c. $2\,MoS_2 + 7\,O_2 \rightarrow 2\,MoO_3 + 4\,SO_2$

$2\,NH_3 + 2\,MoO_3 + H_2O \rightarrow (NH_4)_2Mo_2O_7$

$6\,NH_3 + 7\,MoO_3 + 7\,H_2O \rightarrow (NH_4)_6Mo_7O_{24}\cdot 4\,H_2O$

Coordination Compounds

27. a. pentaamminechlororuthenium(III) ion

b. hexacyanoferrate(II) ion

c. tris(ethylenediamine) manganese(II) ion

d. pentaamminenitrocobalt(III) ion

29. a. hexaamminecobalt(II) chloride

b. hexaaquacobalt(III) iodide

c. potassium tetrachloroplatinate(II)

d. potassium hexachloroplatinate(II)

31. a. $[Co(C_5H_5N)_6]Cl_3$ b. $[Cr(NH_3)_5I]I_2$

c. $[Ni(NH_2CH_2CH_2NH_2)_3]Br_2$ d. $K_2[Ni(CN)_4]$ e. $[Pt(NH_3)_4Cl_2]PtCl_4$

33. a.

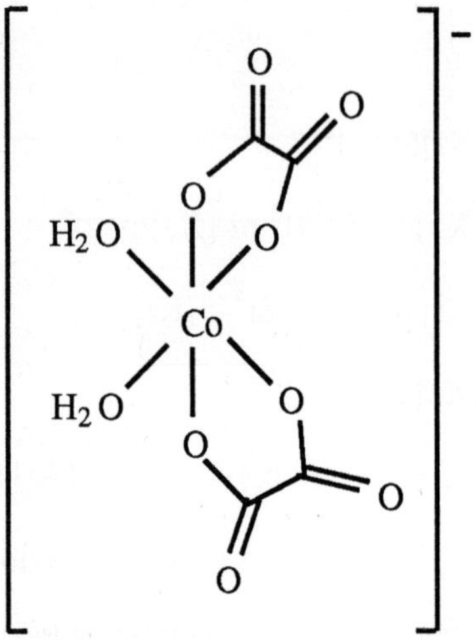

cis 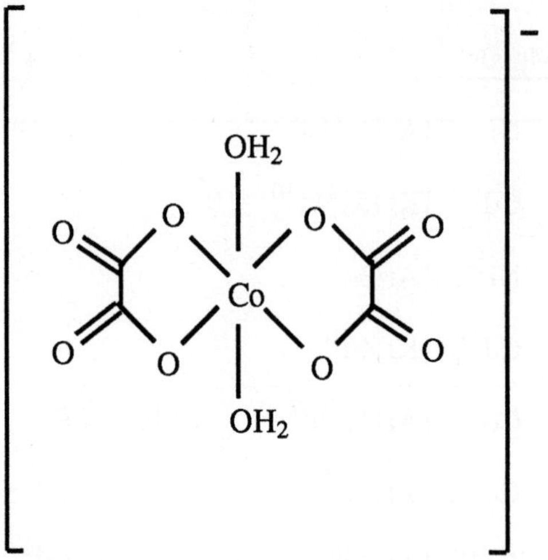 trans

b.

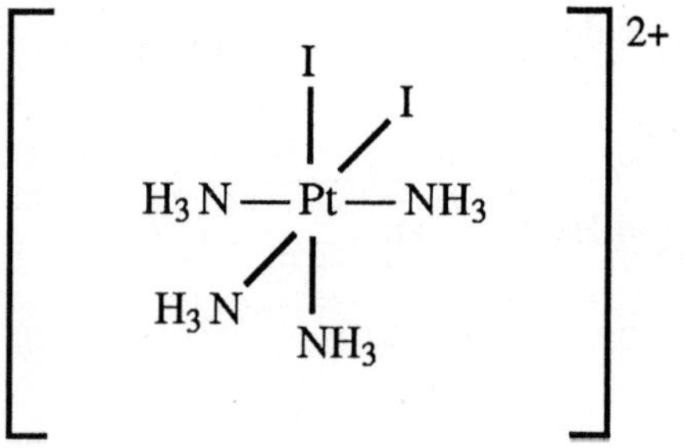

cis 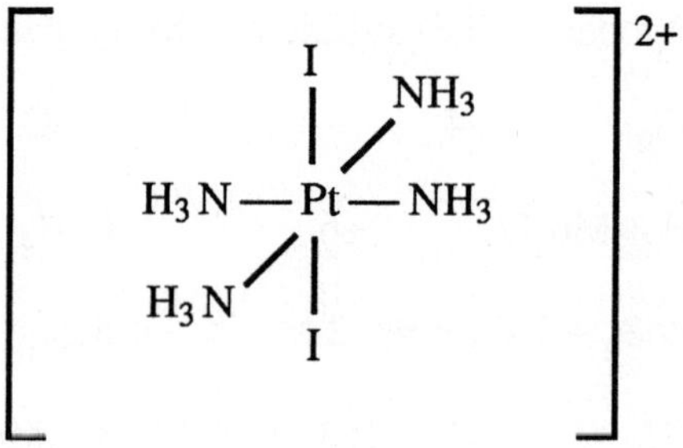 trans

c.

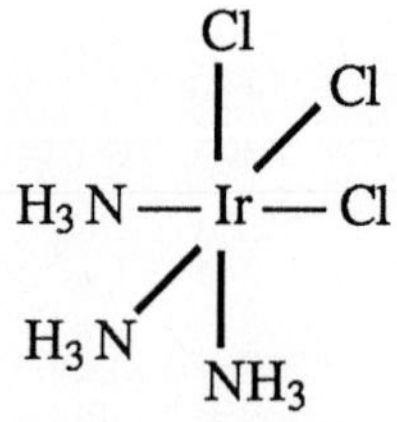

cis 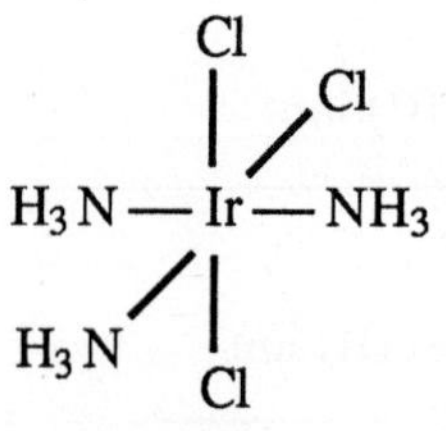 trans

d.

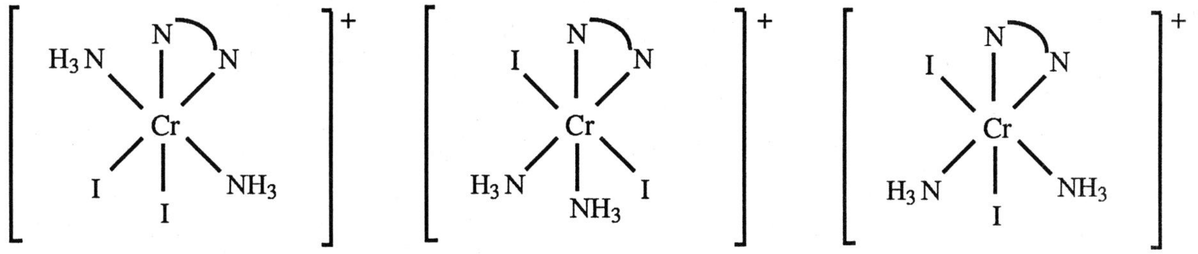

$$[Cr(NH_3)_2(N\text{-}N)I_2]^+$$

H3N N N Cr I I NH3 (+)

I N N Cr H3N NH3 I (+)

I N N Cr H3N I NH3 (+)

35.

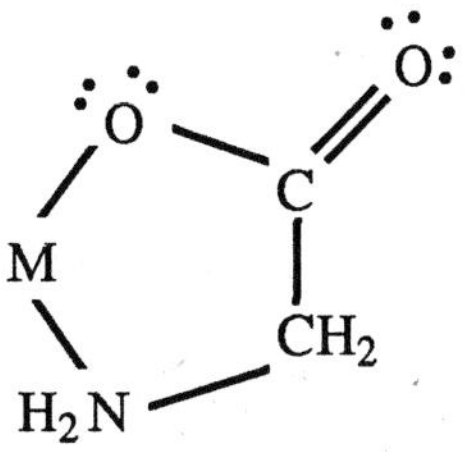

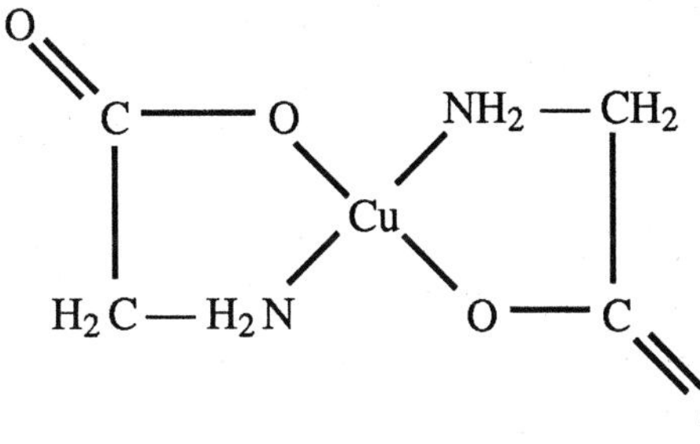

M, O, C, O, CH2, H2N

C O NH2 — CH2 Cu H2C — H2N O — C O

and

C O O — C Cu H2C — H2N NH2 — CH2

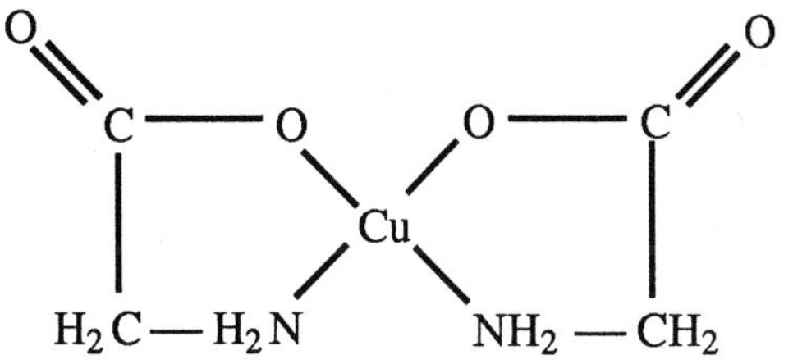

37.

SCN^- ---------- (M-SCN, M-NCS)

NO_2^- ---------- (M-NO_2, M-ONO)

OCN^- ---------- (M-OCN, M-NCO)

N_3^-, en, and I^- are not capable of linkage isomerism.

39.

optically active (mirror image not shown)

optically active (mirror image not shown)

Bonding, Color, and Magnetism in Coordination Compounds

41.

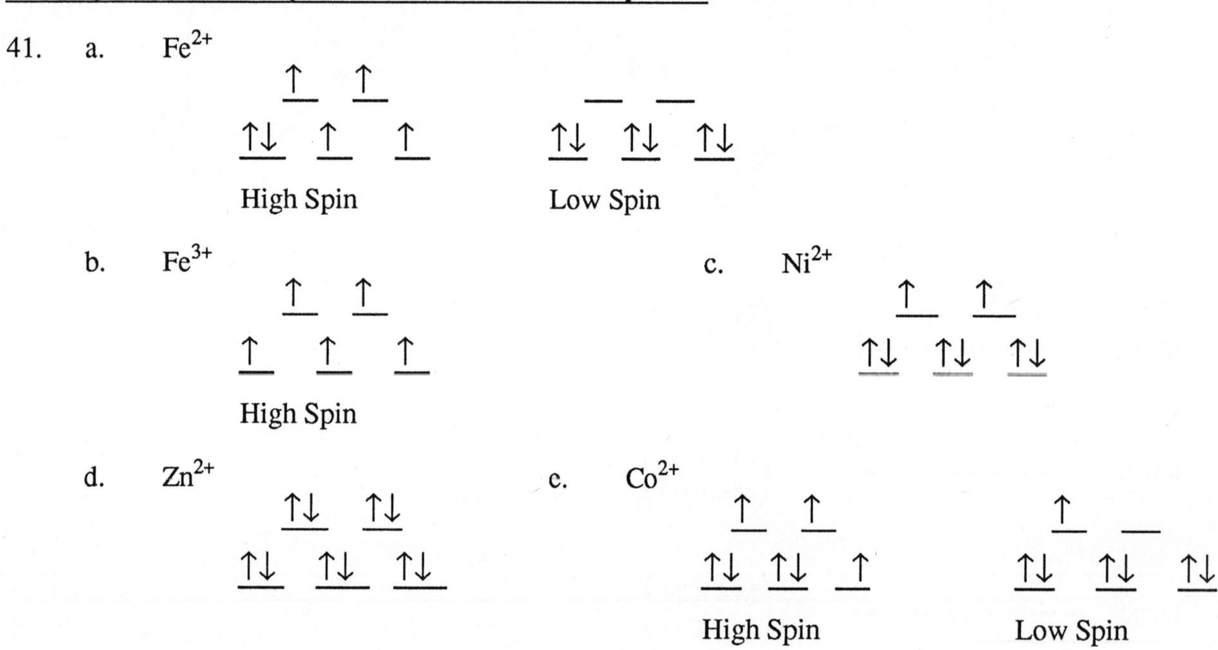

43. $NiCl_4^{2-}$ is tetrahedral and $Ni(CN)_4^{2-}$ is square planar.

$NiCl_4^{2-}$

$Ni(CN)_4^{2-}$

45. Transition compounds exhibit the color complementary to that absorbed. Using Table 20.16, $Ni(H_2O)_6Cl_2$ absorbs red light and $Ni(NH_3)_6Cl_2$ absorbs yellow-green light. $Ni(NH_3)_6Cl_2$ absorbs the higher energy light, therefore, Δ is larger for $Ni(NH_3)_6Cl_2$. NH_3 is a stronger field ligand than H_2O, consistent with the spectrochemical series.

Metallurgy

47. a. $3\ Fe_2O_3 + CO \rightarrow 2\ Fe_3O_4 + CO_2$

$\Delta H° = -393.5 + 2(-1117) - [3(-826) + (-110.5)]$

$\Delta H° = -39$ kJ

$\Delta S° = 214 + 2(146) - [3(90) + 198]$

$\Delta S° = 38$ J/K

b. At 800°C, T = 800 + 273 = 1073 K (carry extra significant figures in T)

$\Delta G° = -39$ kJ - 1073 K(0.038 kJ/K)

= -39 kJ - 41 kJ = -80 kJ

ADDITIONAL EXERCISES

49. a. 4-O on faces × 1/2 O/face = 2-O atoms

2-O atoms inside body, Total: 4-O atoms

8 Ti on corners × 1/8 Ti/corner + 1 Ti/body center = 2 Ti atoms

Formula of unit cell Ti_2O_4: empirical formula TiO_2

b. $\overset{+4\ -2}{2\ TiO_2} + \overset{0}{3\ C} + \overset{0}{4\ Cl_2} \rightarrow \overset{+4\ -1}{2\ TiCl_4} + \overset{+4\ -2}{CO_2} + \overset{+2\ -2}{2\ CO}$

C is being oxidized. Cl_2 is oxidizing agent.

Cl is being reduced. C is reducing agent.

$\overset{+4\ -1}{TiCl_4} + \overset{0}{O_2} \rightarrow \overset{+4\ -2}{TiO_2} + \overset{0}{2\ Cl_2}$

Cl is being oxidized. O_2 is oxidizing agent.

O is being reduced. $TiCl_4$ is reducing agent.

51. a. $2\ CoAs_2(s) + 4\ O_2(g) \rightarrow 2\ CoO(s) + As_4O_6(s)$

As_4O_6 tetraarsenic hexoxide

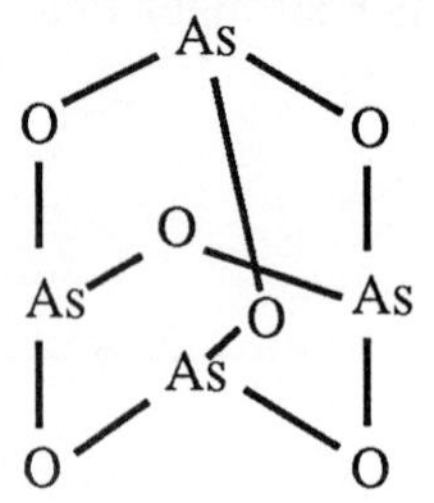

As_4O_6 has a cage structure similar to P_4O_6

b. $Co^{2+}(aq) + OCl^-(aq) \rightarrow Co(OH)_3(s) + Cl^-(aq)$

$(Co^{2+} + 3\ OH^- \rightarrow Co(OH)_3 + e^-) \times 2$

$2\ e^- + H_2O + OCl^- \rightarrow Cl^- + 2\ OH^-$

$2\ Co^{2+}(aq) + 4\ OH^-(aq) + H_2O(l) + OCl^-(aq) \rightarrow 2\ Co(OH)_3(s) + Cl^-(aq)$

c.

	$Co(OH)_3(s)$	$\rightleftharpoons$	$Co^{3+}(aq)$	+	$3\ OH^-(aq)$
Initial	0		0		$1.0 \times 10^{-7}\ M$
	s mol/L $Co(OH)_3$ dissolves to reach equilibrium				
Change	-s	$\rightarrow$	+s		+3s
Equil.			s		$1.0 \times 10^{-7} + 3s$

$2.5 \times 10^{-43} = [Co^{3+}][OH^-]^3 = s(1.0 \times 10^{-7} + 3s)^3$

$\approx s(1.0 \times 10^{-21})$

$s = [Co^{3+}] = 2.5 \times 10^{-22}$ mol/L Assumption good.

d. $Co(OH)_3 \rightleftharpoons Co^{3+} + 3\ OH^-$ pH = 10.00 pOH = 4.00 $[OH^-] = 1.0 \times 10^{-4}\ M$

$2.5 \times 10^{-43} = [Co^{3+}][OH^-]^3 = [Co^{3+}]\ (1.0 \times 10^{-4})^3$

$[Co^{3+}] = 2.5 \times 10^{-31}\ M$

53. $2\ Au(CN)_2^- + 2\ e^- \rightarrow 2\ Au + 4\ CN^-$ $E° = -0.60$ V

$Zn + 4\ CN^- \rightarrow Zn(CN)_4^{2-} + 2\ e^-$ $-E° = 1.26$ V

$2\ Au(CN)_2^- + Zn \rightarrow 2\ Au + Zn(CN)_4^{2-}$ $E°_{cell} = 0.66$ V

$\Delta G° = -nFE° = -(2\ mol\ e^-)\ (96{,}485\ C/mol\ e^-)(0.66\ J/C)$

$= -1.27 \times 10^5\ J = -130\ kJ$

$\log K = \frac{nE°}{0.0592} = \frac{2(0.66)}{0.0592} = 22.30$

$K = 2.0 \times 10^{22}$

55. a. 2 b. 3 c. 4 d. 4

57. a.

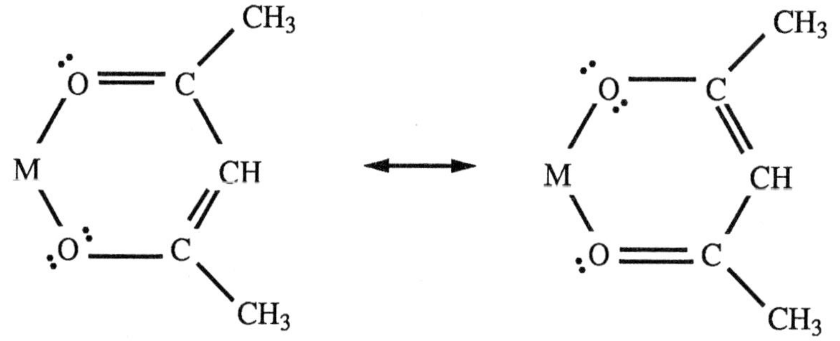

CH3 O C CH M O C CH3 ⟷ CH3 O C CH M O C CH3

b. cis- $Cr(acac)_2(H_2O)_2$ and $Cr(acac)_3$ are optically active.

c. $0.112 \text{ g } Eu_2O_3 \times \frac{304.0 \text{ g Eu}}{352.0 \text{ g } Eu_2O_3} = 0.0967 \text{ g Eu}$

$\% \text{ Eu} = \frac{0.0967 \text{ g}}{0.286 \text{ g}} \times 100 = 33.8\% \text{ Eu}$

$\% \text{ O} = 100 - (33.8 + 40.1 + 4.68) = 21.4\% \text{ O}$

Out of 100 g of compound:

$33.8 \text{ g Eu} \times \frac{1 \text{ mol}}{152.0 \text{ g}} = 0.222 \text{ mol Eu}$

$40.1 \text{ g C} \times \frac{1 \text{ mol}}{12.01 \text{ g}} = 3.34 \text{ mol C}$

$4.68 \text{ g H} \times \frac{1 \text{ mol}}{1.008 \text{ g}} = 4.64 \text{ mol H}$

$21.4 \text{ g O} \times \frac{1 \text{ mol}}{16.00 \text{ g}} = 1.34 \text{ mol O}$

$\frac{3.34}{0.222} = 15.0, \frac{4.64}{0.222} = 20.9, \frac{1.34}{0.222} = 6.04$

Formula is: $EuC_{15}H_{21}O_6$

Each $acac^-$ is $C_5H_7O_2^-$; So Formula is: $Eu(acac)_3$

59. a. $Ru(phen)_3^{2+}$ exhibits optical isomerism.

b. Ru^{2+}: $[Kr]4d^6$, Ru^{2+} is strong field case.

— — e_g

↑↓ ↑↓ ↑↓ t_{2g}

61. CN^- is a weak base.

$Ni^{2+}(aq) + 2\ OH^-(aq) \rightarrow Ni(OH)_2(s)$

$Ni(OH)_2(s) + 4\ CN^-(aq) \rightarrow Ni(CN)_4^{2-}(aq) + 2\ OH^-(aq)$

63. a.

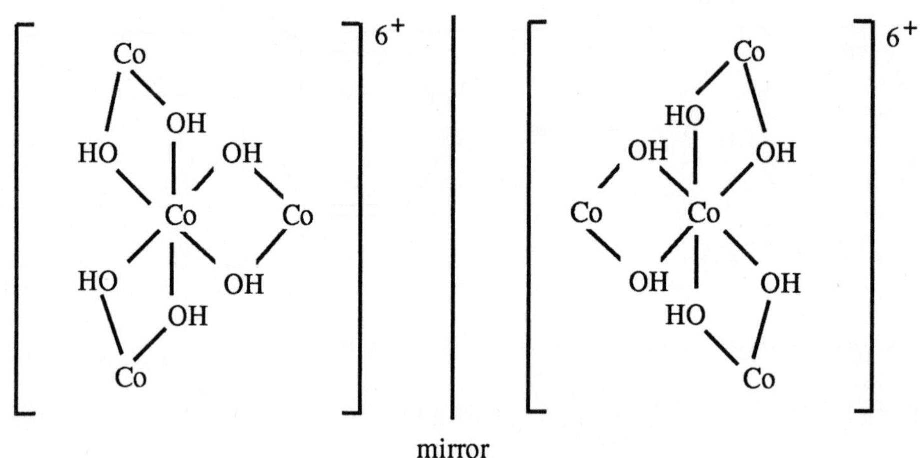

NH_3 molecules are not shown.

b. All are Co(III).

c. none

CHALLENGE PROBLEMS

65. i. $0.0203\ g\ CrO_3 \times \frac{52.00\ g\ Cr}{100.0\ g\ CrO_3} = 0.0106\ g\ Cr$

$\%Cr = \frac{0.0106}{0.105} \times 100 = 10.1\%\ Cr$

ii. $32.92\ mL\ HCl \times \frac{0.100\ mmol\ HCl}{mL} \times \frac{1\ mmol\ NH_3}{mmol\ HCl} \times \frac{17.03\ mg\ NH_3}{mmol} = 56.1\ mg\ NH_3$

$\%\ NH_3 = \frac{56.1\ mg}{341\ mg} \times 100 = 16.5\%\ NH_3$

iii. $73.53 + 16.5 + 10.1 = 100.1$

So the compound is composed of only Cr, NH_3 and I.

Out of 100 g of compound:

$10.1\ g\ Cr \times \frac{1\ mol}{52.00\ g} = 0.194$ $\qquad \frac{0.194}{0.194} = 1$

$16.5\ g\ NH_3 \times \frac{1\ mol}{17.03\ g} = 0.969$ $\qquad \frac{0.969}{0.194} = 5$

$73.53\ g\ I \times \frac{1\ mol}{126.9\ g} = 0.5794$ $\qquad \frac{0.5794}{0.194} = 3$

$Cr(NH_3)_5I_3$ is the empirical formula. Cr(III) forms octahedral complexes.

So compound A is made of $[Cr(NH_3)_5I]^{2+}$ and two I^- ions or $[Cr(NH_3)_5I]I_2$.

iv. $\Delta T_f = iK_fm$, i = 3 ions

$$m = \frac{0.601 \text{ g complex}}{10.0 \text{ g } H_2O} \times \frac{1 \text{ mol complex}}{517.9 \text{ g complex}} \times \frac{1000 \text{ g } H_2O}{\text{kg}} = 0.116 \text{ molal}$$

$\Delta T_f = 3 \times 1.86°C/\text{molal} \times 0.116 \text{ molal} = 0.65°C$ This is close to the measured value.

This is consistent with the formula $[Cr(NH_3)_5I]I_2$.

67. No, since in all three cases six bonds are formed between Ni^{2+} and nitrogen.

$\Delta S°$ for formation of the complex ion is most negative for 6 NH_3 molecules react with a metal ion (7 independent species becomes 1). For penten reacting with a metal ion, 2 independent species become 1, so $\Delta S°$ is less negative. Thus, the chelate effect occurs because the more bonds a chelating agent can form to the metal, the more favorable $\Delta S°$ is for the formation of the complex ion, and the larger the formation constant.

69.

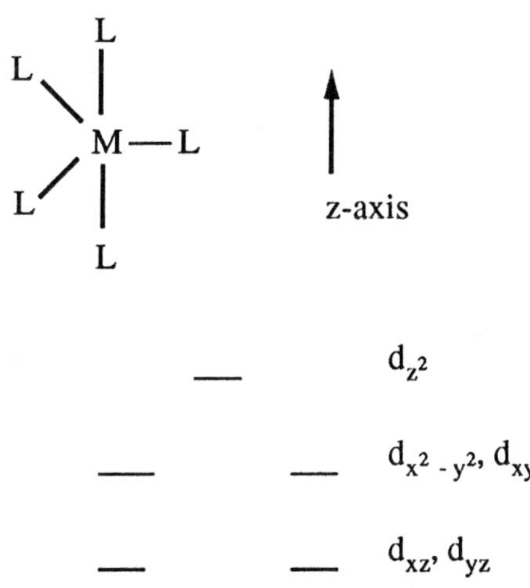

— d_{z^2}

— — $d_{x^2-y^2}, d_{xy}$

— — d_{xz}, d_{yz}

The d_{z^2} will be destabilized much more than in the trigonal planar case. (See Exercise 20.68.)

CHAPTER TWENTY-ONE: THE NUCLEUS - A CHEMIST'S VIEW

QUESTIONS

1. Fission: splitting of a heavy nucleus into two (or more) lighter nuclei.

   Fusion: combining of two light nuclei to form a heavier nucleus.

   Fusion is more likely for elements lighter than Fe; fission is more likely for elements heavier than Fe.

3. ^{14}C levels in the atmosphere are constant or the ^{14}C level at the time the plant died can be calculated. Constant ^{14}C level is a poor assumption and accounting for variation is complicated.

5. No, coal fired power plants also pose risks. A partial list of risks:

Coal	Nuclear
Air pollution	Radiation exposure to workers
Coal mine accidents	Disposal of wastes
Health risks to miners (black lung disease)	Meltdown
	Terrorists
	Public fear

7. For fusion reactions, a collision of sufficient energy must occur between two positively charged particles to initiate the reaction. This requires high temperatures. In fission, an electrically neutral neutron collides with the positively charged nucleus. This has much lower activation energy.

9. Magnetic fields are needed to contain the fusion reaction. Superconductors allow the production of very strong magnetic fields by being able to carry large electric currents. Stronger magnetic fields should be more capable of containing the fusion reaction.

EXERCISES:

Radioactive Decay and Nuclear Transformations

11. a. $^{3}_{1}H \rightarrow ^{0}_{-1}e + ^{3}_{2}He$

    b. $^{8}_{3}Li \rightarrow ^{8}_{4}Be + ^{0}_{-1}e$

    $^{8}_{4}Be \rightarrow 2\,^{4}_{2}He$

    $^{8}_{3}Li \rightarrow 2\,^{4}_{2}He + ^{0}_{-1}e$

    c. $^{7}_{4}Be + ^{0}_{-1}e \rightarrow ^{7}_{3}Li$

    d. $^{8}_{5}B \rightarrow ^{8}_{4}Be + ^{0}_{+1}e$

    e. $^{32}_{15}P \rightarrow ^{32}_{16}S + ^{0}_{-1}e$

13. a. $^{1}_{1}H + ^{14}_{7}N \rightarrow ^{11}_{6}C + ^{4}_{2}He$

    b. $2\,^{3}_{2}He \rightarrow ^{4}_{2}He + 2\,^{1}_{1}H$

    c. $^{1}_{1}H + ^{1}_{1}H \rightarrow ^{2}_{1}H + ^{0}_{+1}e$ (positron)

    d. $^{1}_{1}H + ^{12}_{6}C \rightarrow ^{13}_{7}N$

15. ^{8}B and ^{9}B contain too many protons or too few neutrons.

$^{1}_{1}p \rightarrow ^{1}_{0}n + ^{0}_{+1}e$ (positron emission)

$^{1}_{1}p + ^{0}_{-1}e \rightarrow ^{1}_{0}n$ (electron capture)

^{8}B and ^{9}B might decay by either positron emission or electron capture.

^{12}B and ^{13}B contain too many neutrons or too few protons. Beta emission converts neutrons to protons, so we expect ^{12}B and ^{13}B to be β-emitters.

17. a. Complete decay is 7α and 4β.

$$^{235}_{92}U \rightarrow ^{207}_{82}Pb + 7\,^{4}_{2}He + 4\,^{0}_{-1}e$$

b.

$$^{235}_{92}U \rightarrow ^{231}_{90}Th + ^{4}_{2}He \rightarrow ^{231}_{91}Pa + ^{0}_{-1}e \rightarrow ^{227}_{89}Ac + ^{4}_{2}He$$

$$\rightarrow ^{227}_{90}Th + ^{0}_{-1}e \rightarrow ^{223}_{88}Ra + ^{4}_{2}He \rightarrow ^{219}_{86}Rn + ^{4}_{2}He \rightarrow ^{215}_{84}Po + ^{4}_{2}He$$

$$\rightarrow ^{4}_{2}He + ^{211}_{82}Pb \rightarrow ^{211}_{83}Bi + ^{0}_{-1}e \rightarrow ^{207}_{81}Tl + ^{4}_{2}He \rightarrow ^{207}_{82}Pb + ^{0}_{-1}e$$

Kinetics of Radioactive Decay

19. For $t_{1/2}$ = 12,000 yr

$$k = \frac{\ln(2)}{t_{1/2}} = \frac{0.693}{t_{1/2}} = \frac{0.693}{12{,}000\text{ yr}} \times \frac{1\text{ yr}}{365\text{ d}} \times \frac{1\text{ d}}{24\text{ hr}} \times \frac{1\text{ hr}}{3600\text{ s}}$$

$k = 1.8 \times 10^{-12}\ s^{-1}$

Rate = kN = $1.8 \times 10^{-12}\ s^{-1} \times 6.02 \times 10^{23}$

$= 1.1 \times 10^{12}$ disint./s

For $t_{1/2}$ = 12 hr

$$k = \frac{0.693}{t_{1/2}} = \frac{0.693}{12\text{ hr}} \times \frac{1\text{ hr}}{3600\text{ s}} = 1.6 \times 10^{-5}\ s^{-1}$$

Rate = $1.6 \times 10^{-5}\ s^{-1} \times 6.02 \times 10^{23} = 9.6 \times 10^{18}$ disint./s

For $t_{1/2}$ = 12 min

$$\text{Rate} = \frac{0.693}{12\text{ min}} \times \frac{1\text{ min}}{60\text{ s}} \times 6.02 \times 10^{23} = 5.8 \times 10^{20}\text{ disint./s}$$

For $t_{1/2}$ = 12 s

$$\text{Rate} = \frac{0.693}{12\text{ s}} \times 6.02 \times 10^{23} = 3.5 \times 10^{22}\text{ disint./s}$$

21. $175 \text{ mg } Na_3{}^{32}PO_4 \times \dfrac{32.0 \text{ mg } {}^{32}P}{165.0 \text{ mg } Na_3{}^{32}PO_4} = 33.9 \text{ mg } {}^{32}P, \quad k = \dfrac{\ln(2)}{t_{1/2}}$

$$\ln\left(\frac{N}{N_o}\right) = \frac{-0.69315\, t}{t_{1/2}};\ \ln\left(\frac{m}{33.9}\right) = \frac{-0.69315\,(35.0)}{14.3}$$

$\ln(m) = -1.697 + 3.523 = 1.826;\ m = e^{1.826} = 6.21 \text{ mg } {}^{32}P$ remains

23. $\ln\left(\dfrac{N}{N_o}\right) = -kt = \dfrac{-0.69315\, t}{t_{1/2}}$; If 10.0% decays, then 90.0% is left.

$$\ln\left(\frac{90.0}{100.0}\right) = \frac{-0.69315\, t}{5.26 \text{ yr}};\ t = 0.800 \text{ years}$$

25. $t_{1/2} = 5370$ yr; $\quad k = \dfrac{\ln(2)}{t_{1/2}}, \quad \ln\left(\dfrac{N}{N_o}\right) = -kt$

$$\ln\left(\frac{N}{N_o}\right) = \frac{-0.693\, t}{t_{1/2}} = \frac{-0.693\,(2200 \text{ yr})}{5730 \text{ yr}} = -0.27$$

$$\frac{N}{N_o} = e^{-0.27} = 0.76$$

27. $t_{1/2} = 4.5 \times 10^9$ years

Since 4.5×10^9 years is equal to the half-life, one half of the ^{238}U atoms will have been converted to ^{206}Pb. The numbers of atoms of ^{206}Pb and ^{238}U will be equal. Thus, the mass ratio is:

$$\frac{206}{238} = 0.866$$

Energy Changes in Nuclear Reactions

29. $E = mc^2,\ m = \dfrac{E}{c^2} = \dfrac{3.9 \times 10^{23} \text{ kg m}^2/\text{s}^2}{(3.00 \times 10^8 \text{ m/s})^2} = 4.3 \times 10^6 \text{ kg}$

The sun loses 4.3×10^6 kg of mass each second.

31. $12\,{}^1_1H + 12\,{}^0_1n + 12\,{}^{\ 0}_{-1}e \longrightarrow {}^{24}_{12}Mg$ mass of proton = 1.00728 amu.

$$\Delta m = 23.9850 - [12(1.00728) + 12(1.00866) + 12(5.49 \times 10^{-4})] \text{ amu}$$

$$\Delta m = -0.2129 \text{ amu}$$

$$E = mc^2 = 0.2129 \text{ amu} \times \frac{1 \text{ g}}{6.022 \times 10^{23} \text{ amu}} \times \frac{1 \text{ kg}}{1000 \text{ g}} (2.9979 \times 10^8 \text{ m/s})^2$$

$$E = 3.177 \times 10^{-11} \text{ J}$$

$$\frac{\text{BE}}{\text{nucleon}} = \frac{3.177 \times 10^{-11} \text{ J}}{24} = \frac{1.324 \times 10^{-12} \text{ J}}{\text{nucleon}}$$

For ^{27}Mg

$$12\,^{1}_{1}\text{H} + 15\,^{1}_{0}\text{n} + 12\,^{0}_{-1}\text{e} \longrightarrow\, ^{27}_{12}\text{Mg}$$

$\Delta m = 26.9843 - [12(1.00728) + 15(1.00866) + 12(5.49 \times 10^{-4})]$ amu

$\Delta m = -0.2395$ amu

$$E = mc^2 = 0.2395 \text{ amu} \times \frac{1 \text{ g}}{6.022 \times 10^{23} \text{ amu}} \times \frac{1 \text{ kg}}{1000 \text{ g}} (2.9979 \times 10^8 \text{ m/s})^2$$

$E = 3.574 \times 10^{-11}$ J

$$\frac{\text{BE}}{\text{nucleon}} = \frac{3.574 \times 10^{-11} \text{ J}}{27 \text{ nucleons}} = \frac{1.324 \times 10^{-12} \text{ J}}{\text{nucleon}}$$

33. $^{1}_{1}\text{H} + ^{1}_{0}\text{n} \longrightarrow 2\,^{1}_{1}\text{H} + ^{1}_{0}\text{n} + ^{1}_{-1}\text{H}$ Mass $^{1}_{-1}\text{H}$ = mass $^{1}_{1}\text{H}$

$\Delta m = +2(1.00728) = 2.01456$ amu

$$E = 2.01456 \text{ amu} \times \frac{1 \times 10^{-3} \text{ kg}}{6.02214 \times 10^{23} \text{ amu}} \times (2.997925 \times 10^8 \text{ m/s})^2$$

$= 3.00657 \times 10^{-10}$ J of energy is absorbed per atom or 1.81060×10^{14} J/mol.

The source of energy is the kinetic energy of the proton and the neutron in the particle accelerator.

35. $^{1}_{1}\text{H} + ^{1}_{1}\text{H} \longrightarrow\, ^{2}_{1}\text{H} + ^{0}_{+1}\text{e}$

$\Delta m = (2.01410 - m_e + m_e) - 2(1.00782 - m_e)$

$\Delta m = 2.01410 - 2(1.00782) + 2(0.000549) = -4.4 \times 10^{-4}$ amu

When two mol of $^{1}_{1}$H undergoes fusion, $\Delta m = -4.4 \times 10^{-4}$ g.

$E = 4.4 \times 10^{-7} \text{ kg} \times (3.00 \times 10^8 \text{ m/s})^2 = 4.0 \times 10^{10}$ J

$$\frac{4.0 \times 10^{10} \text{ J}}{2 \text{ mol } ^{1}_{1}\text{H}} \times \frac{1 \text{ mol}}{1.00782 \text{ g}} = 2.0 \times 10^{10} \text{ J/g}$$

Detection, Uses, and Health Effects of Radiation

37. The Geiger-Müller tube has a certain response time. After the gas in the tube ionizes to produce a "count" some time must elapse for the gas to return to an electrically neutral state. The response of the tube levels because at high activities radioactive particles are entering the tube faster than the tube can respond to them.

39. Assuming that (1) the radionuclide is long lived enough such that no significant decay occurs during the time of the experiment, the total counts of radioactivity injected are:

$$0.10\text{ mL} \times \frac{5.0 \times 10^3\text{ cpm}}{\text{mL}} = 5.0 \times 10^2\text{ cpm}$$

Assuming that (2) the total activity is uniformly distributed only in the rats blood, the blood volume is:

$$\frac{48\text{ cpm}}{\text{mL}} \times V = 5.0 \times 10^2;\ V = 10.4\text{ mL} = 10.\text{ mL}$$

41. Release of Sr is probably more harmful. Xe is chemically unreactive, Sr can be easily oxidized to Sr^{2+}. Strontium is in the same family as calcium and could be absorbed and concentrated in the body in a fashion similar to Ca. This puts the radioactive Sr in the bones; red blood cells are produced in bone marrow. Xe would not be readily incorporated in the body.

The chemical properties determine where a radioactive material may be concentrated in the body or how easily it may be excreted. The length of time of exposure and what is exposed to radiation significantly affects the health hazard.

ADDITIONAL EXERCISES

43. Mass of nucleus = mass of atom - mass of electrons

$$= 6.015126 - 3(0.0005486) = 6.013480\text{ amu}$$

$$3\,{}^{1}_{1}H + 3\,{}^{1}_{0}n \longrightarrow {}^{6}_{3}Li$$

$$\Delta m = 6.013480 - [3(1.00728) + 3(1.00866)] = -0.03434\text{ amu}$$

For 1 mol of ^{6}Li, the mass defect is 0.03434 g.

$$E = mc^2 = 3.434 \times 10^{-2}\text{ g} \times \frac{1\text{ kg}}{1000\text{ g}} \times (2.9979 \times 10^8\text{ m/s})^2$$

$$E = 3.086 \times 10^{12}\text{ J/mol}$$

45. mass of nucleus = 2.01410 - 0.000549 = 2.01355 amu

$$u_{rms} = \left(\frac{3\,RT}{M}\right)^{1/2} = \left(\frac{3(8.3145\text{ J/K•mol})\,(4 \times 10^7\text{ K})}{2.01355\text{ g}\,(\frac{1\text{ kg}}{1000\text{ g}})}\right)^{\frac{1}{2}} = 7 \times 10^5\text{ m/s}$$

$$E_K = \frac{1}{2}mu^2 = \frac{1}{2}\left(2.01355\text{ amu} \times \frac{1 \times 10^{-3}\text{ kg}}{6.022 \times 10^{23}\text{ amu}}\right)(7 \times 10^5\text{ m/s})^2$$

$$E_K = 8 \times 10^{-16}\text{ J}$$

47. a. $^{12}_{6}C$: It takes part in the reaction (1st step) but is regenerated in the last step. $^{12}_{6}C$ is not consumed.

b. ^{13}N, ^{13}C, ^{14}N, ^{15}O, and ^{15}N are intermediates.

c. $4\ ^{1}_{1}H \longrightarrow\ ^{4}_{2}He + 2\ ^{0}_{+1}e$

$\Delta m = 4.00260 - 2\ m_e + 2\ m_e - 4(1.00782 - m_e)$

$\Delta m = 4.00260 + 4(0.000549) - 4(1.00782)$; $\Delta m = -0.02648$ amu

For 4 mol $^{1}_{1}H$, $\Delta m = -0.02648$ g

$E = 2.648 \times 10^{-5}\ kg\ (2.9979 \times 10^{8}\ m/s)^2 = 2.380 \times 10^{12}\ J$

$$\frac{2.380 \times 10^{12}\ J}{4\ mol\ H} = \frac{5.950 \times 10^{11}\ J}{mol\ H}$$

CHALLENGE PROBLEMS

49. a. For $2\ H_2O + 2\ e^- \longrightarrow H_2 + 2\ OH^-$, $E° = -0.83$ V

$E°_{H_2O} - E°_{Zr} = +1.53$ V

Yes, the reduction of H_2O to H_2 by Zr is spontaneous, $E° > 0$ (at standard conditions).

b.

$$4\ H_2O + 4\ e^- \longrightarrow 2\ H_2 + 4\ OH^-$$
$$Zr + 4\ OH^- \longrightarrow ZrO_2{\cdot}H_2O + H_2O + 4\ e^-$$

$$3\ H_2O(l) + Zr(s) \longrightarrow 2\ H_2(g) + ZrO_2{\cdot}H_2O(s)$$

c. $E° = -0.83\ V - (-2.36\ V) = +1.53$ V

$\Delta G° = -nFE° = -(4\ mol\ e^-)\ (96{,}485\ C/mol\ e^-)\ (1.53\ J/C)$

$\Delta G° = -5.90 \times 10^{5}\ J = -590.\ kJ$

From the Nernst equation:

$E = E° - \frac{0.0592}{n} \log Q$ At equilibrium, E = 0 and Q = K

$E° = \frac{0.0592}{n} \log K$; $\log K = \frac{4(1.53)}{0.0592} = 103$

$K \approx 10^{103}$

d. $1.00 \times 10^3 \text{ kg Zr} \times \frac{1000 \text{ g}}{\text{kg}} \times \frac{1 \text{ mol Zr}}{91.22 \text{ g Zr}} \times \frac{2 \text{ mol } H_2}{\text{mol Zr}} = 2.19 \times 10^4 \text{ mol } H_2$

$$2.19 \times 10^4 \text{ mol } H_2 \times \frac{2.016 \text{ g } H_2}{\text{mol } H_2} = 4.42 \times 10^4 \text{ g } H_2$$

$$PV = nRT$$

$$V = \frac{nRT}{P} = \frac{(2.19 \times 10^4 \text{ mol})(0.08206 \text{ L atm/mol•K})(1273 \text{ K})}{1 \text{ atm}} = 2 \times 10^6 \text{ L}$$

e. Probably yes, less radioactivity overall was released by venting the H_2, than what would have been released if the H_2 exploded inside the reactor (as happened at Chernobyl). Neither alternative is pleasant, but venting the radioactive hydrogen is the less unpleasant of the two alternatives.

CHAPTER TWENTY-TWO: ORGANIC CHEMISTRY

QUESTIONS

1. There is only one consecutive chain of C-atoms. They are not all in a true straight line since the bond angle at each carbon is a tetrahedral angle of 109.5°.

3. Resonance: All atoms are in the same position. Only the position of π electrons is different.

   Isomerism: Atoms are in different locations in space.

   Isomers are distinctly different substances. Resonance is the use of more than one Lewis structure to describe the bonding in a single compound. Resonance structures are <u>not</u> isomers.

5. a. addition polymer: Polymer formed by adding monomer units to a double bond.

   Teflon: $n\ CF_2{=}CF_2 \rightarrow \left(CF_2 - CF_2\right)_n$

   b. condensation polymer: Polymer that forms when two monomers combine by eliminating a small molecule. Nylon and dacron are examples of condensation polymers.

   c. copolymer: Polymer formed from more than one kind of monomer. Nylon and dacron are also copolymers.

7. A thermoplastic polymer can be remelted; a thermoset polymer cannot be softened once it is formed.

9. Plasticizers make a polymer more flexible. Crosslinking makes a polymer more rigid.

11. a. Replacement of hydrogen with halogen results in fewer H and OH radicals in the flame, reducing flame temperature.

    b. With aromatic groups present it is more difficult to "chip off" pieces of the polymer in the pyrolysis zone.

    c. It is more difficult to "chip off" fragments of the polymer in the pyrolysis zone. The crosslinked polymer chars instead of burns.

13. For a given chain length, there are more hydrogen bonding sites in Nylon-46 than Nylon-6.

EXERCISES

Hydrocarbons

15. $CH_3-CH_2-CH_2-CH_2-CH_2-CH_3$ hexane or n-hexane (highest b.p., least branched)

$$\begin{array}{l} \quad\quad\ \ CH_3 \\ \quad\quad\ \ \ | \\ CH_3-CH-CH_2-CH_2-CH_3 \end{array}$$ 2-methylpentane

```
                 CH3
                  |
CH3 — CH2 — CH — CH2 — CH3
```
3-methylpentane

```
       CH3
        |
CH3 — C — CH2 — CH3
        |
       CH3
```
2,2-dimethylbutane

```
       CH3  CH3
        |    |
CH3 — CH — CH — CH3
```
2,3-dimethylbutane

n-Hexane would have the highest boiling point. It is the least branched and, therefore, hexane will have the strongest dispersion forces between molecules.

17.

a.
```
CH3 — CH — CH2 — CH2 CH3
       |
      CH3
```

b.
```
        CH3
         |
CH3 — C — CH2 — CH — CH3
         |          |
        CH3        CH3
```

c.
```
CH3 — CH — CH2 CH2 CH3
       |
CH3 — C — CH3
       |
      CH3
```

d. The longest chain is 6 carbons long.
```
        3      4      5      6
CH3 — CH — CH2 — CH2 — CH3
        |2
CH3 — C — CH3
        |1
       CH3
```
2,2,3-trimethylhexane

19. a. 1-butene b. 2-methyl-2-butene c. 2,5-dimethyl-3-heptene

21. a. $CH_3—CH_2—CH{=}CH—CH_2—CH_3$ b. $CH_3CH{=}CHCH{=}CHCH_2CH_3$

c.
```
       CH3
        |
CH3 — CH — CH=CHCH2 CH2 CH2 CH3
```

23.

a. 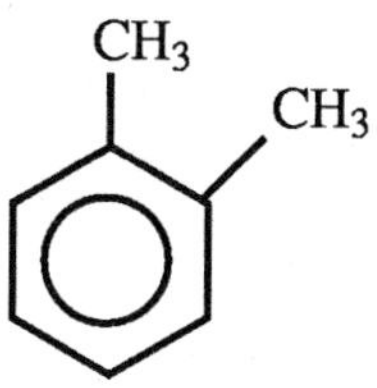 CH_3 CH_3

b. 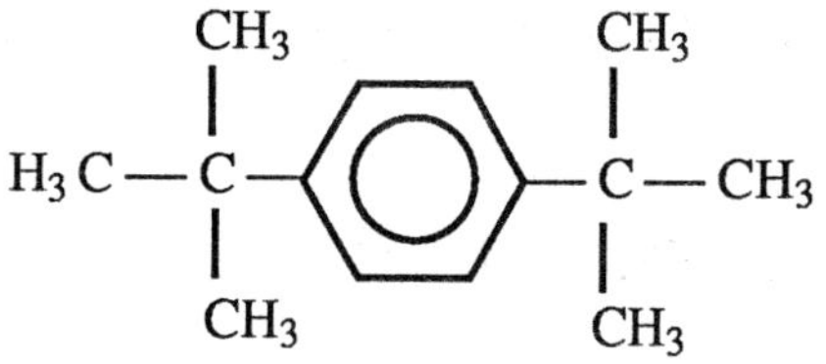 CH_3 CH_3 $H_3C—C—$ $—C—CH_3$ CH_3 CH_3

c. CH_2CH_3 CH_2CH_3

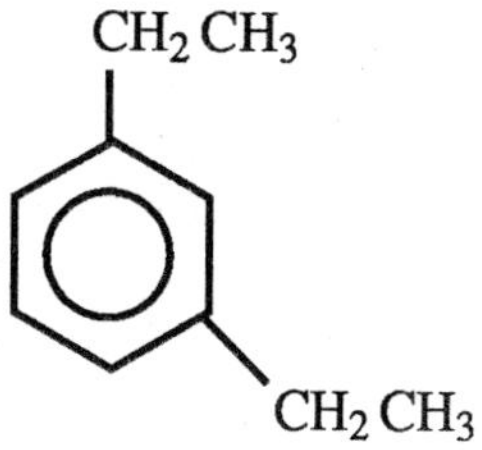

25. a. 1,3-dichlorobutane

b. 1,1,1-trichlorobutane

c. 2,3-dichloro-2,4-dimethylhexane

d. 1,2-difluoroethane

e. chlorobenzene

f. chlorocyclohexane

g. 3-chlorocyclohexene (double bond assumed to be between C_1 and C_2).

26. a. methylcyclopropane

b. t-butylcyclohexane

c. 3,4-dimethylcyclopentene

d. chloroethene (vinyl chloride)

e. 1,2-dimethylcyclopentene

f. 1,1-dichlorocyclohexane

Isomerism

27.

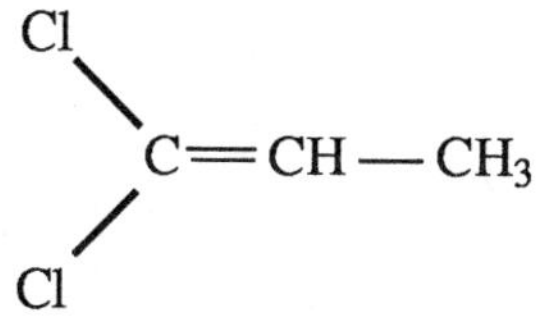

$Cl_2C{=}CH—CH_3$ $CH_2{=}CCl—CH_2Cl$ $CH_2{=}CH—CHCl_2$

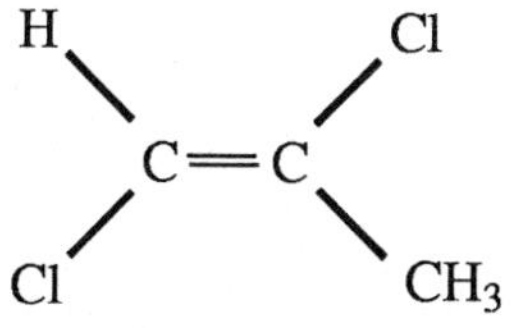

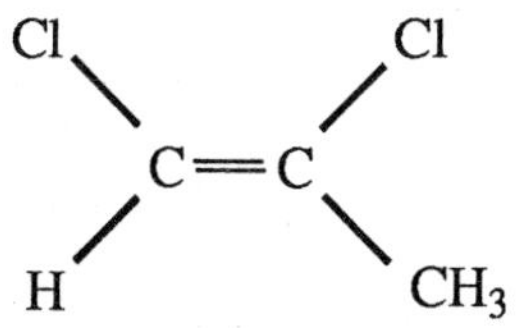

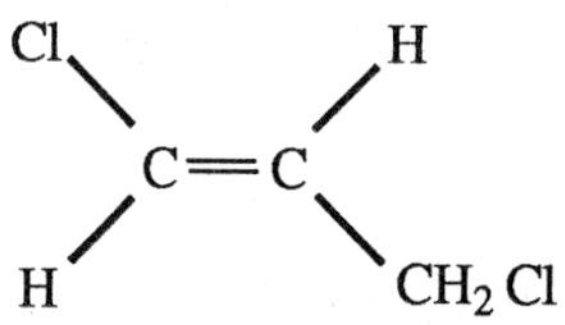

H Cl Cl H $C{=}C$ Cl CH3 Cl Cl H $C{=}C$ H CH3 Cl H $C{=}C$ H CH_2Cl

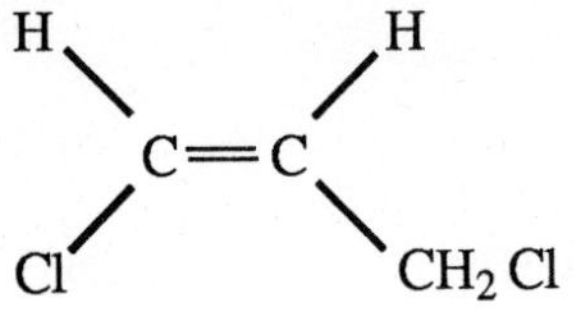

H, H, C=C, Cl, CH_2Cl — Cl, Cl — Cl, Cl, cis — Cl, Cl, trans

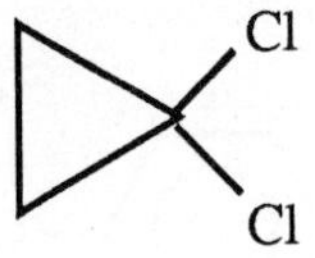

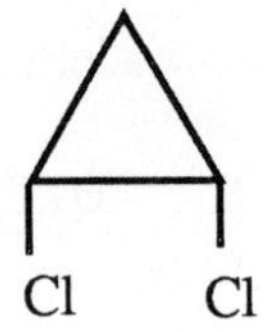

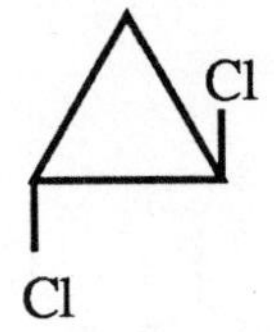

29.

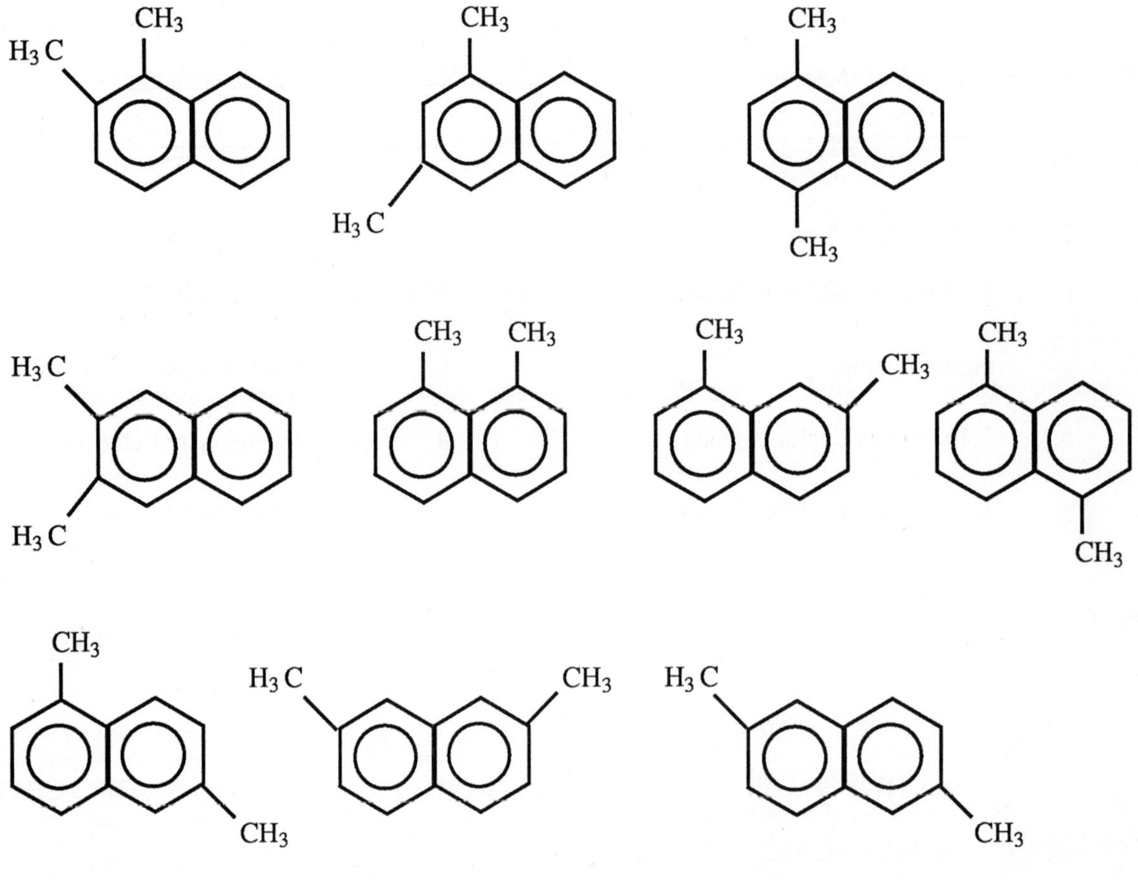

CH_3, H_3C — CH_3, H_3C — CH_3, CH_3

H_3C, H_3C — CH_3, CH_3 — CH_3, CH_3 — CH_3, CH_3

CH_3, CH_3 — H_3C, CH_3 — H_3C, CH_3

31. a. 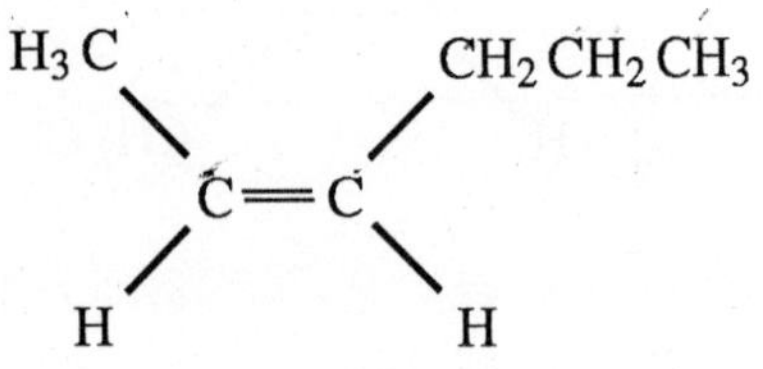 b. 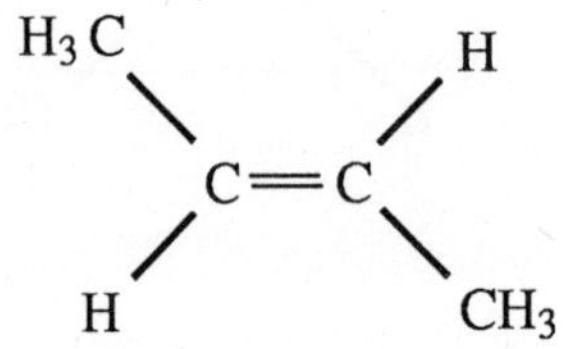 c.

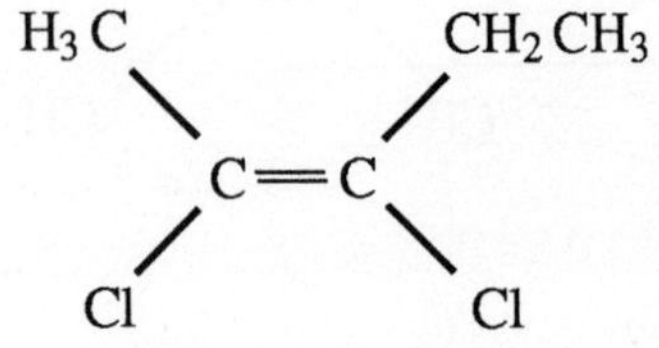

a. H_3C, $CH_2CH_2CH_3$, C=C, H, H

b. H_3C, H, C=C, H, CH_3

c. H_3C, CH_2CH_3, C=C, Cl, Cl

Functional Groups

33. a. ketone b. aldehyde c. ketone d. amine

35. a.

```
         H                 H  H       H
          \               /    \     /
           C ===== C        H    N      primary amine
          /         \       |    |
     O = C           N -- C -- C -- C -- O -- H
                    /     |    |    ||   carboxylic acid
  ketone  \        /      H    H    O
           C ===== C
          /         \
   H -- O   alcohol   H     tertiary amine
```

b. 5 carbons in ring and in $-CO_2H$: sp^2; the other two carbons: sp^3

c. 24 - sigma bonds, 4 - pi bonds

37.

```
        HO — CH — CH2 — NH — CH3
 alcohol ↗    |            ↗ amine
           (benzene ring)
             |        \
            OH         OH
               ↖      ↑
             alcohol (phenol)
```

Reactions of Organic Compounds

39. a. $CH_2{=}CH_2 + Br_2 \rightarrow CH_2Br{-}CH_2Br$

b. $C_6H_6 + Br_2 \xrightarrow{Fe} C_6H_5Br + HBr$

c. $CH_3CO_2H + CH_3OH \rightarrow CH_3CO_2CH_3 + H_2O$

41. a.

$$CH_3-\overset{\overset{\displaystyle O}{\|}}{C}-H$$

b.

$$CH_3-\overset{\overset{\displaystyle O}{\|}}{C}-CH_3$$

c.

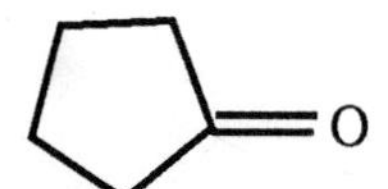

d. No reaction occurs

e.

$$H_3C-\overset{\displaystyle CH_3}{\underset{\displaystyle CH_3}{\overset{|}{\underset{|}{C}}}}-CH_2OH \xrightarrow{[Ox]} H_3C-\overset{\displaystyle CH_3}{\underset{\displaystyle CH_3}{\overset{|}{\underset{|}{C}}}}-\overset{\displaystyle O}{\overset{\|}{C}}H$$

43. a. $CH_3CH{=}CH_2 + Br_2 \rightarrow CH_3CHBrCH_2Br$

b. $CH_3C{\equiv}CH + H_2 \xrightarrow{\text{catalyst}} CH_3CH{=}CH_2 + Br_2 \rightarrow CH_3CHBrCH_2Br$

c. $CH_3CO_2H + HOCH_2CH_2CH_2CH_3 \rightarrow CH_3\overset{\displaystyle O}{\overset{\|}{C}}-O-CH_2CH_2CH_2CH_3 + H_2O$

ethanoic acid butanol butyl acetate or butylethanoate

Polymers

45. a. repeating unit: $\left(CHF-CH_2\right)_n$

monomer: $CHF{=}CH_2$

b. repeating unit:

$$\left(OCH_2CH_2\overset{\displaystyle O}{\overset{\|}{C}}\right)_n$$

monomer: $HO-CH_2CH_2-CO_2H$

c. repeating unit:

$$\left(OCH_2CH_2-O-\overset{\displaystyle O}{\overset{\|}{C}}-CH_2CH_2-\overset{\displaystyle O}{\overset{\|}{C}}\right)_n$$

copolymer of: $HOCH_2CH_2OH$ and $HO_2CCH_2CH_2CO_2H$

d. monomer:

$CH_3-C(C_6H_5)=CH_2$

e. monomer:

$C_6H_5-CH=CHCH_3$

Addition polymers: a, d, and e

Condensation polymers: b anc c

Copolymer: c

47.

$$\left(NH-C_6H_4-NHC(=O)-C_6H_4-C(=O) \right)_n$$

49.

$H_2N-C_6H_4-NH_2$ and $C_6H_2(CO_2H)_4$ (HO_2C, CO_2H, HO_2C, CO_2H)

51. Divinylbenzene crosslinks different chains to each other. The chains cannot move past each other because of the crosslinks; thus, the polymer is more rigid.

ADDITIONAL EXERCISES

53. 2-methyl-1,3-butadiene

55.

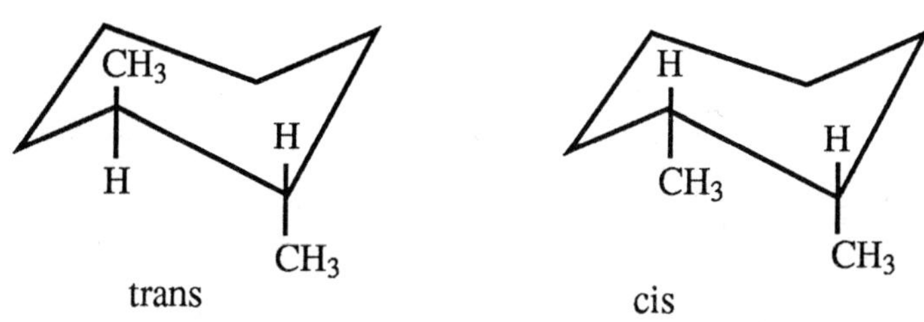

57. a.

$$CH_3-C(OH)=CH_2 \longrightarrow CH_3-C(=O)-CH_3$$

Bonds broken:		Bonds made:	
O—H	467 kJ/mol	C—H	413 kJ/mol
C—O	358 kJ/mol	C=O	799 kJ/mol
C=C	614 kJ/mol	C—C	347 kJ/mol

$\Delta H = 467 + 358 + 614 - (413 + 799 + 347) = 1439 - 1559 = -120.$ kJ

Ketone is more stable; it contains the stronger bonds.

b.

$C_6H_5-C(=O)-NH-CH_3 \longrightarrow C_6H_5-C(OH)=N-CH_3$

Bonds broken:		Bonds made:	
C=O	799 kJ/mol	C—O	358 kJ/mol
C—N	305 kJ/mol	C=N	615 kJ/mol
N—H	391 kJ/mol	O—H	467 kJ/mol

$\Delta H = 799 + 305 + 391 - (358 + 615 + 467) = 55$ kJ

Amide with C=O is more stable; it contains the stronger bonds.

c. For the reaction:

$H_3C-C(=O)-CH_2-C(=O)-CH_3 \longrightarrow H_3C-C(=O)-CH=C(O-H)-CH_3$

Bonds broken:		Bonds made:	
C=O	799 kJ/mol	O—H	467 kJ/mol
C—C	347 kJ/mol	C—O	358 kJ/mol
C—H	413 kJ/mol	C=H	614 kJ/mol

$\Delta H = (799 + 347 + 413) - (467 + 358 + 614) = 1559 - 1439 = 120.$ kJ

The form with two carbonyl groups is more stable because it contains the stronger bonds.

59. The amine group is protonated, making it a positive one charged cation.

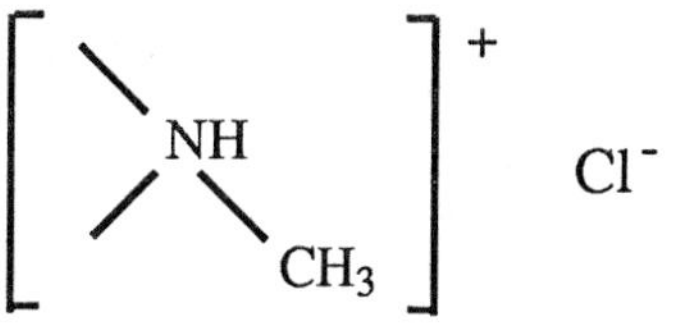

$$\left[\;\diagdown\!\!\!\!\diagup NH - CH_3 \right]^+ \quad Cl^-$$

Morphine hydrochloride is ionic and, hence, more soluble in water.

61. a. The bond angles in the ring are about 60°. VSEPR predicts bond angles close to 109°. The bonding electrons are closer together than they want to be, resulting in stronger electron-electron repulsions. This makes ethylene oxide unstable (reactive).

b. $\left(O - CH_2CH_2 - O - CH_2CH_2 - O - CH_2CH_2 \right)_n$

c.

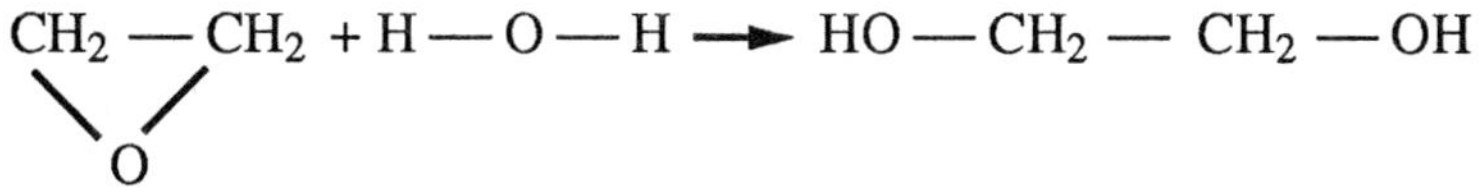

$$CH_2 - CH_2 \text{ (ring through O)} + H - O - H \longrightarrow HO - CH_2 - CH_2 - OH$$

Bonds broken:	Bonds made:
C—O	C—O
H—O	O—H

Since bonds broken = bonds formed, $\Delta H \approx 0$.

$$CH_2 = CH_2 + H - O - H \longrightarrow CH_3CH_2OH$$

Bonds broken:		Bonds made:	
C=C	614 kJ/mol	C—H	413 kJ/mol
H—O	467 kJ/mol	C—C	347 kJ/mol
		C—O	358 kJ/mol

$\Delta H = 614 + 467 - [413 + 347 + 358] = 1081 \text{ kJ} - 1118 \text{ kJ} = -37 \text{ kJ}$

d. $\dfrac{20{,}000 \text{ g}}{\text{mol}} \times \dfrac{1 \text{ ethylene oxide}}{44.05 \text{ g}} = 454 \sim 500 \text{ monomers}$

No. This is just an average. A sample of the polymer will consist of chains with a distribution of lengths.

63.

$$\left(-OCH_2CH_2O\overset{O}{\overset{\|}{C}}NH-C_6H_4-NH\overset{O}{\overset{\|}{C}}OCH_2CH_2O\overset{O}{\overset{\|}{C}}NH-C_6H_4-NH\overset{O}{\overset{\|}{C}}- \right)_n$$

65. a. The polyamide from 1,2-diaminoethane and terephthalic acid is stronger because of the possibilily of hydrogen bonding between chains.

b. The polymer of

$$HO-C_6H_4-CO_2H$$

because the chains are stiffer due to the aromatic ring.

c. polyacetylene is $n\,HC\equiv CH \rightarrow -(CH=CH)_n-$

Polyacetylene is stronger because the double bonds in the chain make the chains stiffer.

67. a. The temperature of the rubber band increases when it is stretched.

b. Exothermic (heat is released).

c. As the chains are stretched they line up more closely resulting in stronger dispersion forces between the chains.

d. Stretching is not spontaneous. ΔG (+)

Since $\Delta G = \Delta H - T\Delta S$, then ΔS must be negative. ΔS (-)
(+) (-)

e.

unstretched

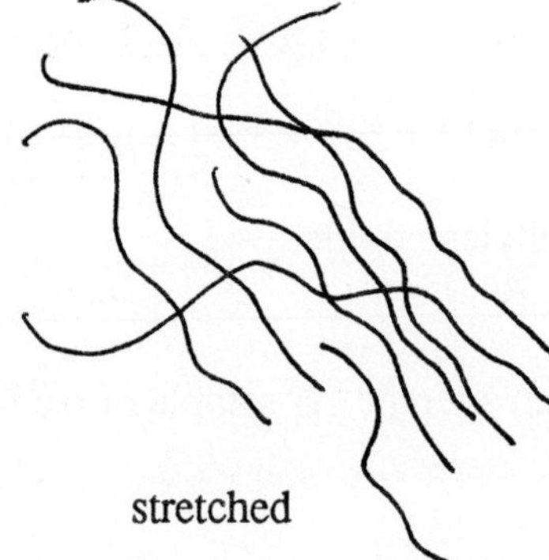

stretched

The structure of the stretched polymer is more ordered (lower S).

69. Compounds are numbered in *The Merck Index.* The number and page for each compound is given for the 10th edition. Uses are listed, look up structure and functional groups present.

a. Dopa, 3427, p. 497; anticholinergic, antiparkinsonian

b. Amphetamine, 606, p. 84; central stimulant

c. Ephedrine, 3558, p. 520; bronchodilator

d. Acetaminophen, 39, p. 7; analgesic, antipyretic

e. Phenacetin, 7064, p. 1035; analgesic, antipyretic

CHALLENGE PROBLEMS

71.

$$\underset{13}{HC}\equiv\underset{12}{C}-\underset{11}{C}\equiv\underset{10}{C}-\underset{9}{CH}=\underset{8}{C}=\underset{7}{CH}-\underset{6}{CH}=\underset{5}{CH}-\underset{4}{CH}=\underset{3}{CH}-\underset{2}{CH_2}-\underset{1}{\overset{\overset{O}{\|}}{C}}-OH$$

73. For the first structure below,

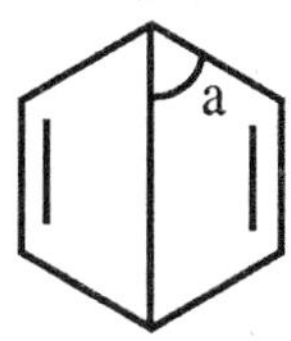

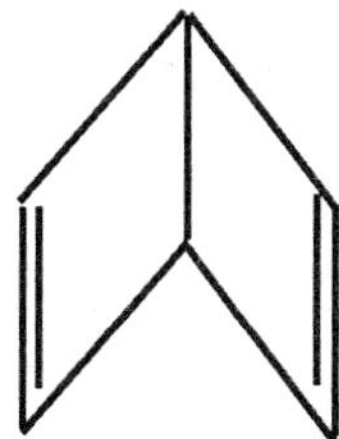

we would predict that the angle (a) should be close to a tetrahedral angle since there are four bonding pairs of electrons around the carbon atom. This would pull some of the C atoms out of the plane. Dewar's benzene has been synthesized. It is a distinct compound with the second structure. Because of the four membered rings, the angle (a) is closer to 90°.

75. a. overall yield = $0.83 \times 0.52 = 0.43$ or 43%

b. $$\text{theoretical yield} = 1000. \times 10^3 \text{ g styrene} \times \frac{1 \text{ mol styrene}}{104 \text{ g}} \times \frac{1 \text{ mol anthraquinone}}{2 \text{ mol styrene}} \times \frac{208 \text{ anthraquinone}}{\text{mol}} = 1.00 \times 10^6 \text{ g or } 1.00 \times 10^3 \text{ kg}$$

Actual yield = $0.43 \times (1.00 \times 10^3 \text{ kg}) = 430$ kg

c. i) Quantities are set up so that 208 lb anthraquinone is the theoretical yield.

$0.9 \times 208 = 187$ lb actual yield.

Note: We will carry extra significant figures and calculate all costs to the nearest cent.

$$\text{Cost} = 148 \text{ lb Phth. Anh.} \times \frac{\$0.35}{\text{lb}} + 293 \text{ lb AlCl}_3 \times \frac{\$0.67}{\text{lb}}$$

$$+ 78 \text{ lb C}_6\text{H}_6 \times \frac{\$0.61}{\text{lb}} = \$295.69$$

$$\frac{\text{cost}}{\text{lb}} = \frac{\$295.69}{187 \text{ lb}} = \$1.58/\text{lb} \text{ anthraquinone produced}$$

ii) $$\frac{208 \text{ lb styrene} \times \frac{\$0.27}{\text{lb}}}{0.43 \times 208 \text{ lb}} = \frac{\$0.63}{\text{lb}}$$

d. The disposal of $AlCl_3$ and H_2SO_4 in i) would present a major problem.

e. iii) To produce 0.8 × 208 = 166 lb of anthraquinone would require:

2 × 78 = 156 lb benzene.

The cost of raw materials per anthraquinone produced is (ignoring the cost of CO):

$$\frac{156 \text{ lb} \times \frac{\$0.61}{\text{lb}}}{166 \text{ lb}} = \frac{\$0.57}{\text{lb}} \quad \text{using benzene}$$

Or to produce 166 lb of anthraquinone would require 182 lb of benzophenone.

$$\text{Cost} = \frac{182 \text{ lb} \times \frac{\$2.40}{\text{lb}}}{166 \text{ lb}} = \frac{\$2.63}{\text{lb}} \quad \text{using benzophenone}$$

iv) To produce 0.9 × 208 = 187 lb of anthraquinone would require 128 lb naphthalene and 54 lb 1,3-butadiene.

$$\text{Cost} = \frac{128 \times \frac{\$0.24}{\text{lb}} + 54 \text{ lb} \times \frac{\$0.12}{\text{lb}}}{187 \text{ lb}} = \frac{\$0.20}{\text{lb}}$$

f. Process (iv) would be best in terms of raw material cost.

CHAPTER TWENTY-THREE: BIOCHEMISTRY

QUESTIONS

1. Primary: amino acid sequence: covalent bonds

   Secondary: features such as α-helix, pleated sheets: H-bonding.

   Tertiary: three dimensional shape: hydrophobic and hydrophillic interactions, salt linkages, hydrogen bonds, disulfide linkages, dispersion forces.

3. Both denaturation and inhibition reduce the catalytic activity of an enzyme. Denaturation changes the structure of an enzyme. Inhibition involves the attachment of an incorrect molecule at the active site, preventing the substrate from interacting with the enzyme.

5. Hydrogen bonding between the —OH groups of the starch and water molecules.

7. Nitrogen atoms with lone pairs of electrons.

9. A deletion may change the entire code for a protein. A substitution will change only one single amino acid in a protein.

11. A polyunsaturated fat contains 2 or more carbon-carbon double bonds.

EXERCISES

Proteins and Amino Acids

13. a.

$$H_2NCH_2CO_2H \rightleftharpoons H^+ + H_2NCH_2CO_2^- \qquad K_a = 4.3 \times 10^{-3}$$

$$H_2NCH_2CO_2^- + H^+ \rightleftharpoons {}^+H_3NCH_2CO_2^- \qquad K_2 = 1/K_a(\text{amino}) = K_b/K_w$$

$$= 6.0 \times 10^{-3}/10^{-14} = 6.0 \times 10^{11}$$

$$H_2NCH_2CO_2H \rightleftharpoons {}^+H_3NCH_2CO_2^- \qquad K = K_aK_2 = 2.6 \times 10^9$$

Equilibrium lies far to the right.

b. ${}^+H_3NCH_2CO_2H$ (1.0 *M* H^+)

$H_2NCH_2CO_2^-$ (1.0 *M* OH^-)

15. a. Aspartic acid and phenylalanine

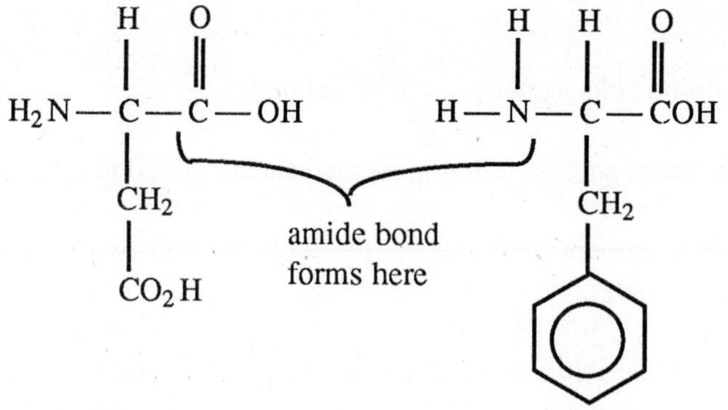

$$H_2N-\underset{\underset{\underset{CO_2H}{|}}{\underset{CH_2}{|}}}{\overset{H}{\overset{|}{C}}}-\overset{O}{\overset{\|}{C}}-OH \qquad H-\overset{H}{\overset{|}{N}}-\underset{\underset{C_6H_5}{\underset{CH_2}{|}}}{\overset{H}{\overset{|}{C}}}-\overset{O}{\overset{\|}{C}}OH$$

amide bond forms here

b. Aspartame contains the methyl ester of phenylalanine. This ester can hydrolyze to form methanol.

$$R-CO_2CH_3 + H_2O \rightleftharpoons RCO_2H + CH_3OH$$

17. a. ionic: Need $-NH_2$ (amine group) on side chain of one amino acid and - CO_2H on side chain of the other.

$-NH_2$ on side chain	$-CO_2H$ on side chain
His	Asp
Lys	Glu
Arg	

b. Hydrogen bonding (X = O or N):

$-X - H \cdots\cdots\cdots O{=}C$ (carbonyl group from peptide bond)

Ser	Asn	Any amino acid
Glu	Thr	
Tyr	Asp	
His	Gln	
Arg	Lys	

c. Covalent: cys . . . cys

d. London dispersion: All non-polar amino acids

e. dipole-dipole: Tyr, Thr and Ser

19.

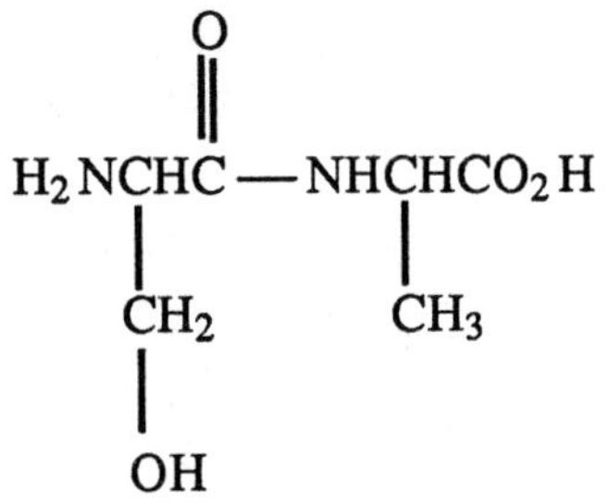

$$\underset{\underset{\text{OH}}{|}}{\underset{|}{\text{CH}_2}} \text{ serine: } H_2NCH(CH_2OH)C(=O)—NHCH(CH_3)CO_2H$$

$$H_2NCH(CH_3)C(=O)—NHCH(CH_2OH)CO_2H$$

ser - ala 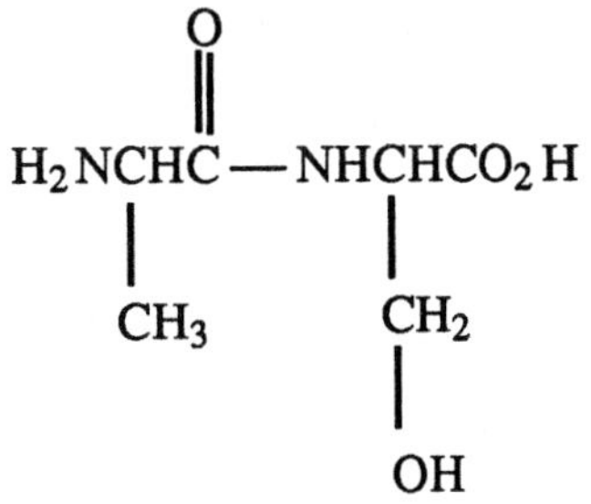 ala - ser

21. From —NH_2 to —CO_2H end:

phe-phe-gly-gly, gly-gly-phe-phe, gly-phe-phe-gly

phe-gly-gly-phe, phe-gly-phe-gly, gly-phe-gly-phe

Six tetrapeptides are possible.

23. Glutamic acid: R = —$CH_2CH_2CO_2H$

Valine: R = —$CH(CH_3)_2$

A polar side chain of the amino acid is replaced by a non-polar group. This could affect the tertiary structure and the ability to bind oxygen.

Carbohydrates

25.

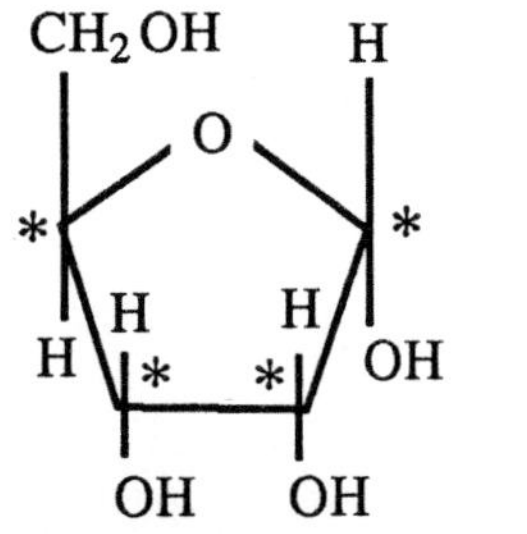

D-ribose

D-mannose

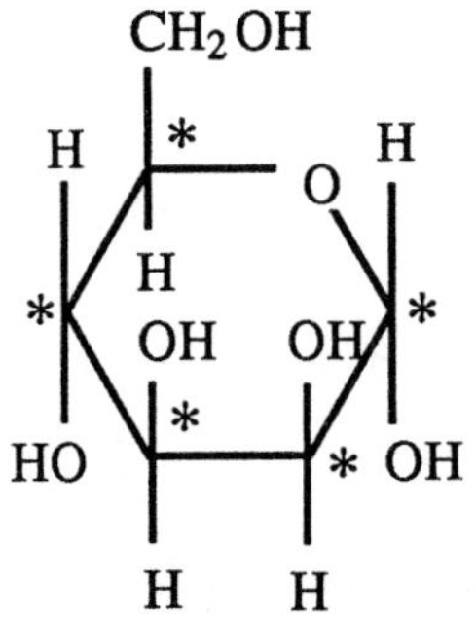

Optical Isomerism and Chiral Carbon Atoms

27. A chiral carbon has four different groups attached to it. A compound with a chiral carbon is optically active.

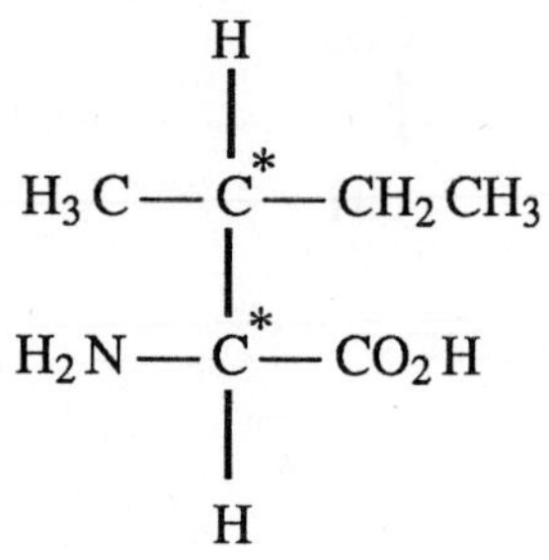

isoleucine: $H_3C—\overset{*}{C}H—CH_2CH_3$ attached to $H_2N—\overset{*}{C}H—CO_2H$

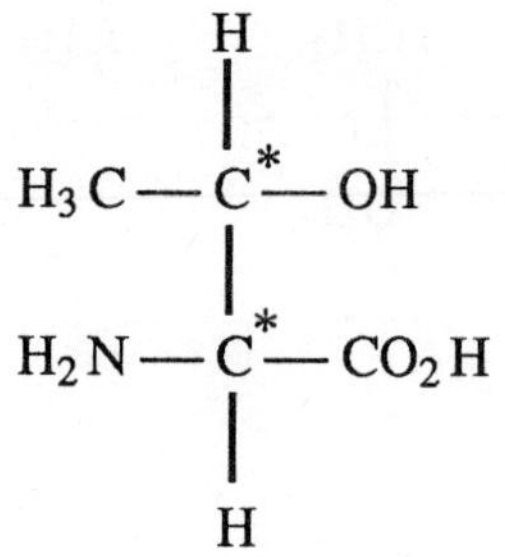

threonine: $H_3C—\overset{*}{C}H—OH$ attached to $H_2N—\overset{*}{C}H—CO_2H$

29.

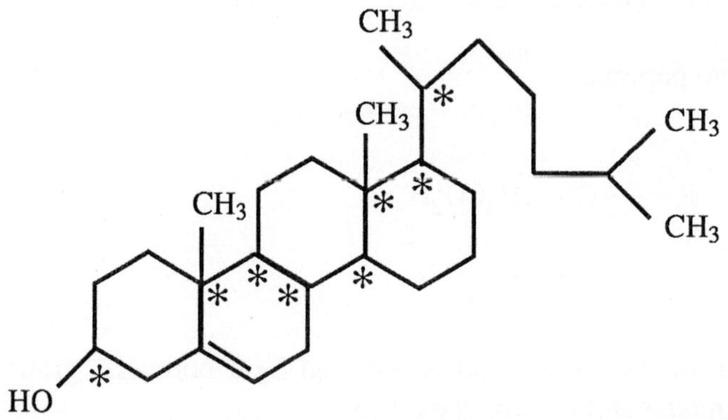

Eight chiral carbon atoms (marked with *).
Note: Some hydrogen atoms are omitted from the structure.

31.

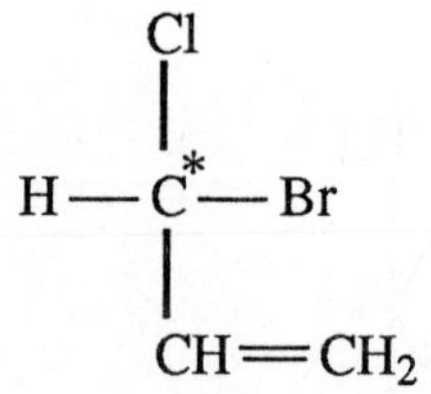

$H—\overset{*}{C}(Cl)(CH{=}CH_2)—Br$ is optically active. The chiral carbon is marked with *.

Nucleic Acids

33. T-A-C-G-C-C-G-T-A

35. Uracil: will H-bond to adenine.

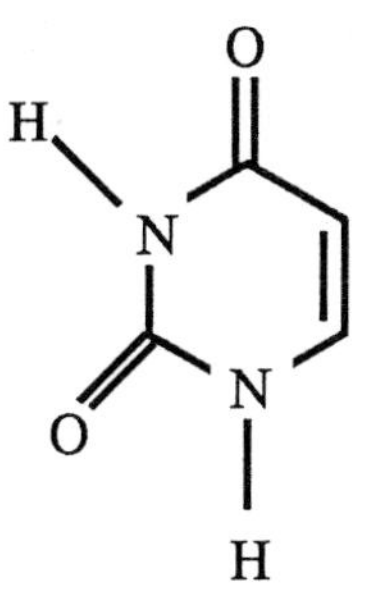

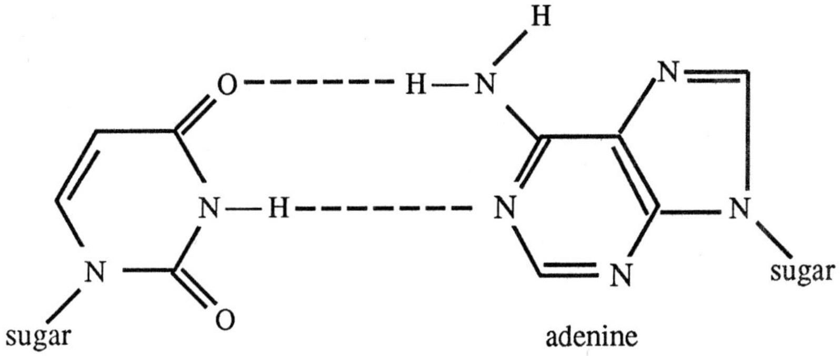

37. Base pair:

RNA		DNA
A	. . . . .	T
G	. . . . .	C
C	. . . . .	G
U	. . . . .	A

a.

Glu:	CTT, CTC
Val:	CAA, CAG, CAT, CAC
Met:	TAC
Trp:	ACC
Phe:	AAA, AAG
Asp:	CTA, CTG

b. DNA sequence for Met - Met - Phe - Asp - Trp:

TAC - TAC - AAA - CTA - ACC
or or
AAG CTG

c. four

d. C - T - T - A - C - C - A - A - A

Glu - Trp - Phe

e. C - T - C - A - C - C - A - A - A

C - T - T - A - C - C - A - A - G

C - T - C - A - C - C - A - A - G

Lipids and Steroids

39.

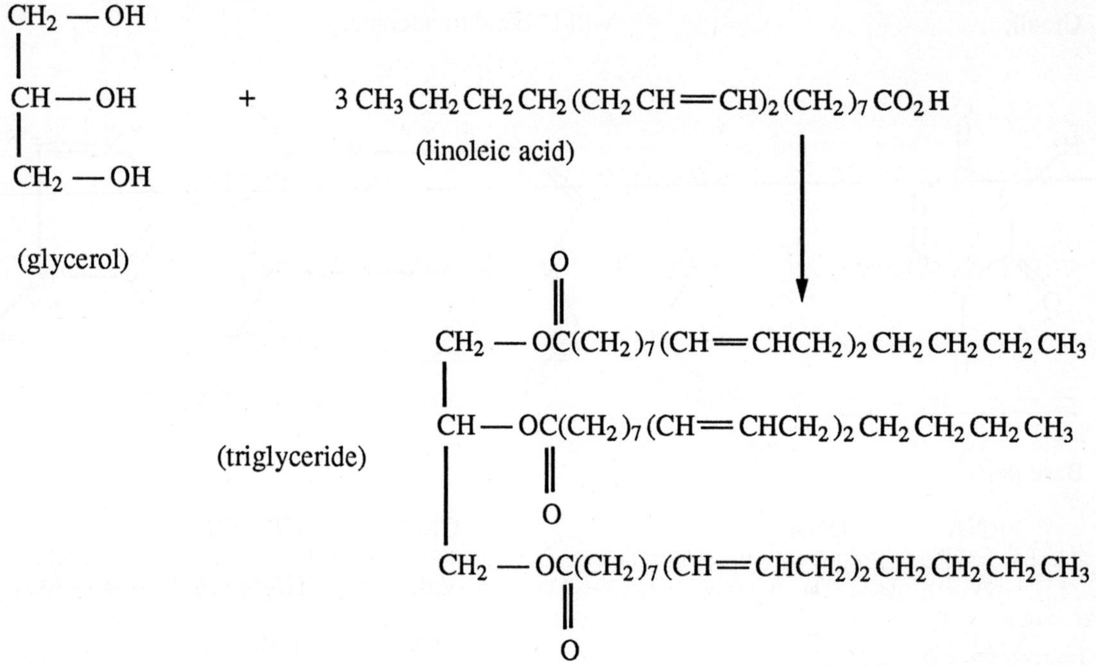

$$\begin{array}{l} CH_2-OH \\ | \\ CH-OH \\ | \\ CH_2-OH \end{array} \quad + \quad 3\ CH_3CH_2CH_2CH_2(CH_2CH{=}CH)_2(CH_2)_7CO_2H$$

(glycerol) (linoleic acid)

$\downarrow$

(triglyceride)

$$\begin{array}{l} CH_2-OC(=O)(CH_2)_7(CH{=}CHCH_2)_2CH_2CH_2CH_2CH_3 \\ | \\ CH-OC(=O)(CH_2)_7(CH{=}CHCH_2)_2CH_2CH_2CH_2CH_3 \\ | \\ CH_2-OC(=O)(CH_2)_7(CH{=}CHCH_2)_2CH_2CH_2CH_2CH_3 \end{array}$$

41.

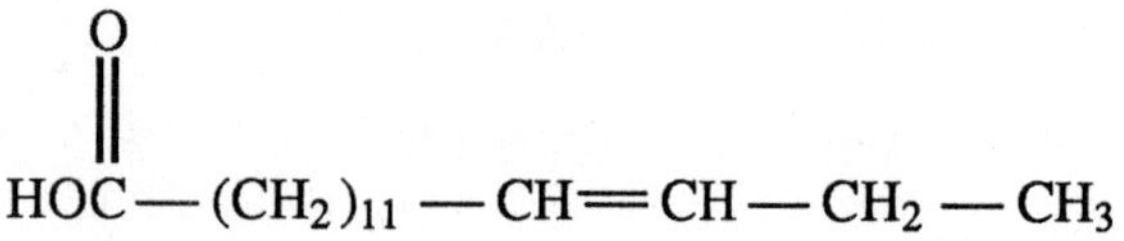

$HOC(=O)-(CH_2)_{11}-CH{=}CH-CH_2-CH_3$ 16 carbon omega-3 fatty acid

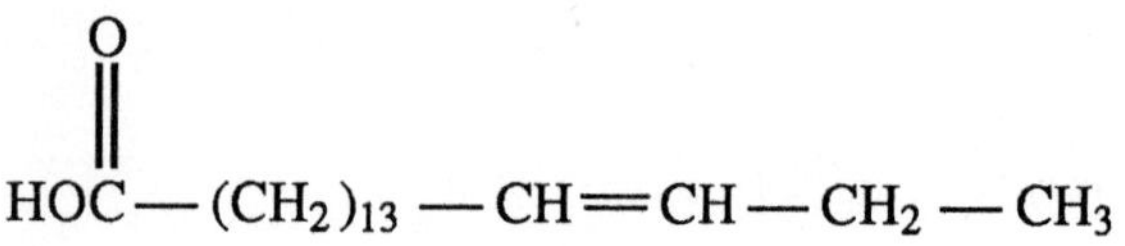

$HOC(=O)-(CH_2)_{13}-CH{=}CH-CH_2-CH_3$ 18 carbon omega-3 fatty acid

43. See #1986; p. 282; *The Merck Index, 10th Ed.*

ADDITIONAL EXERCISES

45. Glutamic acid:

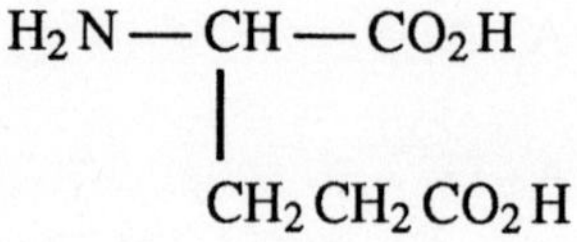

$$\begin{array}{l} H_2N-CH-CO_2H \\ \qquad\ \ | \\ \qquad\ CH_2CH_2CO_2H \end{array}$$

Monosodium glutamate: one of the acidic protons is lost.

$$\begin{array}{c} H_2N — CH — CO_2H \\ | \\ CH_2CH_2CO_2^- Na^+ \end{array}$$

47. $5 \times 10^9 \text{ pairs} \times \dfrac{340 \times 10^{-12} \text{ m}}{\text{pair}} = 1.7 \text{ m } (5'7'')$ (Ignoring significant figures.)

49. Structures can be found on pp. 1434-1439 of *The Merck Index, 10th Ed.*

a. A - fat soluble

b. E - fat soluble

c. K_5 - water soluble

d. K_6 - water soluble

51. $\Delta G = \Delta H - T\Delta S$

For the reaction, we break a P—O and O—H bond and make a P—O and O—H bond. Thus, $\Delta H \approx 0$. $\Delta S < 0$, since 2 molecules going to one molecule. Thus, $\Delta G > 0$, not spontaneous.

CHALLENGE PROBLEMS:

53. For a dipepetide there are $5 \times 5 = 25$ different cases.

For a tripeptide: $5 \times 5 \times 5 = 125$ different cases

So for 25 amino acids in the polypeptide there are $5^{25} = 2.98 \times 10^{17}$ different polypeptides.

55. We can use the reaction:

$$^+H_3NCH_2CO_2H \rightleftharpoons 2\,H^+ + H_2NCH_2CO_2^- \qquad K_{eq} = 7.3 \times 10^{-15}$$

$$7.3 \times 10^{-15} = \frac{[H^+]^2[H_2NCH_2CO_2^-]}{[^+H_3NCH_2CO_2H]} = [H^+]^2$$

$[H^+] = 8.5 \times 10^{-8}$ pH = 7.07 = isoelectric point